Recent Developments in Applied Biostratigraphy

The Micropalaeontological Society Special Publications

This collection of papers on the role of micropalaeontology in hydrocarbon exploration is the first in a series of Special Publications on a range of topics that will promote the aims of The Micropalaeontological Society, and communicate our subject to a wider audience. Some will appear as thematic collections of papers while others will continue the publication of stratigraphical atlases of a series of microfossil groups (including revisions of previously published volumes).

This volume is published by the Geological Society. Previous volumes still in print are published by Kluwer.

Calcareous Nannofossil Biostratigraphy. Paul R. Bown. ISBN 0-412-78970-1, August 1998.
Early Evolutionary History of Planktonic Foraminifera. M. K. BouDagher-Fadel, F. T. Banner, J. E. Whittaker & M. D. Simmons. ISBN 0-412-75820-2, November 1997.
Micropalaeontology and Hydrocarbon Exploration in the Middle East. M. D. Simmons. ISBN 0-412-42770-2, June 1994.
A Stratigraphic Index of Dinoflagellate Cysts. A. J. Powell. ISBN 0-412-36280-5, November 1991.
Ostracoda and Global Events. R. Whatley & C. Maybury. ISBN 0-412-36300-3, July 1990.

It is recommended that reference to all or part of this book should be made in one of the following ways:

POWELL, A. J. & RIDING, J. B. (eds) 2005. *Recent Developments in Applied Biostratigraphy.* The Micropalaeontological Society, Special Publications. The Geological Society, London.

BUTLER, N., CHARNOCK, M. A., HAGER, K. O. & WATKINS, C. A. 2005. The Ravenscar Group: a coeval analogue for the Middle Jurassic reservoirs of the North Sea and offshore Mid Norway. *In*: POWELL, A. J. & RIDING, J. B. (eds) *Recent Developments in Applied Biostratigraphy.* The Micropalaeontological Society, Special Publications. The Geological Society, London, 43–53.

THE MICROPALAEONTOLOGICAL SOCIETY SPECIAL PUBLICATIONS

Recent Developments in Applied Biostratigraphy

EDITED BY

A. J. POWELL
Consultant Biostratigrapher, Le Muids, Switzerland

and

J. B. RIDING
British Geological Survey, Nottingham, UK

2005
Published for The Micropalaeontological Society
by
The Geological Society
London

THE MICROPALAEONTOLOGICAL SOCIETY

The Micropalaeontological Society (TMS) exists 'to advance the education of the public in the study of Micropalaeontology' and is operated 'exclusively for scientific and educational purposes and not for profit'. It was founded as The British Micropalaeontological Group (BMG) in 1970, following a proposal by Professor Leslie Moore of the University of Sheffield and several colleagues who wished to organize a group of palaeontologists with a mutual interest in the study of the microfossils present in British type sections and the provision of a forum for the communication of their results.

Under the guidance of Dr Bob Cummings the Group became the British Micropalaeontological Society (BMS) in 1975 during a period of rapid expansion and the development of the science, particularly in the field of hydrocarbon exploration.

The geographical development of micropalaeontology has resulted in a growth in the international membership of the Society, such that the name was changed to The Micropalaeontological Society in 2001. It is a registered charity in the UK (No. 284013). The Society publishes *The Journal of Micropalaeontology* and from 1973 to 1999 the BMS also published *A Stereo-Atlas of Ostracod Shells*.

The Society comprises six specialist groups which study Foraminifera, Microinvertebrates, Nannofossils, Ostracods, Palynology and Silicofossils. Membership of any of these specialist groups is by personal choice of individual members. The Society organizes an annual general meeting and a number of other scientific meetings and field excursions throughout the year. With the Palaeontological Association, Palaeontographical Society and The Geological Society of London, the TMS assists in the planning of the annual Lyell Meeting of The Geological Society. The Society also manages a website at http://www.nhm.ac.uk/hosted_sites/tms.htm

Published by The Geological Society from:
The Geological Society Publishing House
Unit 7, Brassmill Enterprise Centre
Brassmill Lane
Bath BA1 3JN,
UK
(*Orders*: Tel. +44 (0)1225 445046
Fax +44 (0)1225 442836)
Online bookshop: www.geolsoc.org.uk/bookshop

British Library Cataloguing in Publication Data
A catalogue record for this book is available from the British Library.

ISBN 1-86239-187-4

Typeset by Type Study, Scarborough, UK
Printed by MPG Books Ltd, Bodmin, UK

Distributors
USA
AAPG Bookstore
PO Box 979
Tulsa
OK 74101-0979
USA
Orders: Tel. +1 918 584-2555
Fax +1 918 560-2652
E-mail bookstore@aapg.org

India
Affiliated East–West Press Private Ltd
Marketing Division
G-1/16 Ansari Road, Darya Ganj
New Delhi 110 002
India
Orders: Tel. +91 11 2327-9113/2326-4180
Fax +91 11 2326-0538
E-mail affiliat@vsnl.com

Japan
Kanda Book Trading Company
Cityhouse Tama 204
Tsurumaki 1-3-10
Tama-shi, Tokyo 206-0034
Japan
Orders: Tel. +81 (0)423 57-7650
Fax +81 (0)423 57-7651
E-mail geokanda@ma.kcom.ne.jp

Contents

Recent developments in applied biostratigraphy: an introduction

A. J. POWELL[1] & J. B. RIDING[2]
[1]*Rue du Village 20, CH-1273 Le Muids, Switzerland*
(e-mail: jamespowell@bluewin.ch)
[2]*British Geological Survey, Keyworth, Nottingham NG12 5GG, UK*
(e-mail: jbri@bgs.ac.uk)

This volume in the Special Publication Series of The Micropalaeontological Society (TMS) is the result of an extremely successful joint meeting of TMS, the American Association of Stratigraphic Palynologists (AASP) and the North American Micropaleontology Section (NAMS) of SEPM that took place at University College London between 11 and 13 September 2002. The main theme of this international meeting was 'Recent Developments in Applied Biostratigraphy' and the vision was to encourage trans-Atlantic exchange of ideas, ultimately to seed new research initiatives. In particular, the aim was to develop an integrated multidisciplinary approach in both the academic and industrial realms. The editors hope that through publication of this volume, this goal will have been realized. The conveners of the meeting were:

- James Powell (Dinosystems) acting as TMS Secretary at the time;
- James Riding (British Geological Survey) acting as both TMS Treasurer and AASP President-Elect at the time;
- Chris Denison (ChevronTexaco) representing AASP;
- Tom Dignes (ExxonMobil) representing NAMS;
- Rachel Preece (ChevronTexaco) representing TMS in the USA;
- Alan Lord (UCL) acting as Local Secretary;
- Sue Mathews (UCL) providing Local Support.

Over 200 delegates registered for the meeting from 25 countries (Argentina, Austria, Australia, Belgium, Brazil, Canada, Colombia, Denmark, France, Germany, India, Ireland, Italy, Jamaica, Mexico, Netherlands, Norway, Oman, Saudi Arabia, Spain, Switzerland, Trinidad, UK, USA and Venezuela). A group photograph of the delegates was published in the AASP journal *Palynology* (vol. 27, pp. 270, 271) in 2003. Over the three days of the meeting, there were two parallel sessions comprising 78 oral presentations. Sessions were devoted to:

- quantitative methods;
- Caribbean/South America biostratigraphy;
- Sultanate of Oman biostratigraphy;
- sequence biostratigraphy;
- reservoir biostratigraphy;
- faeces facies;
- open micropalaeontology session;
- open palynology session.

There were also over 40 posters on display. The abstracts for all the oral presentations and posters given at this meeting were published in *Palynology* (vol. 27, pp. 217–268) in 2003. On the evening of 12 September a special session was arranged at The Geological Society at which two talks were delivered; John Marshall spoke on 'Arthur Raistrick: Britain's Premier Palynologist' and Martin Farley on 'Forging a Path for Biostratigraphy'. Following the meeting, Iain Prince, Bruce Tocher and Ian Jarvis led a field trip to the Isle of Wight, and Garry Jones and Brian O'Neill ran a workshop on the biostratigraphy and sequence stratigraphy of the Gulf of Mexico.

The papers in this volume have been grouped into four broad themes.

- Northwest European development (Monteil, Fleming Field; and Payne *et al.*, Mungo Field);
- outcrop analogues (Butler *et al.*, Ravenscar Group; and Jones *et al.*, Ainsa System);
- international exploration and development (Hughes, Saudi Arabia; Ilyina *et al.*, Siberia; Jaramillo *et al.*, Colombia; and Stead & Awad, Sudan);
- methodologies and techniques (Dale *et al.*, statistical modeling; Gary *et al.*, modeling biostratigraphic data; and Williams *et al.*, non-acid wellsite palynology).

From: POWELL, A. J. & RIDING, J. B. (eds) 2005. *Recent Developments in Applied Biostratigraphy.* The Micropalaeontological Society, Special Publications, 1–3. 1747-602X/$15.00

Monteil presents a non-taxonomic biostratigraphical technique ('morphostratigraphy') aimed at providing refined reservoir correlations. Morphostratigraphy is based upon the detection and recognition of morphological characters that may be correlatable; taxa are broken down into a spectrum of variable characters regardless of their generic or specific attribution. He uses as an example data from the Paleocene Maureen Formation in the Fleming Field, UK Continental Shelf (UKCS). Monteil concludes that this technique provides finer biostratigraphical resolution than using the approach of conventional bioevents. This is particularly important to reservoir compartmentalization modelling and to understanding production behaviour. Also using an example from the Paleocene succession of the UKCS, **Payne *et al.*** describe the difficulties of reservoir-scale correlation in a salt diapir flank structure within the Mungo Field. By eschewing the 'golden spike' approach, they attempt to unravel the complexity of the system by using integrated biostratigraphical evidence in parallel with heavy mineral analysis.

The Ravenscar Group of the Cleveland Basin, northeast England is a coeval outcrop analogue for the Middle Jurassic (Aalenian to Bathonian) reservoirs of the North Sea and offshore Mid Norway and provides **Butler *et al.*** material for their contribution. The integration of miospore and microplankton data with sedimentary facies has resulted in improvements in the stratigraphical resolution for the Brent Province of the East Shetland Basin and Mid Norway. **Jones *et al.*** utilize the submarine fan/channel complex of the Eocene Ainsa System, located in the Spanish Pyrenees, to contribute to a hydrocarbon reservoir analogue model. Utilizing various micropalaeontological parameters, they investigate the possibilities of differentiating sub-environments within the system.

Hughes provides two papers based on material coming from Saudi Arabia. The first discusses the Aptian Shu'aiba Formation, the main reservoir in the Shaybah Field. Because carbonate reservoirs are typically too thin for their geometries to be determined by seismic or wireline log evidence, micropalaeontological analysis and biofacies interpretation is crucial. Hughes subdivides the reservoir into a number of biofacies within four main depositional units. The Permo-Triassic Khuff Formation is the subject of the second paper by **Hughes**. Through the recognition and application of biofacies tiering, he has identified significant deepening and shoaling trends that can be used to help determine depositional cycle boundaries and maximum flooding events. This paper is of particular significance to those interested in the Permian extinction event and the Permian/Triassic boundary in general.

The Tyumenskaya superdeep well SDW-6 provides an important section of the Middle and Upper Jurassic succession of West Siberia **Ilyina *et al.*** have used samples from this Callovian to Volgian stratigraphical reference section to study a broad range of microfossil groups (including palynomorphs). They have established a detailed dinoflagellate cyst biozonation for the succession as well as establishing transgressive–regressive trends on the basis of microbenthos and microplankton.

In recent years, hydrocarbon exploration in Colombia has moved to structurally complex regions where seismic resolution is poor and palynology is the only useful micropalaeontological discipline. **Jaramillo *et al.*** describe the continental palynology of the Paleocene–Eocene succession of the Cuervos Formation of Llanos and the Llanos foothills. Testing their informal biozonation in a sidetrack to a well drilled in the Llanos foothills, they suggest that such an approach is not only reliable, but can also be successfully applied to exploration in such settings.

If Siberia and Colombia have at least some reliable biostratigraphical information available in the public domain, the same cannot be said of Sudan. It is the Muglad Basin of Sudan where **Stead & Awad** have studied the non-marine Cenozoic succession. These authors have developed a palynological biozonation for a virtually complete Cenozoic succession, which should enhance future hydrocarbon exploration in the region.

Dale *et al.* report on the application of correspondence analysis to an industrial palynological dataset based upon methods developed for ecological signals from modern dinoflagellate cysts. Applying their technique to a dataset from the Norwegian North Sea, Dale *et al.* believe that the two downhole shifts in ecological trends, that they identified in four wells, correspond to major flooding surfaces, and suggest orientation of the well sites to a paleoshoreline. However, it should be pointed out that these results are potentially highly data-set dependent. Nevertheless, their study demonstrates that, on occasion, integrating all data may not work as well as treating spatially partitioned sections of the dataset individually.

Gary *et al.* present an innovative method that uses a sample-by-sample multivariate similarity matrix of biostratigraphical data to define high-resolution correlation between two wells.

They utilize three different similarity/dissimilarity metrics to calculate the matrix for the raw data, presence–absence data and percent-range transformation. The resultant correlation path may be constrained by other information, such as maximum flooding surfaces. The authors use data from two wells from the Lower Cretaceous Blake Field of the UKCS to demonstrate their methodology.

In the final paper, **Williams *et al*.** describe a new palynological processing technique applicable to wellsite situations. The method minimizes the use of hazardous chemicals by employing a primarily mechanical extraction technique. Using examples from Mid-Norway (Upper Cretaceous), the Norwegian North Sea (Middle Jurassic to Upper Cretaceous), West of Shetland and the Forties Field, UKCS (Upper Paleocene), the authors show how the method has been applied to the horizontal biosteering of wells, and the maximization of mature field recoverable reserves. Importantly, the technique virtually eliminates health and safety risks.

The editors are grateful to the following individuals for acting as referees for the papers in this volume: Andrew Alderson (Ichron Ltd, Northwich, Cheshire), Graham Booth (Millennia Ltd, Alton, Hampshire), Paul Britton (StrataData Ltd, Ottershaw, Surrey), Marcel BouDagher-Fadel (University College, London), Paul Cornick (Petrostrat Ltd, Llandudno, Conwy), Thomas Demchuk (ConocoPhillips, Houston, Texas), Stanley Duxbury (Duxbury Stratigraphic Consultants, Aberdeen), Susanne Feist-Burkhardt (The Natural History Museum, London), John Gregory (Kronos Consultants, St Albans, Hertfordshire and The Natural History Museum, London), Nicholas Holmes (Ichron Ltd, Northwich, Cheshire), Richard Howarth (University College, London), David Jolley (University of Sheffield), Robert Wyn Jones (BP Exploration Co. Ltd, Sunbury-on-Thames, Middlesex), Simon Payne (BP Exploration Operating Co. Ltd, Aberdeen), James Powell (Le Muids, Switzerland), Roberto Rettori (Università degli Studi di Perugia, Italy), Richard Reyment (Naturhistoriska riksmuseet, Stockholm, Sweden), Keith Richards (KrA Stratigraphic, Deganwy, Conwy), James Riding (British Geological Survey, Nottingham), Michael Simmons (Neftex Petroleum Consultants Ltd, Abingdon, Oxfordshire), David Smith (Truro, Cornwall). The editors also wish to thank TMS Special Publications Editor, Professor Malcolm Hart (University of Plymouth), for his efforts in bringing this volume to fruition. J. B. Riding publishes with the permission of the Executive Director, British Geological Survey (NERC).

Morphostratigraphy: a new non-taxonomic biostratigraphical technique applied to a turbiditic deep-sea reservoir (Paleocene Maureen Formation, Fleming Field, UKCS)

E. MONTEIL

BG Group, 100 Thames Valley Park Drive, Reading RG6 1PT, UK

Current address: Geoscience Australia, GPO Box 378, Canberra, ACT 2601, Australia

(e-mail: eric.monteil@ga.gov.au)

Abstract: This paper presents the results of a study that assesses a new non-taxonomic biostratigraphical technique as a tool for refining reservoir correlations. In this example from the Paleocene Maureen Formation, Fleming Field, UKCS, morphostratigraphy provides greater biostratigraphical resolution than a more conventional biozonation or bioevent approach and significantly improves reservoir correlation. As a result, a different reservoir sand connectivity model is proposed. This model explains production anomalies experienced in the field. The study also clearly demonstrates the applicability of morphostratigraphy to refine correlations within deep marine turbidite settings.

Palynology is widely used in exploration for dating and correlation. However, since key markers may be rare or absent, it is sometimes difficult to identify sufficient bioevents to correlate wells, particularly at the reservoir scale. This is especially crucial to compartmentalization modelling and understanding production behaviour, as many reservoirs occur within a single biozone. In an attempt to solve this problem, a new technique called morphostratigraphy is tested to assess whether this method can be effectively applied. For technical (data availability and suitability) and strategic reasons (sand connectivity reservoir, volume uncertainty) a turbiditic deep-sea reservoir, the Paleocene Maureen Formation in the Fleming Field is selected for this pilot study. Specific aims of the study are threefold: to establish a morphostratigraphical zonation scheme for the Fleming Field; to test morphostratigraphy as a means of refining chronostratigraphical framework and reservoir correlations; to evaluate the potential of morphostratigraphy for improved reservoir correlation in deep marine turbidite settings.

Area of study and material

Dataset selection

Over 2000 palynological slides from three geographical areas were reviewed to determine the most suitable dataset for morphostratigraphical analysis. The Paleocene Maureen Formation in the Fleming Field was identified as the best dataset available. Even though poor palynomorph recovery seemed to be a common problem, historical slides from the Fleming Field wells generally had relatively good preservation and diversity. Reprocessing the samples from selected wells provided slides with much better assemblages in terms of both species diversity and absolute abundance.

The 16/29a-9 and 22/5a-A3 wells were selected for this study because both wells have a relatively long cored interval and historical biostratigraphical reports (data generated by Jonathan Bujak) indicated that the cored intervals were coeval. The biostratigraphical scheme of Mudge & Bujak (1996*a*, *b*) provided the necessary 'broad-brush' correlation framework. Wells without cored intervals were not included when building the morphostratigraphical scheme.

Area of study

The Fleming Field is located on the eastern flank of the Central Graben, UKCS, and is bound to the east by the Jaeren High (Fig. 1). The Maureen Formation reservoir sands onlap onto the Jaeren High. The reservoir comprises stacked Paleocene turbidite sandstones sourced from the East Shetland Platform to the north and northwest, which combined to form a low stand fan/wedge complex (Stewart 1987; Vining *et al.* 1993).

Material

Closely spaced mudstone samples from the two study wells were selected, in part based upon

From: Powell, A. J. & Riding, J. B. (eds) 2005. *Recent Developments in Applied Biostratigraphy*. The Micropalaeontological Society, Special Publications, 5–21. 1747-602X/$15.00

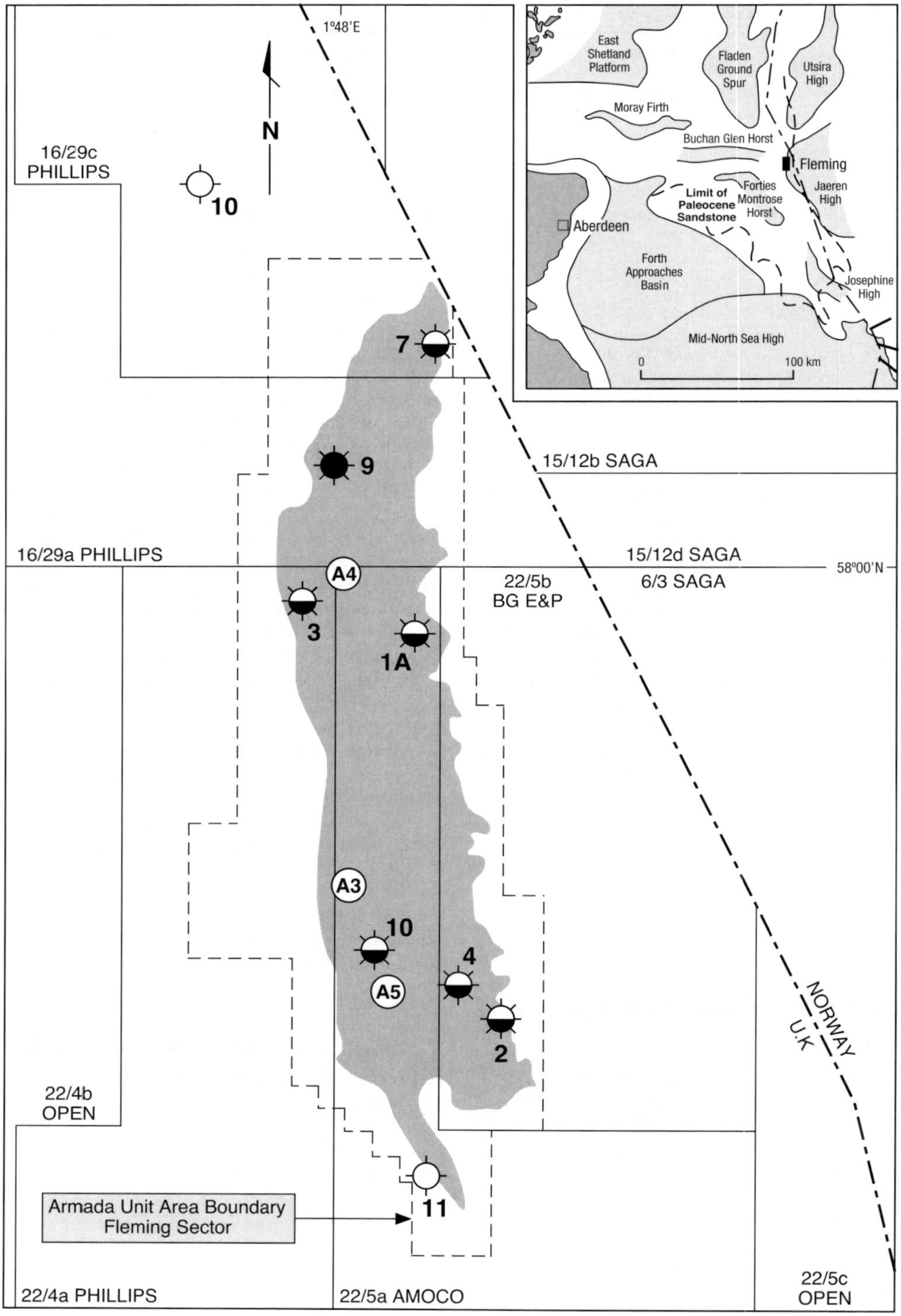

Fig. 1. Location map.

Table 1. *Depth intervals and thicknesses for the studied sections*

	Interval (ft KB)	Thickness (ft)	Thickness (m)
Well 16/29a-9	9112.33–9171.13	58.8	17.92
Well 22/5a-A3	14881.23–14914.82	33.6	10.24

detailed sedimentological logging undertaken by Brian Cullen (Reservoir Associates North Sea).

Core depths are given as measured depths in feet below kelly bushing. The sample coverage averages one sample per 1.8 feet for the 16/29a-9 well and one sample per 2 feet for the 22/5a-A3. Depth intervals and thickness of the studied sections are listed in Table 1.

Technical issues

This study illustrated the importance of three key issues in morphostratigraphical analysis: sample quality, sample density and microscope performance. As highlighted by Wrenn (1998), high-quality palynological analyses cannot be derived from poorly processed samples. This is particularly true for morphostratigraphy. Routine sample analyses from wells in the Fleming Field proved inadequate for establishing the morphozonation, but were sufficient, once the scheme was developed, to enable the morphozonation scheme to be applied to the other Fleming Field wells. In order to establish a morphostratigraphical zonation, sample density must be sufficient to capture the thinnest morphozones. Once established, zone boundaries may be identified using lower sampling density. Lastly, a good microscope equipped with an interferential contrast capability is an absolute necessity.

The concept of morphostratigraphy *sensu* Monteil (1990)

This concept was developed using dinoflagellate cysts belonging to the Ceratioid group (Monteil 1985), the *Wallodinium/Palaeotetradinium* plexus (Monteil 1989) and the genus *Amphorula* (Monteil 1990). It is based upon the recognition of morphological characters (morpho-characters) that may be directly correlatable. Morpho-characters range in complexity from simple cyst body size variations to minute changes in distal process terminations. Morphostratigraphy offers the potential for the identification and quantification (event-maximization) that have applications in various fields of geology, including biostratigraphy, ecostratigraphy, and graphic correlation. Morphostratigraphy differs from routine biostratigraphy in the degree of taxon subdivision and event classification.

A comparison between morphostratigraphy and conventional biostratigraphy is illustrated in Figure 2. In standard event-based biostratigraphy, the distribution of taxa is plotted through time (Fig. 2a). Species are defined by a unique set of morphological features, herein represented as variously numbered pins (Fig. 2b). However, many ambiguous species descriptions in the literature force an element of subjectivity into the species identification process, which may lead to misinterpretation. This problem is overcome by morphostratigraphy, which breaks down taxonomic units into a spectrum of variable morpho-characters, regardless of their taxonomic generic or specific attribution (Fig. 2b). Selected morpho-characters are plotted singly or in combination through time (Fig. 2c).

Sometimes, as in the example illustrated in this study, each individual morpho-character cannot be considered independently of its parent taxon. It can be perceived to be the process of splitting formae into smaller taxonomic units based on subtle morphological variations. As a result, in the initial event-detection phase, it is a microscopy-intensive procedure. Once morpho-events are established, however, recognition of these events can be readily carried out during routine biostratigraphical analysis.

Morphostratigraphical application in the Paleocene Maureen Formation, Fleming Field, UKCS

Morphostratigraphical analysis for this study required five successive steps:

(1) identify two candidate wells with long cored intervals previously interpreted to be coeval;
(2) recognize correlatable bioevents (FADs, LADs and acmes);
(3) differentiate random variants from systematic variant morpho-characters;

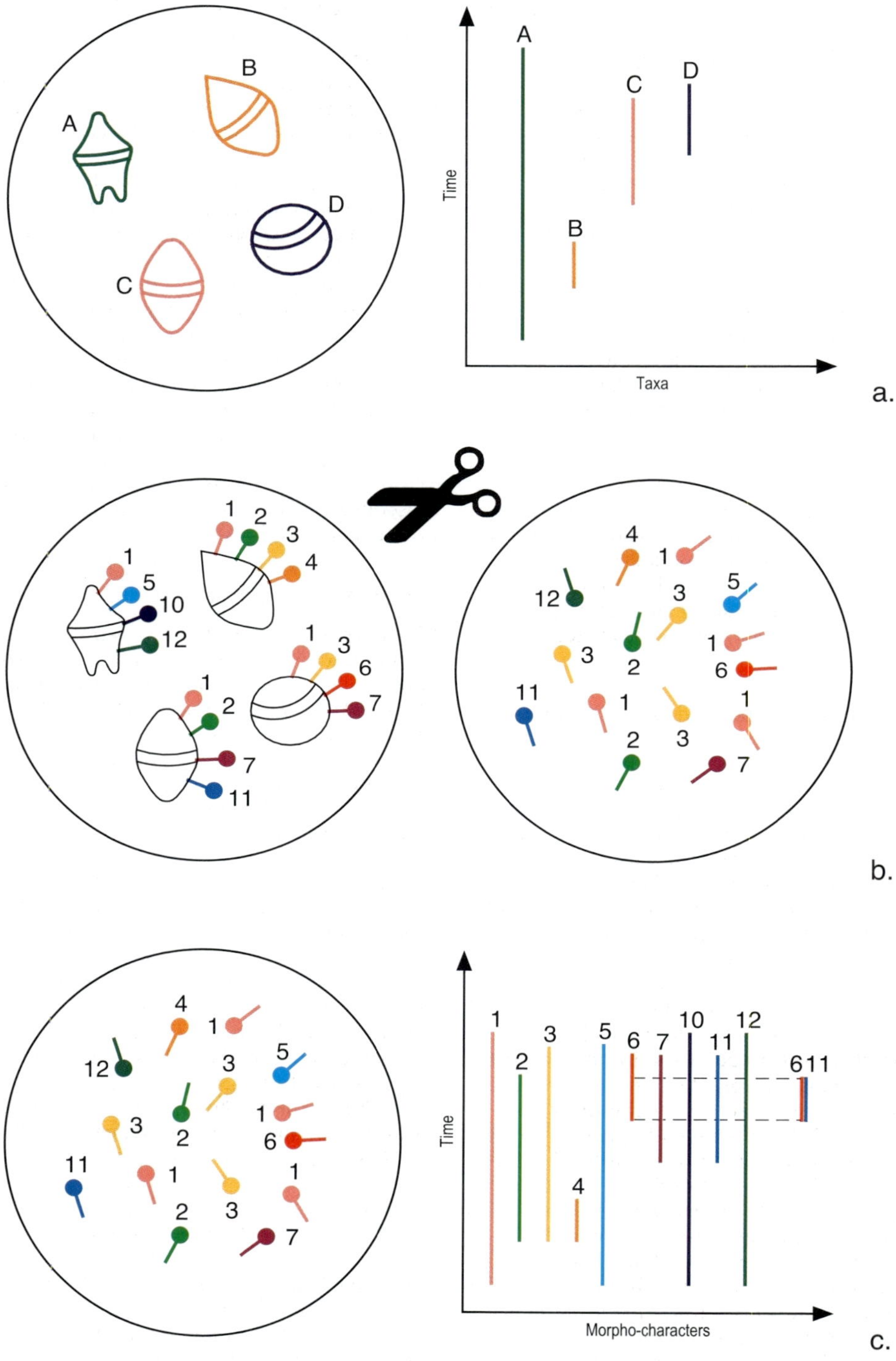

Fig. 2. Concept of morphostratigraphy (Monteil 1990). (**a**) Conventional biostratigraphy: taxa are plotted against time. (**b**) Taxonomic breakage. (**c**) Morphostratigraphy: selected morpho-characters are plotted against time.

(4) calibrate morpho-characters against bioevents;
(5) build the morphostratigraphical zonation;

Two types of features are considered in morphostratigraphy. Those assumed to be 'phyletic', such as archeopyle type or horn type, and those considered to be ecophenotypic, such as ornamentation development type, relative size of the cyst, or relationships between processes. The methodology involves a number of steps, which can be summarized as follows. After plotting a comprehensive set of morphological variants on a range chart for each well, the analyst must distinguish random variants (noise) from systematic variants (signal). Parent taxa bearing selected morpho-characters are identified through an open nomenclatural system, e.g. parent taxon A, B or C. According to their perceived correlative value, morpho-characters are assigned to one of four groups ranging from unusable, questionably usable, and potentially usable to usable. Stratigraphical ranges of usable morpho-characters are then calibrated against acme events, first appearance datums (FADs) and last appearance datums (LADs).

Selection of morphological variants in the Paleocene Maureen Formation

Dinoflagellate cyst assemblages from the Paleocene Maureen Formation *p.p.* were studied for this investigation. The usability of morphological variants was assessed and categorized from unusable to usable, as described above.

Three examples of unusable morpho-characters are mentioned here, though more exist. Dissimilar distal process terminations were noticed within the *Oligosphaeridium complex* plexus where distal terminations are largely and strongly perforated or, alternatively, minutely and weakly perforated. Variation in the length of processes within the *Oligosphaeridium complex* plexus, as well as variation in overall cyst size of *Palaeoperidinium pyrophorum* was determined to be random variants for this study.

Questionably usable morpho-variants include those which exhibit numerous morphological variations over a very short section, but do not have value in the interval studied, such as the three morphotypes identified in the *Spiniferites* group. The following variations in the connection or non-connection of the processes were observed. Parent taxon A specimens have processes with long distal extremities, which are never linked; parent taxon B have processes with long distal extremities, which are sometimes linked with distal extremities of neighbour processes; and parent taxon C have processes with long distal extremities, which are always linked, and form a reticulum with relatively narrow meshes. In other localities or over a greater section these morpho-variants may have considerable value.

Other morphotypes that have a high potential to be usable morpho-characters, but were not investigated in sufficient detail to be used in this study, have been assigned to the potentially usable group. They include: *Spiniferites 'palmatus', Spiniferites 'sweet', Achomosphaera 'ancoriformis', Impagidinium* sp. 2 (folded) etc. These few examples of parent taxon names included in this group, which are not discussed further, give some insight into the potential increase in resolution of the morphostratigraphical approach still available.

The seven usable morpho-characters employed to build the morphozonation scheme are defined and discussed below.

Morpho-character definitions

Six morpho-characters were identified in the *Spiniferites 'magnificus'*, *Spiniferites 'post-alcicornu'* and *Thalassiphora 'delicata'* plexus. A seventh was based on the acme of the algae *incertae sedis* type I. To facilitate the discussion, each individual morpho-character is presented in relation to its parent taxon. Identification criteria of morpho-characters are summarized and illustrated in Figure 3.

(1) *Spiniferites 'magnificus'* plexus. *Spiniferites 'magnificus'* is an informal species named by Robertson Research International (RRI). Though never formally described, this taxon has been extensively used in consultant reports because of its very peculiar morphology. Monteil & Williams (1998) demonstrated that 12 distinct morpho-characters of *S. 'magnificus'* might be used to zone the early Danian of the Ekofisk Formation. This species is also referred to as *Spiniferites* n. sp. by Stewart (1987) and *Spiniferites* cf. *supparus* by Powell (1992).

In the Paleocene Maureen Formation, six new morpho-characters (A, B, C, D, E) of *S. 'magnificus'* were identified. Only morpho-characters A and B were used to build the morphozonation described below. *S. 'magnificus'* A and B are both characterized by their large size (*c.* 60–70 μm) with A having short, flexuous and thin processes, and B having long, rigid and thick processes. The medium-sized (*c.* 40–45 μm) *S. 'magnificus'* C and E are differentiated by structure and distal extremities. *S. 'magnificus'*

MORPHO-CHARACTER	PARENT TAXON	DEFINITION	EXAMPLE
A	*Spiniferites 'magnificus' A*	processes short, flexuous and thin	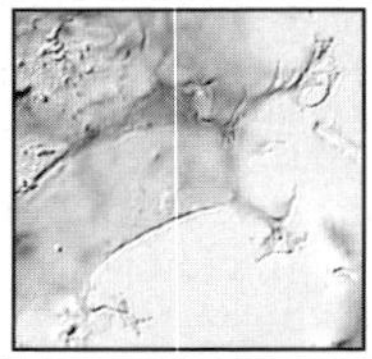
B	*Spiniferites 'magnificus' B*	processes long, rigid and thick	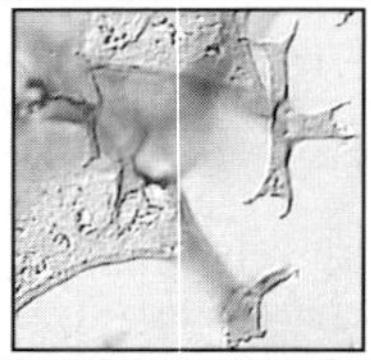
C	*Thalassiphora 'delicata' A*	single fenestrae	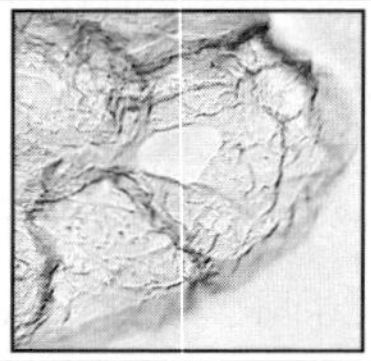
D	*Thalassiphora 'delicata' B*	multi fenestra	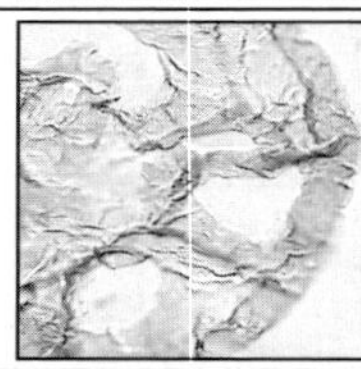
E	*Spiniferites 'post-alcicornu' A*	low septa	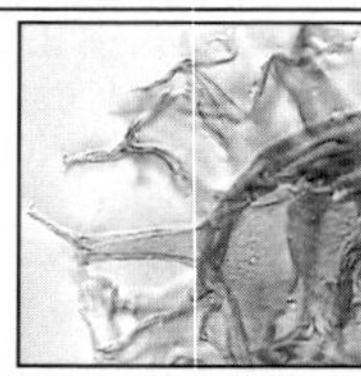
F	*Spiniferites 'post-alcicornu' B*	low septa + membrane linking processes together	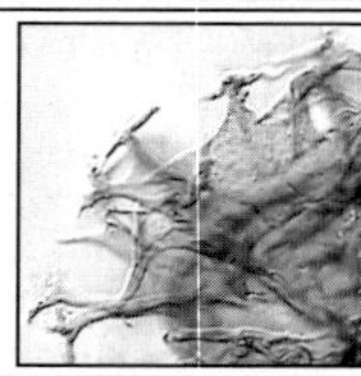
G	*Algae 'incertae sedis' type 1*	large size globular	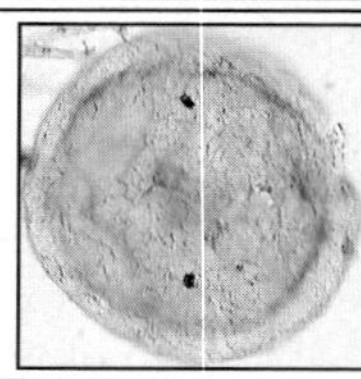

Fig. 3. Main identification criteria of the seven selected morpho-characters. All photos are at the same magnification.

C has a soft structure with short and simple distal extremities whilst E has a stiff structure with long, complex distal process extremities. *S. 'magnificus'* D is characterized by its very small size (*c.* 25 μm).

Interestingly, early Danian Ekofisk Field morpho-characters of *S. 'magnificus'* are clearly different from those encountered in the late Danian from the Fleming Field. For instance, in 22/5a-A3 at 14897.75' it was possible to identify a reworked early Danian morphotype of *S. 'magnificus'*.

(2) *Thalassiphora 'delicata'* plexus. *Thalassiphora delicata* was first described by Williams & Downie in Davey *et al.* (1966) from the early Eocene of the London Clay at Enborne Valley (Berkshire, UK). Based upon current observations, Paleocene specimens appear to be morphologically different from Eocene specimens and have been referred to as *Thalassiphora* cf. *delicata* by Stewart (1987) and Powell (1992). Like *S. 'magnificus'*, this taxon is widely used by consultants as *Thalassiphora* cf. *delicata* (e.g. RRI) or *T. 'archa'* (e.g. Ichron).

Pending a taxonomic revision of this complex, and rather than use an invalid name to differentiate these forms, the species *'delicata'* is herein maintained but used in quotations for Paleocene specimens. *Thalassiphora 'delicata'* A differs *from Thalassiphora 'delicata'* B in having one single fenestra rather than multiple fenestrae. The first stratigraphical occurrences of *T. 'delicata' A* and *T. 'delicata' B* are always marked by the appearances of small-sized forms. Rare specimens of *Thalassiphora inflata* are often associated with T *'delicata'* A but never with *T. 'delicata' B.*

(3) *Spiniferites 'post-alcicornu'* plexus. *Spiniferites 'post-alcicornu'* A and B are very similar to *Achomosphaera alcicornu sensu* Monteil & Williams 1998 from the Lower Danian of the Ekofisk Formation. *S. 'post-alcicornu'* A differs from A. *alcicornu* sensu Monteil & Williams 1998 by the presence of low parasutural septa. *S. 'post-alcicornu* 'B differs from S. *'post-alcicornu'* A in having high proximal membranes linking the processes together.

(4) Algae *incertae sedis* type I. This probable freshwater algae appears in abundance at the top of the core in 16/29a-9.

Morphostratigraphical analysis results

The morphostratigraphical scheme established using pilot wells 16/29a-9 and 22/5a-A3 is shown in Figures 4 and 5, respectively. Ranges of up to seven morphological characters were plotted against depth and calibrated to conventional bioevents, producing seven single-character morphozones. Morphozone boundaries are indicated by a horizontal line. An oblique line, as at 9126.82' in 16/29a-9 (Fig. 4), indicates that only one or two parent taxa bearing the morpho-character C or D were observed, as opposed to more than 20 in the overlying sample. This remark supports the need for quantitative data recommended in the conclusions of this study. Combined-character morphozones MR 1 to MR 7 were defined by the overlap of long-ranging single-character morphozones.

An additional zone and sub-zone were subsequently defined. Zone MR8 corresponds to an interval rich in globular *incertae sedis* algae type 1. Sub-zone MR6' corresponds to the top of zone MR6, and is marked by an acme of the species *Palaeoperidinium pyrophorum*.

Results are summarized as follows:

(1) The MR7/MR8 boundary is marked by both strong increases in *Thalassiphora 'delicata'* A and algae *incertae sedis type* 1, and the incoming of rare specimens of *Thalassiphora inflata.*
(2) Bioevent calibration against morphozones demonstrates unequivocally that the first downhole occurrences of *Cerodinium diebelii* and *Alisocysta reticulata* in well 16/29a-9 are too low compared to well 22/5a-A3 (Fig. 6). Palynofacies dissimilarity between the two wells indicates that the occurrences are clearly environmentally controlled.
(3) The two wells display similar, homogeneous patterns in the succession of the morphozones through the interval interpreted here to be coeval. In fact, the similarity of morphostratigraphical development in the two wells is striking and seems to indicate that this scheme could be extended to adjacent wells.

Morphostratigraphical correlation between wells 16/29a-9 and 22/5-A3

Comparison of this morphozonation scheme to the biozonation of Mudge & Bujak (1996*a*, *b*) is shown in Figure 6. Nine morphozones and subzones were defined in the interval of the three original biozones, resulting in a threefold increase in biostratigraphical resolution. The morphostratigraphical correlation also indicates that the two intervals are not entirely coeval as

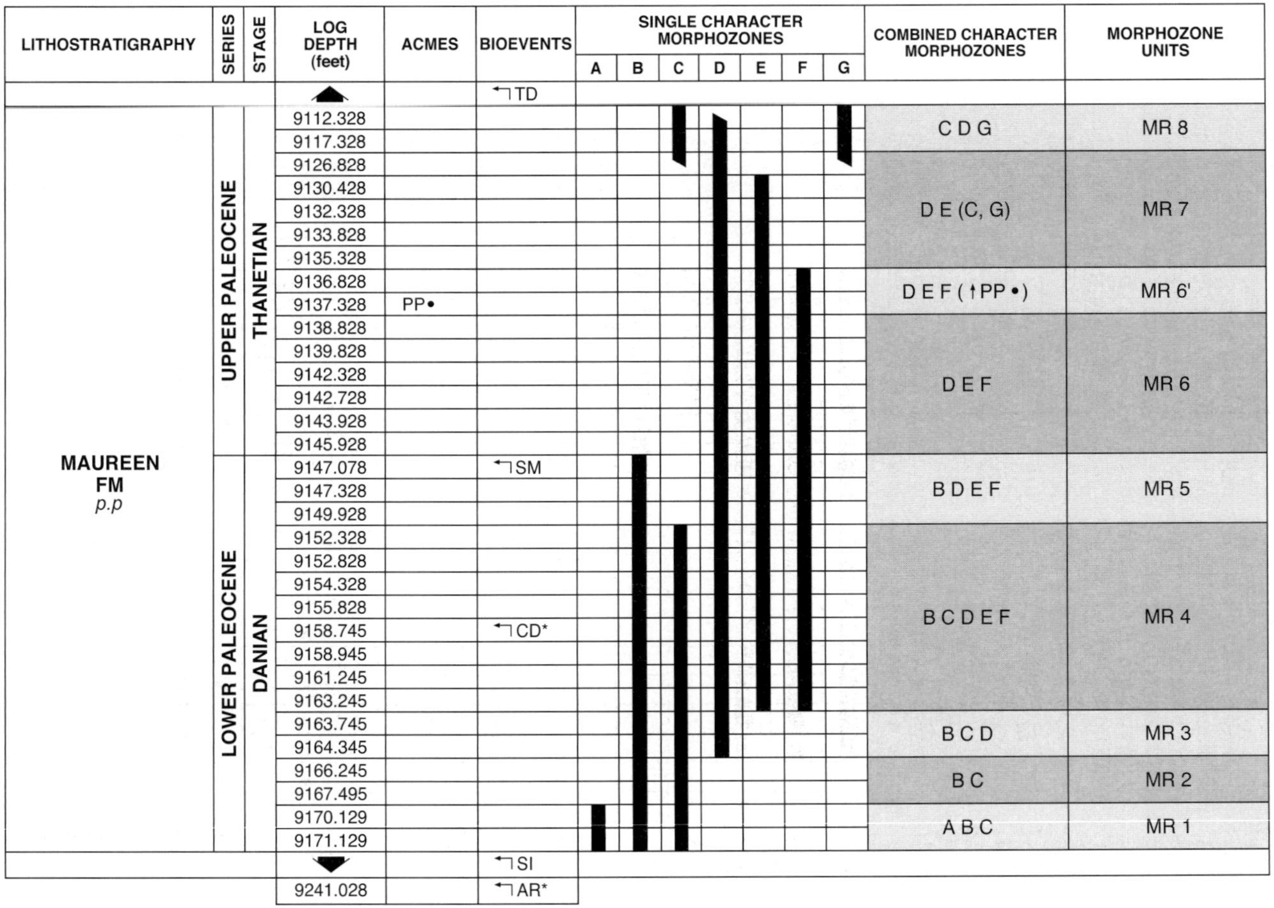

Fig. 4. Morphostratigraphical analyses of well 16/29a-9. Morpho-characters key: **A** *Spiniferites 'magnificus'* A; **B** *Spiniferites 'magnificus'* B; **C** *Thalassiphora 'delicata'* A; **D** *Thalassiphora 'delicata'* B; **E** *Spiniferites 'post-alcicornu'* A; **F** *Spiniferites 'post-alcicornu'* B; **G** Algae *incertae sedis* Type I. Bioevent key: SI, *Senoniasphaera inornata*; AR, *Alisocysta reticulata*; SM, *Spiniferites 'magnificus'*; CD, *Cerodinium diebelii*; TD, *Thalassiphora 'delicata'*. Acme key: PP, *Palaeoperidinium pyrophorum*. Note: CD* and AR* LADs are too low in this well compared to well 22/5a-A3.

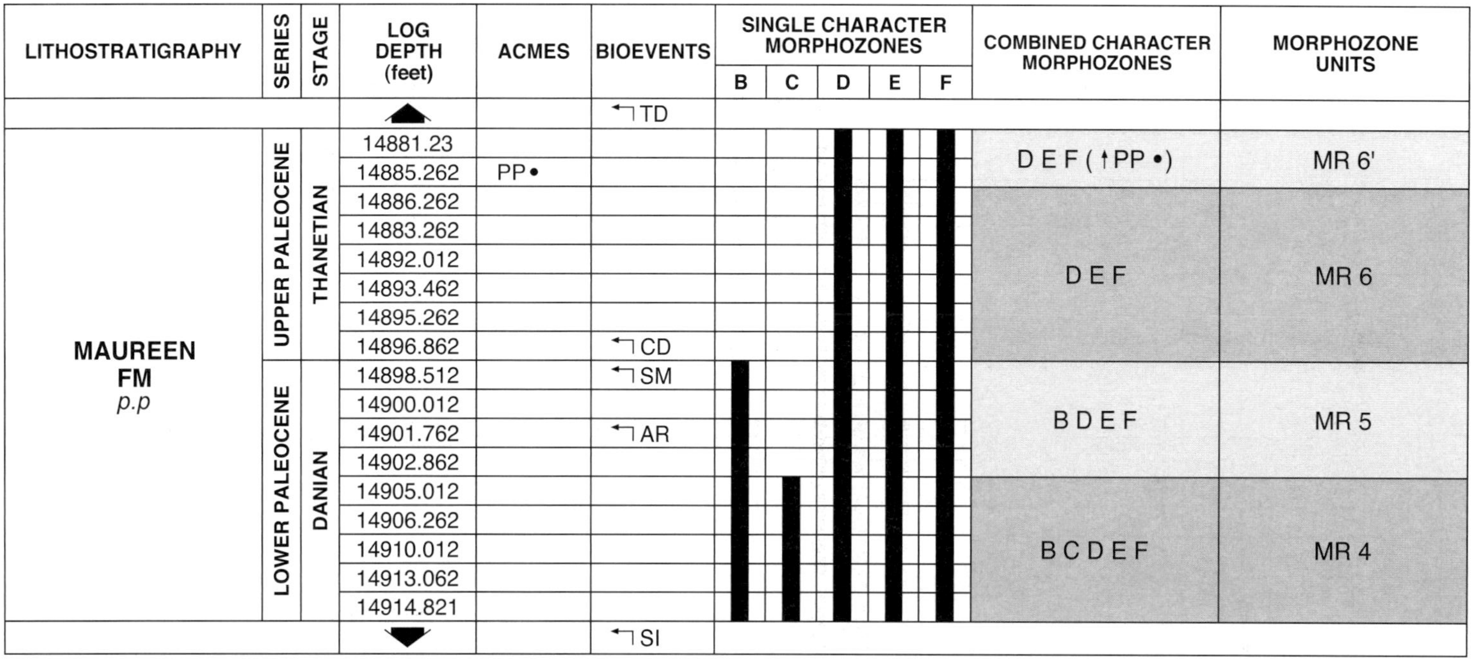

Fig. 5. Morphostratigraphical analyses of well 22/5a-A3. Morpho-characters key: **A** *Spiniferites 'magnificus'* A; **B** *Spiniferites 'magnificus'* B; **C** *Thalassiphora 'delicata'* A; **D** *Thalassiphora 'delicata'* B; **E** *Spiniferites 'post-alcicornu'* A; **F** *Spiniferites 'post-alcicornu'* B; **G** Algae *incertae sedis* Type I. Bioevent key: SI, *Senoniasphaera inornata*; AR, *Alisocysta reticulata*; SM, *Spiniferites 'magnificus'*; CD, *Cerodinium diebelii*; TD, *Thalassiphora delicata*. Acme key: PP, *Palaeoperidinium pyrophorum*.

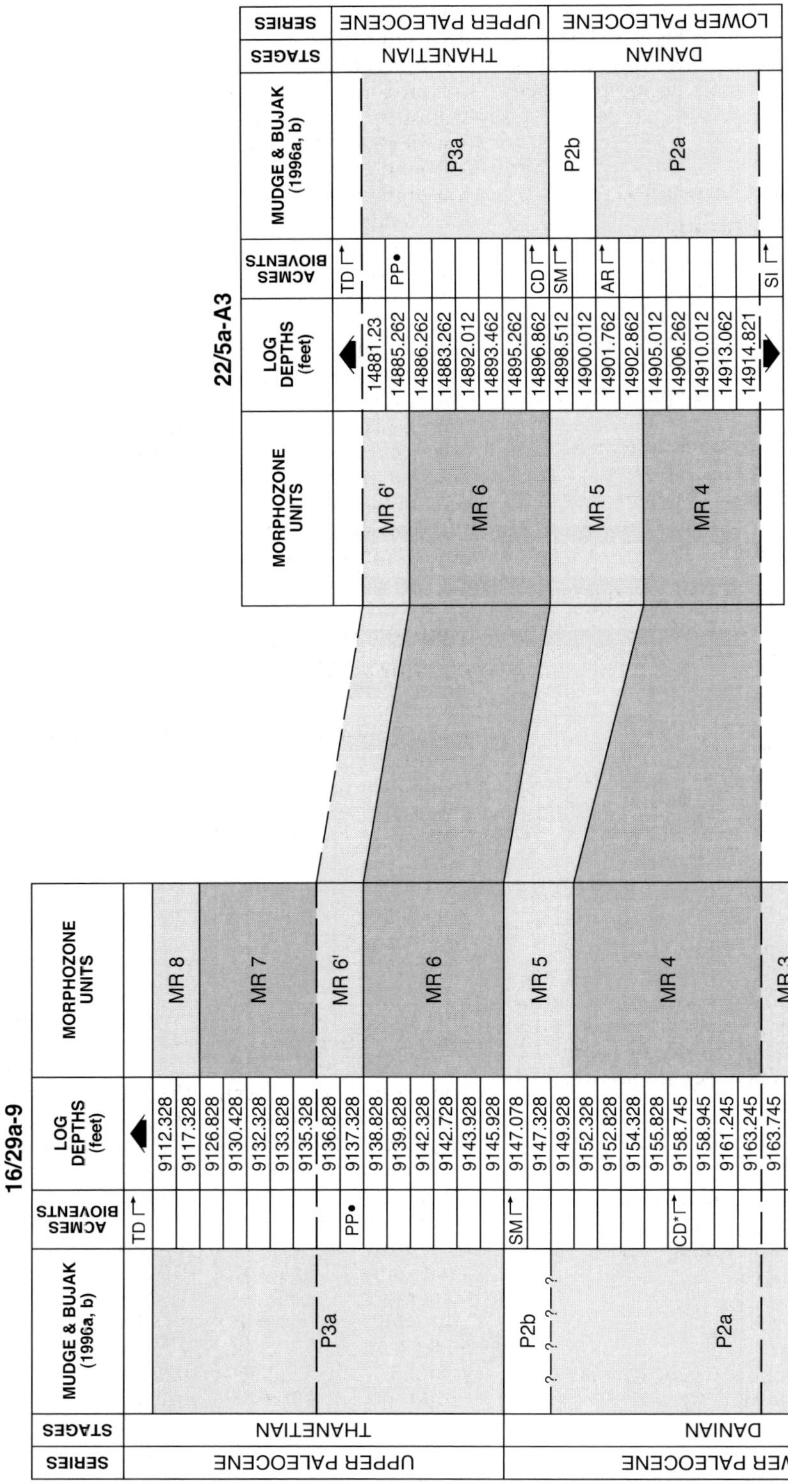

Fig. 6. Morphozonation and correlations of wells 16/29a-9 and 22/5a-A3. Comparison with the biozonation of Mudge & Bujak (1996*a*, *b*)

previously thought. Furthermore, the LDO of *Alisocysta reticulata* in well 16/29a-9 is shown to be too deep, this is important as this taxon was key to the original reservoir correlation (see below).

Morphostratigraphical correlations in relation to the gamma-ray log

Morphozonations established for the two study wells were plotted against the gamma-ray log curves (Fig. 7). Boundaries between morphozones were placed arbitrarily in the gap between samples bounding a morphozone (plain line), except where moved within sampling gap intervals (dashed lines) to a better match. Since gamma log signatures for the Maureen Formation appear to display a consistent lateral development in relation to defined morphostratigraphical zones, a gamma-ray log correlation conditioned using the morphozones is suggested.

Log characterizations of MR4, MR5, MR6 and MR6' morphozones recognized in both wells, are discussed below from bottom to top:

MR4

Morphozone MR4 seems to be associated with a retrograde signature followed by a prograding profile of the gamma-ray log. The top boundary of MR4 may be better placed in both wells at the overlying gamma-ray peak.

MR5

This morphozone appears to comprise a prograding profile followed by a retrogradational profile of the gamma-ray log. Based on the MR5 top boundary position in 22/5a-A3, the top boundary of MR5 should probably be moved to a slightly shallower depth (dashed line) in well 16/29a-9.

MR6

Morphozone MR6 seems to coincide with a complex gamma-ray log profile; a retrograde profile followed by a prograding profile with a final retrogradational unit.

MR6′

Morphozone MR6′ seems to correspond to a prograding profile followed by a retrograding profile of the gamma-ray log. In both wells, the acme of *Palaeoperidinium pyrophorum* first occurs during the gamma-ray increase trend with the acme marking the boundary between MR6 and MR6′.

The positive correlation observed between morphozones and gamma-ray log curves suggest a possible causative link that may explain changes in dinoflagellate cyst morphology, namely that they may be mediated by environmental or hydrological factors that also affect sediment composition. In modern dinoflagellates, acme events occur in cosmopolitan species, which thrive in stressed environments or upwelling areas. Similar observations have been made in early Danian carbonates of the Ekofisk Field.

If these observations are confirmed in other wells of the Fleming Field, a morphozone/gamma-ray log model could further understanding of reservoir compartmentalization. Likewise, if this model is correct it should be possible to predict the depth of the FDO of A. *reticulata and C. diebelii* in other wells, assuming similar field-wide environmental conditions.

Implications for the understanding of reservoir architecture

One of the major uncertainties in turbiditic depositional environments is lateral and vertical sandstone connectivity. During the course of this study, production data from 22/5a-A3 presented some anomalies, which put into question the understanding of the reservoir architecture, or more pertinently reservoir connectivity. Until then the reservoir compartmentalization model was based on an internal company study, which proposed a correlation of the sandstones based on two probabilistic biostratigraphical techniques: graphic correlation and ranking and scaling (RASC). This study identified three possibly regional mudstones named from shallowest to deepest, the *Alisocysta reticulata* (Ar), the *Globorotalia planocompressa* (Gpl), and the *Spiniferites 'magnificus'* (Sma) mudstones, and concluded that these mudstones may vertically compartmentalize the reservoir (Fig. 8).

In an attempt to help resolve these problems and to extract operational value from the established morphostratigraphical scheme, morphostratigraphical analyses were undertaken on eight additional wells from the Fleming Field: 16/29c-7, 22/5a-A4, 22/4a-3, 22/5a-A3, 22/5a-10, 22/5a-A5, 22/5b-2, 22/5b-4 and 22/5a-11.

Most of the additional palynological slides studied were from ditch cutting or sidewall core samples and the identification of morphozones was based solely on top occurrences. Only

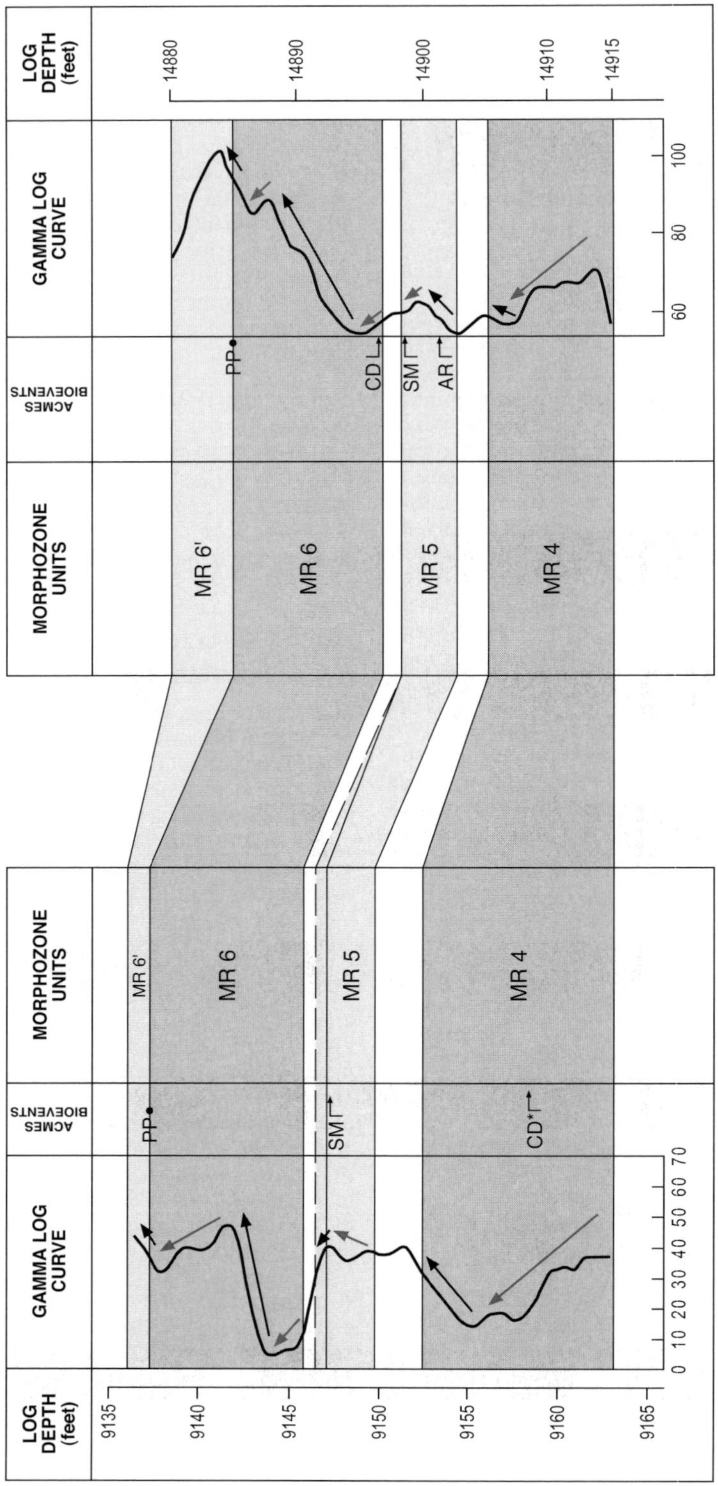

Fig. 7. Suggested gamma-ray log correlations based on morphozones. *Indicates that the observed occurrence is too low.

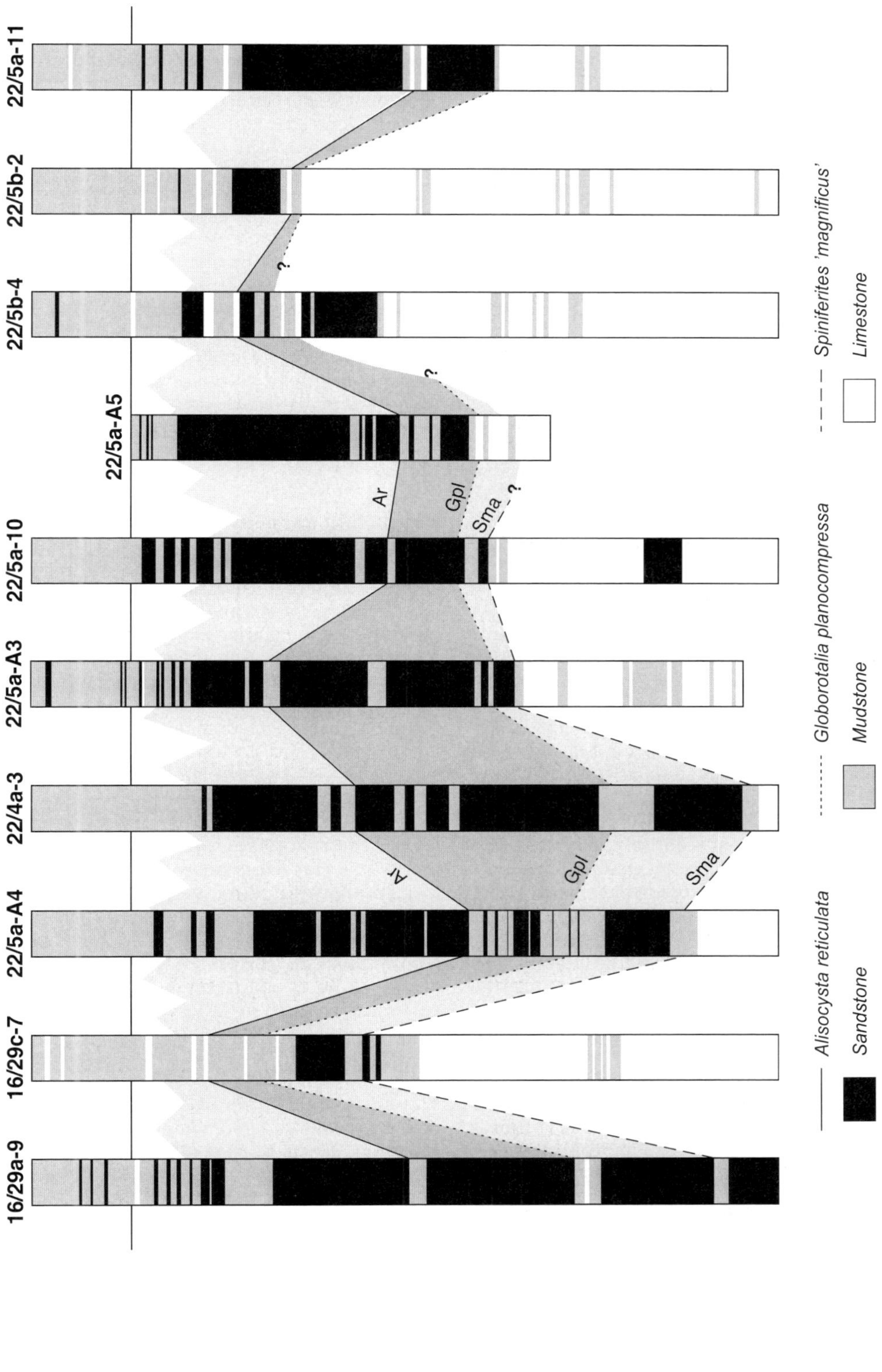

Fig. 8. Original reservoir compartmentalization in the Fleming Field (Maureen Formation, UKCS).

four tops, MR 6, MR 5, MR 4 and MR 1, representing the top of morpho-characters F (parent taxon *Spiniferites 'post-alcicornu'* B), B (parent taxon *Spiniferites 'magnificus'* B), C (parent taxon *Thalassiphora 'delicata'* A) and A (parent taxon *Spiniferites 'magnificus'* A), respectively, could be used. Differentiation between the MR6 zone and the MR 6′ subzone (marked by an acme of *Palaeoperidinium pyrophorum*) was not possible due to inadequate sample spacing.

The morphostratigraphical characterization of the mudstones suggests a different partitioning of the sandstone units. The four sandstone units identified in well 16/29a-9, where all morphozones are present, are named as follows: S4 indicates the sandstone unit above MR6 top; S3 indicates the sandstone unit between MR6 and MR5 tops; S2 indicates the sandstone unit directly below MR1 top. The sandstone unit below S2 has been logically named S1. This new sandstone correlation is illustrated in Figure 9.

Simultaneously, an in-house agglutinated foraminiferal biofacies analysis was undertaken using a fuzzy logic approach and is detailed in Wakefield *et al.* (2001). This analysis confirmed the presence of three laterally extensive mudstones, which may be seals within the Fleming Field. It also demonstrated that the Ar event is not as laterally extensive as originally thought, and that the overlying *Eoglobigerina trivialis/Pullenia americana* (Et/Pam) bioevent should been used instead to identify a regional mudstone within the field. This is in agreement with the new sandstone correlation based on morphostratigraphy, as summarized below:

(1) The dating of the mudstone named after the Et/Pam event is entirely consistent with the morphostratigraphy characterization.
(2) The dating of the mudstone named after the *Globorotalia planocompressa* (Gpl) event matches at around 70% with dating inferred from morphostratigraphy. The remaining 30% does not contradict the existing correlation. This discrepancy is probably due to the poor quality of the palynological data for wells 16/29c-7, 22/5a-10 and 22/5b-2.
(3) The dating of the mudstone named after the *Spiniferites 'magnificus'* (Sma) event is entirely consistent with morphostratigraphy results.
(4) However, it must be noted that this study identifies *Spiniferites 'magnificus'* consistently higher in the reservoir than did the original biostratigraphical analysis. This discrepancy most likely results from recognition of different morphotypes of *Spiniferites 'magnificus'* and does not change the interpretation.

It is noteworthy that similar reservoir correlations were derived by these two distinct approaches. The agreement of these disparate studies increases confidence in the proposed morphostratigraphical correlation.

Morphostratigraphy provides the potential for a more refined correlation than the agglutinated foraminiferal technique for the study interval where data is available, and may provide a better means for evaluation of mudstone continuity. It produces an ultra-high resolution zonation for reservoir modelling without recourse to additional data. This is in contrast to the three-stage method of data generation, graphic correlation and agglutinated foraminifera biofacies analysis on which was based the compartmentalization model of Wakefield *et al.* (2001).

Production anomaly resolution: an example

Morphostratigraphical analysis provides a clue to the cause of anomalous production data observed in the Fleming Field. During the maintenance program, well 22/5a-A3 was 'shut in' while production continued from adjacent wells 22/5a-A1 and 22/5a-A5. When production was resumed, the expected change in pressure was not observed. This suggested that the producing sand in 22/5a-A3 was isolated from the remaining production wells.

The reservoir correlation model inferred from morphostratigraphical dating of mudstones (Fig. 9b) shows sandstone unit S4 located above the three tops MR 6, MR5 and MR 4 in 22/5a-A3 while the sand unit ?S2 is located below MR 4 in 22/5a-A5. This demonstrates that the perforated sand in 22/5a-A3 (sand unit S4) is not in communication with the perforated interval in 22/5a-A5 (sand unit ?S2).

Conclusions

This study greatly improves the stratigraphical resolution and understanding of the Paleocene Maureen Formation reservoir architecture. Morphostratigraphy provides a resolution three times greater than that provided by a conventional biostratigraphical approach, improving our knowledge of the timing of geological events and significantly impacting on our understanding of the reservoir connectivity.

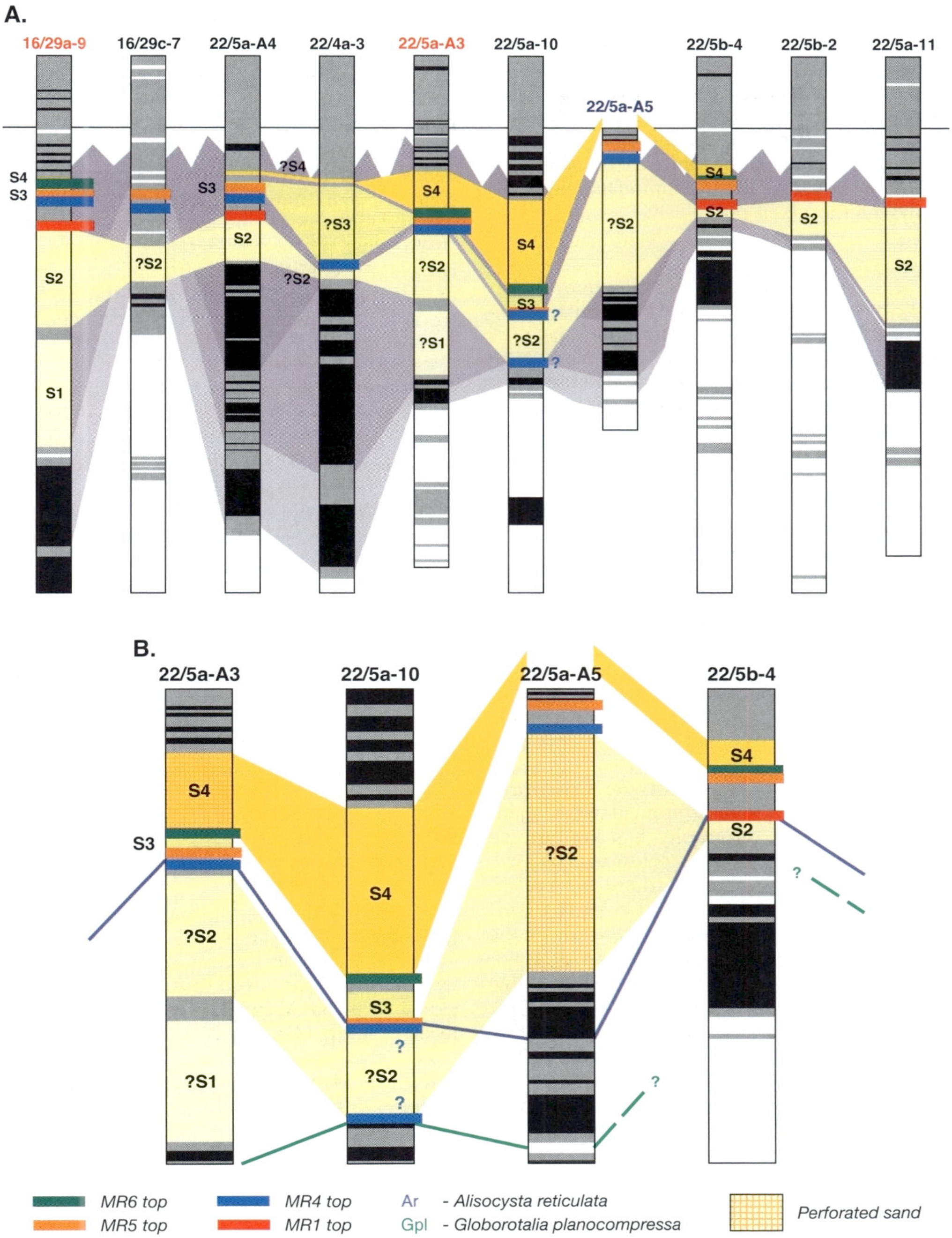

Fig. 9. **A**. Morphostratigraphical dating of mudstones and subsequent suggested reservoir compartmentalization in the Fleming Field (Maureen Formation, UKCS). Inferred identification of sands S1, S2, S3, S4, is based on 16/29a-9 well where all morphozones were defined; **B**. Production problem solving.

It confirms the presence of three laterally extensive mudstones that may act as hydrocarbon seals. In summary, morphostratigraphy produced an ultra-high resolution zonation in a single step, without recourse to additional data. This was in contrast to the three-stage method of data generation, graphic correlation and RASC, and agglutinated foraminiferal biofacies analysis undertaken previously.

Study recommendations

Integrating additional morpho-characters would refine the morphostratigraphical model described in this paper. Although only seven morpho-characters were used to establish this morphozonation scheme, approximatively ten additional morpho-characters were recognized in the samples. It therefore seems likely that integration of additional morpho-characters would increase event resolution and further improve reservoir correlation.

Calibration of the gamma-ray log curve against the defined morphozones should be further investigated. Correlation between the morphozonation scheme and gamma-ray curves in the Maureen Formation suggests that it would be possible to create an integrated morphozonation/gamma log reservoir model.

Business impact

Morphostratigraphy may constitute an interesting alternative biostratigraphical tool where conventional biostratigraphy has not proved beneficial, or where better resolution is needed. This is especially applicable where the precise placement of wells within a particular reservoir interval is important. In such circumstances, the stratigraphical resolution provided by morphostratigraphy may prove more applicable than conventional biostratigraphical approaches.

As with routine micropalaeontology and palynology, morphostratigraphy can be rapidly applied to well cuttings, sidewall cores, conventional cores, once key morpho-characters have been identified, and the morphostratigraphical scheme established.

Application of the morphostratigraphical zonation to wells from the Fleming Field indicates that morphostratigraphy may also have potential in the biosteering of wells. However, this application remains untested.

This study has clearly demonstrated the potential of morphostratigraphy for improved correlations within deep marine turbidites where sand architecture is particularly complex and may be below the resolution of seismic.

Recommendations for future applications

Future applications of morphostratigraphy could involve the regional, palaeoenvironmental, and further testing at reservoir-scale in different lithologies. In order to best apply morphostratigraphy to practical problems, work needs to be undertaken to define its limits in terms of age, material preservation, sampling density and depositional environment.

Applying morphostratigraphy to basin-scale correlation could significantly impact regional understanding and sequence stratigraphy. More work on regional applications, particularly in parts of the section where extinctions are scarce, is recommended.

There is great potential in the use of morpho-character assemblages as proxies of depositional environment. The method might be applied to equivalent outcrop and subsurface sections to evaluate the impact of lateral facies changes on morphozone assemblages and how morpho-characters vary from proximal to distal settings. Determining what factors affect morpho-character variation would contribute greatly to the use of palynomorphs for environmental reconstruction.

Until now, morphostratigraphy has only been tested in clay-rich deep marine depositional settings. The author has noticed that morphozones can be established in very thin turbidite mudstones, and believes the method could be useful in sand-rich systems as well. It should be applied to other deep marine settings and tested in carbonate reservoirs. Monteil & Williams (unpublished) have demonstrated that in the Ekofisk Field, where it is often difficult to follow the path of seismic marker, morphostratigraphical resolution is three times better than correlations based on nannofossils.

For whatever application, the further development of morphostratigraphy should involve the use of statistical methods to increase the objectivity of the interpretation.

The author would like to thank BG Group and other members of the Armada Group (BP, Centrica, Eni, ConocoPhillips, and Total) for permission to publish this manuscript. Special thanks are also extended to S. Burley, R. Garden, and G. Oakes (BG-group) for their support of the study. I would like to thank P. Highton, A. Racey (BG-group) and R. Williams (Norwegian Petroleum Directorate) for constructive comments of an earlier draft of the manuscript. M. Wakefield (BG-group) is warmly thanked for invaluable assistance and improvements to the manuscript. Grateful acknowledgement is also due for the critical and thorough review by I. Rex (Geoscience Australia). P. Cornick (Petrostrat Ltd) and an unknown referee are thanked for their constructive criticisms and suggestions. Figures and plates were produced with the help of L. Emerton and S. Mezzomo.

References

DAVEY, R.J., DOWNIE, C., SARJEANT, W.A.S. & WILLIAMS, G.L. 1966. Studies on Mesozoic and Cainozoic dinoflagellate cysts. *Bulletin British Museum (Natural History) Geology*, Supplement **3**, 1–243.

MONTEIL, E. 1985. *Les dinokystes du Valanginien du Sud-Est (Ardèche; France)*. Unpublished PhD thesis, Université Pierre et Marie Curie, Paris.

MONTEIL, E. 1989. The genus *Palaeotetradinium* Deflandre 1936 emend.: Systematics, phylogeny, morphostratigraphy from Late Jurassic to Early Tertiary. *Program and Abstracts, 4th International Conference on Modern and fossil Dinoflagellates, April 16–22, Woods Hole, United-States*, 80.

MONTEIL, E. 1990. Revision and emendation of dinocyst genus *Amphorula* Dodekova 1969. The concept of morphostratigraphy. *Bulletin Centres Recherche Exploration–Production Elf-Aquitaine*, **14**, 2, 597–609.

MONTEIL, E. & WILLIAMS, R.W. 1998. Detailed morphology of the *Spiniferites* complex in the Danian of Mid-Norway. *Norges Teknisk-naturvitenskapelige Universitet, Vitenskapsmuseet, Rapport Botanisk, Serie* 1998–1, Abstracts from the Sixth International Conference on Modern and Fossil Dinoflagellates, 'Dino6', Trondheim, June 1998, 112–113.

MUDGE, D.C. & BUJAK, J.P. 1996*a*. An integrated stratigraphy for the Paleocene and Eocene of the North Sea. *In*: KNOX, R.W. O'B., CORFIELD, R. M. & DUNAY, R.E. (eds) *Correlation of the Early Paleogene in Northwest Europe*. Geological Society, London, Special Publications, **101**, 91–113.

MUDGE, D.C. & BUJAK, J.P. 1996*b*. Paleocene biostratigraphy and sequence stratigraphy of the UK Central North Sea. *Marine and Petroleum Geology*, **13**, 295–312.

POWELL, A.J. 1992. Dinoflagellate cysts of the Tertiary system. *In*: POWELL, A.J. (ed.) *A Stratigraphic Index of Dinoflagellate Cysts*. British Micropalaeontological Society Publication series, 155–251.

STEWART, I.J. 1987. A revised stratigraphic interpretation of the early Paleogene of the Central North sea. *In*: BROOKS, J. & GLENNIE, K. (eds) *Petroleum Geology of North West Europe*. Graham & Trotman, London, 557–576.

VINING, B.A., IOANNIDES, N.S. & PICKERING, K.T. 1993. Stratigraphic relationships of some Tertiary lowstand depositional systems in the Central North Sea. *In*: PARKER, J.R. (ed.) *Petroleum Geology of Northwest Europe: Proceedings of the 4th Conference*. Geological Society, London, 17–30.

WAKEFIELD, M.I., COOK, R.J., JACKSON, H. & THOMPSON, P. 2001. Interpreting biostratigraphical data using fuzzy logic: the identification of regional mudstones within the Fleming Field, UK North Sea. *Journal of Petroleum Geology*, **24**, 417–440.

WRENN, J.H. 1998. The importance of palynologic sample processing to the oil industry. *In*: BRYANT, V.M. & WRENN, J.H. (eds) *New Developments in Palynomorph Sampling, Extraction, and Analysis*. American Association Stratigraphic Palynologist Contributions Series, **33**, 7–2 1.

Mungo Field UK North Sea 22/20, 23/16a: stratigraphy, salt diapirs and reservoir development (or 'The Riddle of the Sands')

S. N. J. PAYNE[1], P. A. CORNICK[2], L. F. DRAPER[1], H. NICHOLSON[3], A. C. MORTON[4], P. HUGGINS[1] & R. ANDERTON[1*]

[1]*BP Exploration Operating Company Ltd, Burnside Road, Farburn Industrial Estate, Dyce, Aberdeen AB21 7PB, UK (e-mail: paynesnj@bp.com)*

[2]*Petrostrat Ltd, Tyldesley House, 48 Clarence Road, Craig-y-Don, Llandudno LL30 1TW, UK*

[3]*BP Norge AS, Godesetdalen 8, 4065 Stavanger, Norway*

[4]*HM Research Associates Ltd, 100 Main Street, Woodhouse Eaves, Leicestershire LE12 8RZ, UK*

[*]*Current address: RA Geo Consulting Ltd, Home Knowe, Trossachs Road, Aberfoyle, Stirling FK8 3SX, UK*

Abstract: Mungo is an oil and gas field located within the Eastern Trough of the UK central North Sea. It comprises a salt diapir flank structure reservoired within Paleocene turbidite sandstones. Biostratigraphical data from early wells indicated extreme geological complexity; reworking, caving and injection were all invoked. Using the principle that the dinocyst genus *Apectodinium* did not migrate into the North Sea until Unit S1a of the Forties Sandstone Member, interpretation centred on the last downhole occurrence (LDO) of this genus. Once established, this 'golden spike' became the key to interpreting well stratigraphy. The complexity thereby imposed fitted notional models of a 'Forties mélange' reservoir, with older sediments slumped off the growing diapir during the time of Forties deposition. Inter-well correlation was rendered doubtful and the role of biostratigraphy downgraded. Continued development drilling induced a further attempt to unravel this complexity. Emphasis was changed, downplaying the *Apectodinium* driven model, seeking instead any evidence of an *in situ* stratigraphy. Re-examination of biostratigraphical and core material, combined with heavy mineral analysis, has recognized a full 'normal' Paleocene succession in which five layers can be correlated. The prime reservoir is now assigned to the Maureen Formation, albeit with significant volumes of the Forties Sandstone Member structurally emplaced, presumably by injection, entraining fossiliferous Forties muds. Possible mechanisms for injection are discussed.

The Mungo hydrocarbon accumulation is focused around a salt diapir that lies approximately 250 km east of Aberdeen in 88 m of water depth in the Eastern Trough of the North Sea Central Graben (Fig. 1). It straddles the boundary between the BP-operated blocks 22/20 and 23/16a, which together form the P059 licence. Mungo is located in a prolific hydrocarbon province sourced from the Upper Jurassic Kimmeridge Clay Formation. The accumulation is reservoired predominantly in Paleocene turbidite sandstones with some potential in the Cretaceous chalks (as yet not fully appraised). The reservoir is sealed both by overlying Eocene mudstones and the updip pinch-out of sands against the Permian Zechstein salt stock (Fig. 2).

The Mungo discovery well, 23/16a-3, was drilled high up on structure in 1989. The well tested both gas and condensate, predominantly from Cretaceous chalk and diapiric caprock; Paleocene sandstones encountered in this well were very thin. Subsequent seismic interpretation and appraisal drilling demonstrated considerable potential in the Paleocene sandstones downdip of the discovery well on the flanks of the diapir. These sandstones are now the principal reservoir horizon. Seismic imaging and interpretation on Mungo is challenging due to a combination of steep dips (45–65 degrees), complex faulting and a poorly defined salt/reservoir interface. Currently no reliable intra-reservoir horizons are seismically mappable between the top of the Sele Formation and the top of the Ekofisk chalk. The Mungo diapir has a total structural relief at top reservoir level of over 1600 m, with an oil column of almost 1000 m and a 300 m gas leg.

From: POWELL, A. J. & RIDING, J. B. (eds) 2005. *Recent Developments in Applied Biostratigraphy*. The Micropalaeontological Society, Special Publications, 23–42. 1747-602X/$15.00

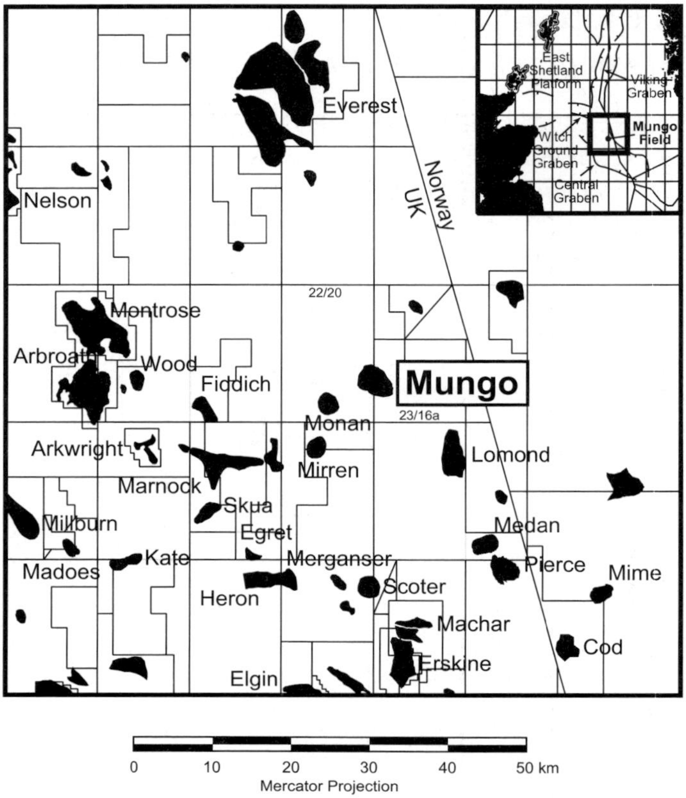

Fig. 1. Location of Mungo Field, blocks 22/20 and 23/16a, central North Sea, with adjacent hydrocarbon accumulations. Inset map shows regional context.

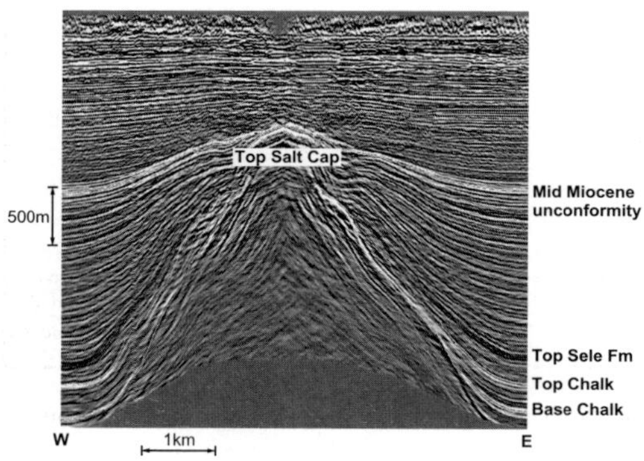

Fig. 2. Seismic section west to east across Mungo diapir. Vertical exaggeration × 2.

The salt stock is approximately 1.4 by 0.8 km wide and the reservoir covers an area some 6 km in diameter. Recoverable reserves are estimated at 155 mb of oil and 79 BCF of gas. Production commenced in 1998, reached plateau at 60 000 BOPD, and is currently on decline at 55 000 BOPD and 50 MMSCF of gas.

North Sea Paleocene fan stratigraphy

Towards the end of the Early Paleocene, the development of the North Atlantic thermal plume led to the uplift of northern Britain, resulting in the influx of large volumes of clastic sediment into the North Sea basin, primarily through the Moray Firth. Pelagic carbonate sedimentation of the Ekofisk Formation was abruptly terminated and superseded by deep-water clastics. In three main phases, turbidity currents deposited the Montrose Group Maureen and Lista, and the Moray Group Forties, submarine fans.

Discriminating between these three Paleocene sand-prone packages in the central North Sea is readily accomplished. Through this

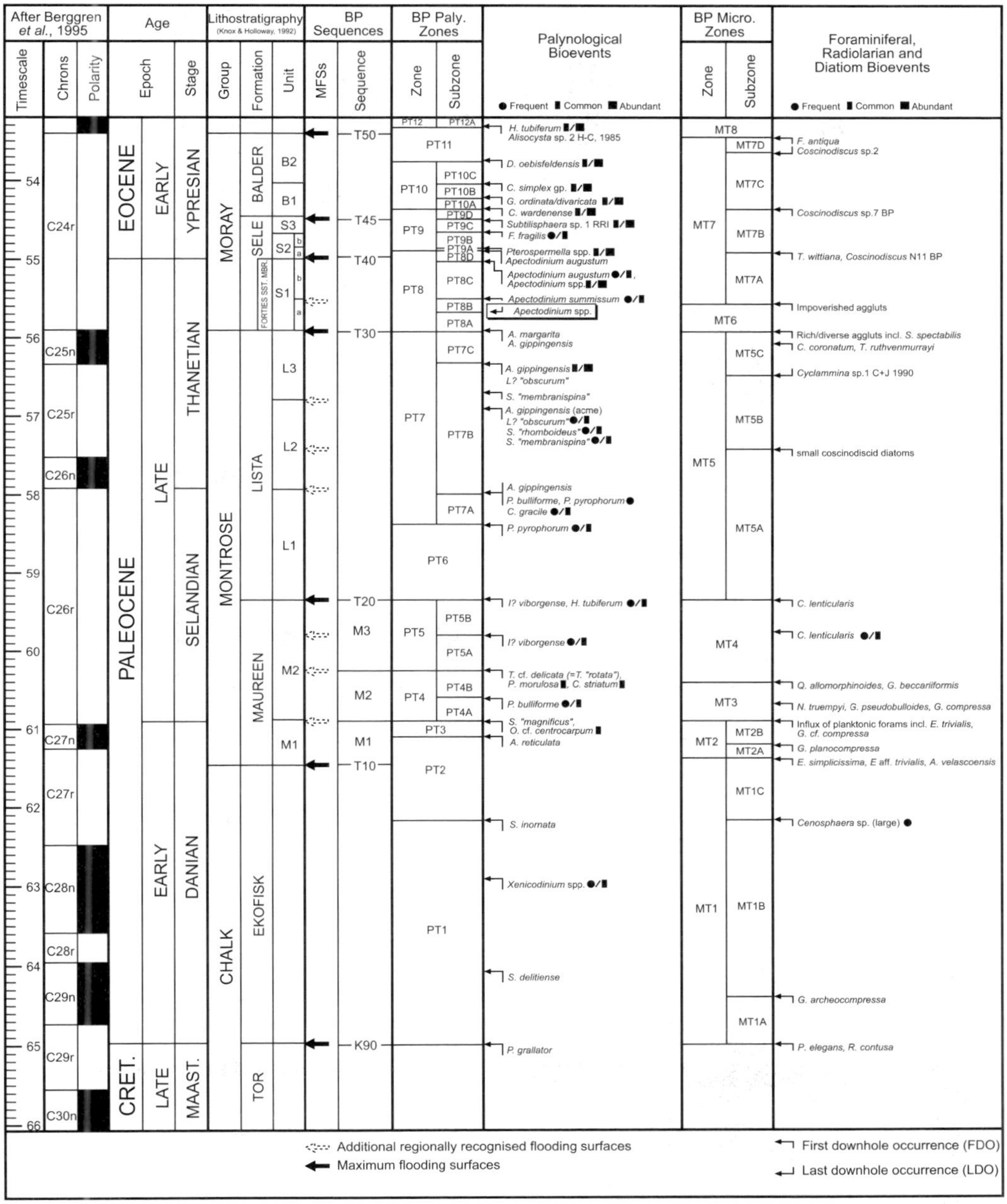

Fig. 3. Central North Sea stratigraphic framework with reference to Mungo Field with definition of BP biozones. Note the 'golden spike', the LDO of *Apectodinium* spp. within Unit S1a of the Forties Sandstone Member.

interval, biostratigraphy has a peerless track record for stratigraphical subdivision, correlation and palaeoenvironmental determination. Differentiation is based primarily on the biotas of the mudstones that envelop the sandstone packages, in addition to log character and lithology (Fig. 3). Boundaries between conformable units are characteristically seen as gamma ray maxima and microfossil abundance and diversity peaks, representative of flooding surfaces. Mungo lies within the mid- to outer fan portions of the Maureen, Lista and Forties systems, 200 to 250 km from the contemporary delta slopes and 100 to 150 km from the downfan limit of sand deposition (Reynolds 1994).

The oldest sand-prone package in the Paleocene of the North Sea, the Maureen Formation, is a heterolithic unit predominantly composed of gravity-flow sandstone deposits interbedded with finer clastics, and both autochthonous basinal and allochthonous carbonates (Ahmadi *et al.* 2003). The geometry of the Maureen fans was controlled by the pre-existing basinal topography that in turn reflects the Jurassic structural grain (Anderton 1997). The Maureen Formation is the most sand-prone depositional system in the Paleocene, with a 41% net sand-grain volume (Liu 1995). Mudstones within the Maureen Formation are typically grey or greenish-grey and bioturbated; marls are also commonly present. The Maureen Formation is regionally subdivided into units M1 and M2 (Knox & Holloway 1992). A further subdivision can be readily made on the basis of biostratigraphy and wireline logs, and this is used on Mungo and herein to subdivide unit M2 into M2 and M3.

Unit M1 is the oldest clastic package in the Paleocene and directly overlies Ekofisk Formation chalks and marls. This interval is characterized by planktonic and calcareous benthic foraminifera (BP biozone MT2), the prevalence of these groups becoming markedly reduced at the top of M1, a response to decreasing oxygenation. Of the former, the first downhole occurrences (FDOs) or extinctions of *Eoglobigerina trivialis*, *Globorotalia* cf. *compressa* and *G. planocompressa* are diagnostic. Palynologically, the gamma maximum at the top of M1 is defined by the FDO of dinocysts *Alisocysta reticulata* (BP biozone PT2) and *Spiniferites 'magnificus'* (biozone PT3), and is closely associated with a downhole increase in *Operculodinium* cf. *centrocarpum* (large, robust forms).

The top of Unit M2 is defined by the FDO of the dinocyst *Thalissiphora* cf. *delicata* (biozone PT4) and the FDOs of calcareous benthic foraminifera including *Quadrimorphina allomorphinoides* and *Gavelinella beccariiformis* (biozone MT3). Within this unit the last vestiges of a planktonic foraminiferal fauna, typically comprising *Globorotalia pseudobulloides* and *G. compressa,* disappear, suggesting increasing water-mass restriction and basin isolation. Additional palynological constraint for this unit is provided by downhole numerical increases in the dinocysts *Paleocystodinium bulliforme*, *Cerodinium striatum* and the colonial alga *Palambages morulosa*.

The top of the youngest, informal, Maureen unit, M3, is defined by the FDO of the dinocyst *Isabelladinium*? *viborgense* (biozone PT5) and the FDO, then acme, of the radiolarian *Cenodiscus lenticularis* (biozone MT4). A downhole numerical increase in the dinocyst *Hystrichosphaeridium tubiferum* is also seen at this level.

The succeeding Lista Formation contains the thickest and most extensive fan system in the North Sea Tertiary and probably corresponds to the period of maximum elevation of the thermally uplifted source areas (Anderton 1997). It has an estimated net sand-grain volume of 35%, making it the second most sand-prone system in the Paleocene after the Maureen Formation (Liu 1995). Sediments comprise non-calcareous, blocky mudstones, characteristically greenish-grey in colour, with subordinate sandstones. Rich, diverse and complex agglutinated foraminiferal faunas are recovered from the intensely bioturbated mudstones, dominated by *Zoophycos*, indicative of viable bottom-water oxygenation levels. Knox & Holloway (1992) established a threefold regional subdivision of the Lista Formation into L1, L2 and L3. The lower unit, L1, is defined by the FDO of rare *Paleocystodinium bulliforme* (biozone PT7a), common *Cordosphaeridium gracile* and the FDO of common *Paleoperidinium pyrophorum* (biozone PT6). Influxes of coscinodiscid diatoms (biozone MT5a) approximate to Unit L1.

The top of unit L2 is defined by the acme of *Areoligera gippingensis* (biozone PT7b), together with rare to common occurrences of *Spiniferites 'rhomboideus'*, *S. 'membranispina'* and *Leptodinium? 'obscurum'*. The upper unit, L3, is readily discerned by the FDO of the dinocysts *Alisocysta margarita* and *Areoligera gippingensis* (biozone PT7c), together with an agglutinated foraminiferal association of high diversity, abundance and, in part, complexity including *Spiroplectammina spectabilis* (biozone MT5).

The youngest sand-prone package in the North Sea Paleocene, the Forties Sandstone Member of the Sele Formation, is present across much of the Central Graben and Outer Moray Firth, and differs markedly from the Lista Formation, both biostratigraphically and lithologically. Sele mudstones are typically medium to dark grey, carbonaceous, pyritic, fissile and commonly laminated. The lower S1a unit is characterized by a low diversity assemblage of simple agglutinated foraminifera (biozone MT6) reflecting deposition in a dysaerobic benthic environment. Palynofloras are dominated by the dinocysts *Areoligera* spp. (excluding *A. gippingensis*), *Spiniferites ramosus* group, *Impletosphaeridium* spp.

together with reworked contemporaneous spores including *Cicatricosisporites* spp., *Deltoidospora* spp. and *Camerozonosporites* spp., and superabundant bisaccate pollen. In the upper part of Unit S1a the dinocyst genus *Apectodinium* appeared in the North Sea basin (seen as a last downhole occurrence) and rapidly increases in diversity and abundance.

The bulk of the Forties Sandstone Member, the S1b unit, is characterized by a distinctive palynofacies with a proliferation of various species of *Apectodinium* (biozones PT8d–c) and amorphous organic matter. A downhole increase in numbers of *Apectodinium* spp., including common *A. augustum*, is a reliable marker for the top of the Forties Sandstone Member. The S1b unit is further characterized by several species of pollen which are rarely seen within the underlying S1a unit, these including *Interpollis supplingensis*, *Labrapollis labraferoides*, and, in the upper part of Unit S1b, a major influx of *Caryapollenites veripites*. A low-diversity diatom assemblage (biozone MT7a) is recorded in this unit. No benthos is recorded in S1b, an indication of palaeoenvironmental deterioration to anaerobia.

The appearance of members of the dinocyst genus *Apectodinium* in the lower part of the Sele Formation has proven a robust, pragmatic and distinctive tool for defining the Forties Sandstone Member in the North Sea. Powell (1988) refers to the Base *Apectodinium* Datum and this last downhole occurrence (LDO) is widely used as a definitive 'golden spike' for the recognition of the lower part of the Forties Sandstone Member, within Unit S1a. This datum is known to be diachronous across Europe and North Africa. Brinkhuis *et al.* (1994) suggests that the genus migrated northwards during the Late Paleocene, possibly indicating climatic warming. However, in the North Sea basin, the use of this datum to differentiate the Forties Sandstone Member from the Lista Formation, reinforced by the co-occurrence of dark grey, laminated mudstones, is considered to be highly reliable. To place samples yielding *Apectodinium* spp. in a formation older than the Sele Formation requires special pleading.

Well 23/16a-4: 'the well from hell' and the 'Forties mélange'

In 1990 well 23/16a-4 was drilled into the prognosed area of thickest Paleocene clastics on the northeast flank of the Mungo diapir. This well found well-developed Paleocene turbidite sands that flowed oil at up to 4633 BOPD. Both clastics and the underlying chalk were cored although the latter contained no shows.

Biostratigraphical analysis of well 23/16a-4 revealed an exceptionally complex, seemingly chaotic, geological and stratigraphical succession. A profound mixing of taxa from the Paleocene clastic packages, particularly the Forties Sandstone Member and the Maureen Formation, was evident; indeed discerning the boundaries between the Forties Sandstone Member, and the Lista and the Maureen Formations proved exceptionally difficult both on microfossil and lithological criteria (Fig. 4).

The model that seemed most apposite to the biostratigraphical data from 23/16a-4, given the diapir flank scenario, was one of conventional, if extreme, reworking. The pervasive presence of mudstones equivalent to the Forties Sandstone Member was identified by the presence of *Apectodinium* spp. (with the application of the Base *Apectodinium* 'golden spike' rule) and grey-black, laminated mudstones; the deepest credible recovery of members of this genus (in core) was regarded as approximating to the base of the Forties Sandstone Member. It was thereby considered that the Forties Member, the youngest unit identified from the microfossils, represented *in situ* sedimentation. Older reworked material came predominantly from the Maureen Formation, a common feature of the Forties Sandstone Member regionally. Minor reworking of Lista, Ekofisk and Tor formations were also evident. The emplacement of older sediments into younger was considered to have occurred largely by slumping, sliding and reworking off the diapir itself in Forties time. Thereby arose the model of a 'Forties mélange'.

The thorough level of mixing was, however, a cause for concern and a range of possible explanatory mechanisms, or combinations thereof, were invoked. All could be accommodated within a broader mélange model. These possibilities were:

- geological reworking, mixing older taxa into younger horizons – the 'Forties mélange';
- injection or intrusion of older sediments into younger;
- drilling artefact of downhole contamination through caving in cuttings samples;
- artefact of sample contamination including mud-cake contamination (particularly relevant to sidewall core samples) or mislabelling at rigsite or in the laboratory.

A note of disquiet was sounded by the fact that some intervals within the 'Forties mélange'

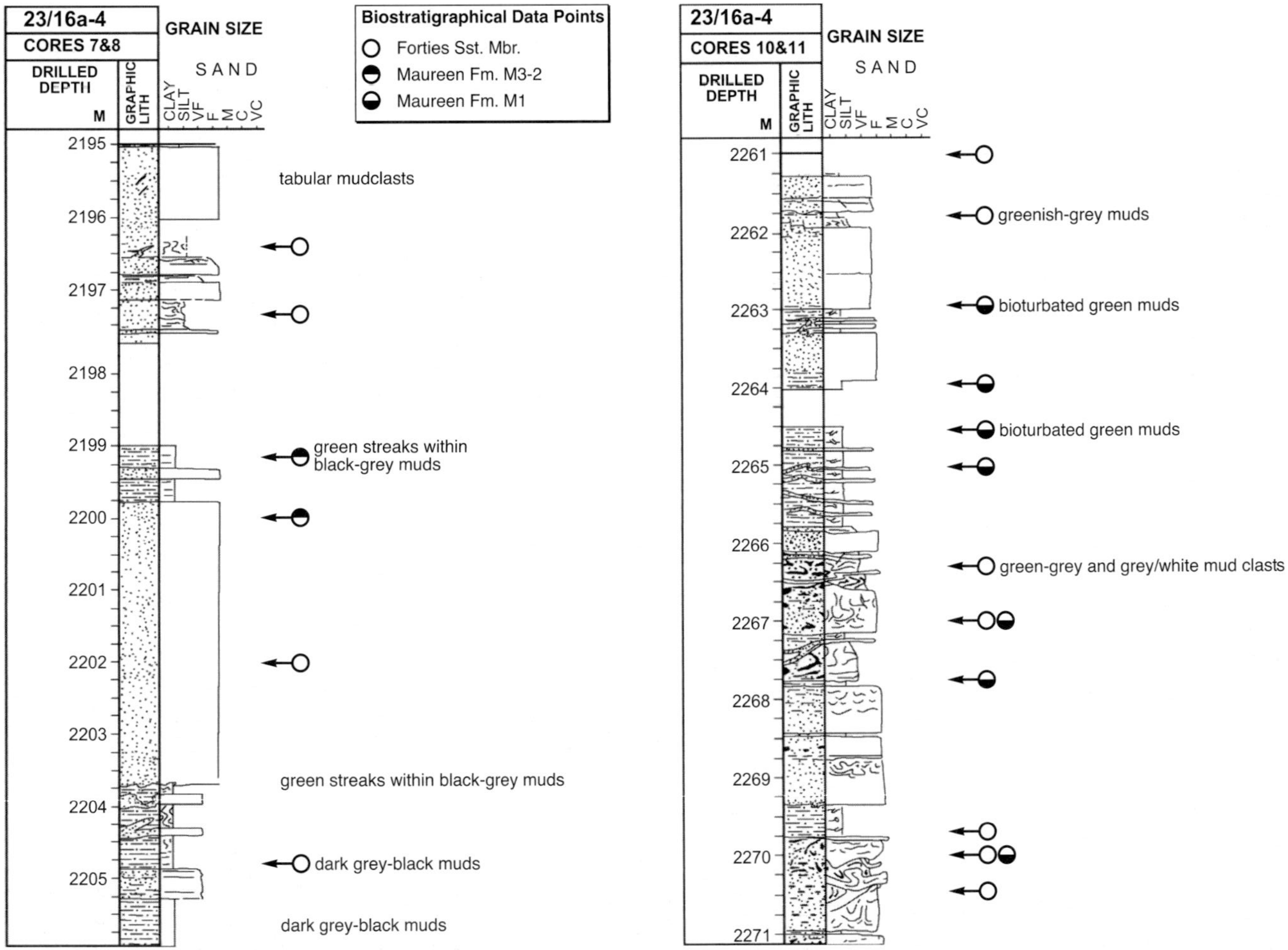

Fig. 4. Lithological logs for selected cored intervals from well 23/16a-4 with biostratigraphic data points and observations on mudstone colouration; depths in metres. Complexity is apparent from the mixing of both biostratigraphic assemblages and mudstone lithotypes.

yielded an entirely non-Forties biota, but were nonetheless enveloped by Forties material. This was reconciled as the result of slumping of coherent olistoliths of older material off the growing diapir during the time of Forties deposition, into a background of Forties sediments. Concern was reinforced with the recognition that supposedly reworked intervals seemed to show a 'ghosting' of a coherent stratigraphical sequence, atypical of an unroofing model of reworking (where progressive denudation should entail older reworked material being found in the youngest sediments). Likewise, the presence of Forties sands and mudstones within the Ekofisk chalks in well 23/16a-4, recognized biostratigraphically and lithologically from core, was difficult to explain. No mechanism could be envisaged for implanting younger, incompetent muds into older, presumably largely lithified, carbonates.

Recurrent stratigraphical complexity

Through the early 1990s, appraisal drilling continued apace on Mungo Field. Well 22/20–3 and its sidetrack 22/20–3Z established the oil-water contact and proved an oil column of nearly 1000 m and a 300 m gas leg. Geological complexity and unpredictability continued to be widespread and not seismically mappable. Stratigraphical correlation, even between main bores and adjacent side-tracks, was often fraught with difficulty, with major lateral facies changes and net to gross variations (0.25 to 0.55) implied.

Core from well 23/16a-5 showed a large slump fold (around 5 m wavelength) and soft-sediment deformation entirely within a well-defined Forties package. This gave considerable comfort to the model of slumping and sliding off the diapir in the time of Forties deposition. This concept propagated into the seismic interpretation and recumbent folds began to be postulated in what hitherto had been a seismically opaque interval. Also from cores in this well, injection was clearly seen and considered to have been driven by pressure release, older sediments injected upwards into younger strata. No mechanism could be seen for implanting younger muds downwards into older sediments. Biostratigraphical correlation of reservoir sands, even formations, between 23/16a-5 and its two sidetracks, one updip, one downdip, seemed unrealistic.

The imposition of a model of chaotic mélange with its inherent unpredictability, coupled with the lack of seismic definition, removed the expectation for biostratigraphy to help layer the reservoir. Nonetheless, microfossil data continued to be gathered and interpreted on all new wells. Thereafter, sand distribution around the field was modelled stochastically by lithofacies categorization, established in core and extrapolated to uncored wells using wireline logs. For calculations of the volume of oil in place it was assumed that both 'good' and 'poor' areas would be found in the oil leg, and reservoir models spanning a range of possibilities were investigated. Uncertainty about the continuity, connectivity and compartmentalization of the reservoir remained, giving concerns about the distribution of hydrocarbons around the diapir. It was, however, considered that the Mungo reservoir had sufficiently high mean net to gross at 0.40 (ranging from 0.25 to 0.55) to provide hydrocarbon offtake with aquifer support. Overall, deformation was considered to have increased connectivity but increased tortuosity.

New role for biostratigraphy

A change of emphasis emerged for biostratigraphy, the discipline assuming a rigsite operational role. The main task, from the latter stages of appraisal drilling onwards, was to aid the picking of a casing point in the Sele Formation mudstones immediately above the topmost Forties sandstones, based primarily on the FDO of *Apectodinium augustum*. Placing the casing shoe too high would expose the open well-bore to the mechanically unstable Moray Group mudstones while drilling the reservoir section; too low and a proportion of reservoir pay could be lost behind casing. Rigsite nannofossil analysis was used to ensure that the well terminated within *in situ* Ekofisk chalk, ensuring that the wellbore did not reach total depth in an allochthonous chalk 'raft' within the Montrose Group, with the risk of leaving clastic reservoir beneath unexploited. Additionally, this discipline was also used to ensure that wells terminated above the potentially fractured Cretaceous Tor Formation chalks, with scope for major drilling mud losses or gas 'kicks'.

Wellsite palynological coverage over the Miocene and Oligocene was also used in an effort to define, and thereby predict, an interval prone to intense caving, considered to be an effect of pressure release above and around the intruded salt stock.

On well 22/20-A03Z the top of the Forties Sandstone Member was conclusively picked on palynological criteria some 200 m higher than seismically prognosed. This led to a remigration and reinterpretation of seismic data and the validation of steeper dips on the west side of the diapir, previously discarded as being too steep to be geologically realistic. Increasing the dip of the reservoir units reduced the bulk rock volume and lowered estimates of oil in place in this segment of the field, an important caution on the economic materiality of Mungo Field.

Towards a new model

Through 1999 and 2000 the Mungo biostratigraphical dataset was revisited in a further attempt to layer an 'unlayered' reservoir and predict reservoir performance. A profound change of emphasis was contemplated, downplaying the significance of the Base *Apectodinium* 'golden spike' and the reworking model thereby imposed. Instead the hitherto anomalous 'ghosting' of sequential Lista and Maureen data, and the possibility of extensive sediment injection stratigraphically 'downward', younger into older, was to be further investigated.

Whilst biostratigraphy could not resolve a correlation of mudstones around and within the Mungo reservoir, it was decided to investigate the applicability of a complementary tool that focussed instead on the sandstones. The spur for this was a re-examination of core from the first development well, 22/20-A01, on the southwestern flank of the field. In this core three significant types of sediment were evident (see Fig. 7 for lithological overview):

- greenish-grey bioturbated mudstones;
- coarse grained, bedded, massive oil-stained sandstones with apparently concordant relationships with the greenish-grey mudstones;
- finer grained, tighter sandstones (generally not oil-stained), often with irregular contacts with enveloping sediments and a disturbed and brecciated fabric, containing disrupted intraclasts of grey-black mudstones, and a smaller proportion of greenish-grey mudstones.

From a lithostratigraphical angle it was considered that the greenish-grey bioturbated mudstones might be background hemipelagic Maureen sedimentation intercalated with coarse Maureen sandstones; these two lithotypes comprise the bulk of the cored interval. The sandstones, in particular, show a layering consistent with depositional dips expected on the diapir flank when structurally restored.

The relationship between these units and the finer, tighter sandstones appeared to be markedly different. The latter, typically of the order of 1 m in thickness, often have a seemingly irregular and discordant relationship with the other lithotypes. Typically, grey-black, Forties-type mudstones are present within these apparently disturbed sands as brecciated, angular intraclasts with no trace of bedding; some disrupted greenish-grey muds typical of the Maureen Formation are also seen within these units. The inference was made that the finer, apparently disturbed sands might be allochthonous packages emplaced within the Maureen Formation.

Biostratigraphical analysis of selected mudstones over this section in 22/20-A01 conforms neatly to the lithostratigraphical interpretation. Greenish-grey mudstones yield Maureen assemblages that span several biozones in the correct stratigraphical order. In stark contrast, grey black mudstones taken from within the finer grained sandstones have yielded uniquely Forties palynofloras, comprising not only species of *Apectodinium*, but also associations of pollen diagnostic of the Forties Sandstone Member.

One explanation for this complex association would invoke the remobilization and intrusion of younger sediments into older strata (stratigraphically 'downward'). This is the antithesis of the 'Forties mélange' model, with Maureen reworked into Forties. Whilst it honours the data it runs somewhat counter-intuitive. Intrusion or injection could readily explain the association of Forties and Maureen biota: Forties palynomorphs being recovered from the brecciated grey-black mudstone intraclasts entrained with injected Forties sands, emplaced into the Maureen Formation. A dyke breccia could be interpreted from the 22/20-A01 core, with a mixing of Forties sand and muds, together with a minor Maureen mudstone component. The latter could be explained by injection into an uncompacted Maureen host body. This bears comparison with remobilized Eocene sandstones in the North Sea Alba Field in which Duranti *et al.* (2002) record shale clast breccias in proximity to host shale units, with host shale fragments becoming increasingly mixed with the remobilized sandstones to form breccias. O. Parize (pers. comm.) has observed a similar texture in injected sand bodies in the Upper Aptian Saint-Andre-de-Rosans section in the Vocontian Basin of southeast France. Here shaly intraclasts are present at the dyke margins, typically over an interval of 1–5 cm.

The concept of remobilization allows a possible explanation for the persistent anomaly (in core and cuttings samples) of Forties Sandstone Member palynomorph assemblages recovered within the Maureen Formation on Mungo Field, and less commonly in the Lista and Ekofisk formations. An injection theory elegantly obviates concerns over the apparent sequential reworking of older units in the correct stratigraphical order into the Forties. The implication now made was that the 'ghosting' of sequential stratigraphy was valid, but was significantly disrupted by the emplacement of Forties material.

Heavy mineral analysis

To test this developing hypothesis, heavy mineral analysis was initiated, focused on the cored section of well 22/20-A01. It was hoped that mineralogical differences might discriminate between the sand bodies on Mungo Field and confirm or refute the possibility of injection as a significant feature of Mungo Field. A secondary objective was to assess whether there is any evidence for variations in sediment provenance beyond the regional input from the northwest via the Moray Firth. An additional, more local source, potentially the Jaeren High to the northeast, might entail a different model of sand distribution around the diapir. The sample set also included two well-dated control wells from the same sector of the field; 23/16a-5 has indisputable, biostratigraphically defined Forties Sandstone Member sands, while 23/16a-5Z has similar for the Maureen Formation.

In terms of overall mineralogy there is considerable overlap between the Maureen sandstones in 23/16a-5Z, the Forties section in 23/16a-5 and the interval in 22/20-A01. Assemblages are relatively diverse with eleven detrital species including apatite, anatase, chrome spinel, garnet, gahnite, monazite, rutile, staurolite, tourmaline and zircon. As a consequence of this similarity, reliance has to be placed on more sensitive provenance criteria such as heavy mineral ratio and garnet geochemical data (see Morton & Hallsworth 1994). Of the ratio criteria, rutile:zircon (RuZi), monazite:zircon (MZi), and chrome spinel:zircon (CZi) show only subtle differences between the Forties and Maureen control sections.

In contrast, however, there are distinct differences between garnet:zircon (GZi) and apatite:tourmaline (ATi) between the sucessions in the control wells (Fig. 5). Both ratios are lower in the Maureen section than in the Forties, with only limited overlap between the two. These differences suggested that these ratios could be used to test the remobilization hypothesis in 22/20-A01. The differences between the the control wells are further emphasized by crossplots of GZi and ATi against the zircon:apatite ratio (ZAi) (Fig. 6). ZAi is not a reliable provenance discriminator, because it ratios two minerals with contrasting hydraulic behaviour, but it is sensitive to differences in hydraulic conditions during sedimentation. As a result it is useful for discerning Forties from Maureen on a local scale.

The sequence from 22/20-A01 is interpreted to include sandstones belonging to the Maureen Formation together with subordinate material incorporated from the Forties Sandstone Member (see Fig. 7). Downhole variations in ratio data show systematic differences between *in situ*, layered, coarser sandstones and finer sandstones with evidence for deformation. This suggests that the two sandstone types are not genetically related, supporting the injection concept. The sandstones that appear to be layered have generally lower GZi and ZAi than the disturbed units, although ATi is more variable. On the various crossplots the *in situ* Maureen sandstones show a high degree of overlap with Maureen sandstones from 23/16a-5Z, providing a strong support for a genetic relationship.

Conversely, the sandstones that display evidence for disturbance plot in different parts of the various crossplots, away from the areas defined by the Maureen sandstones. There is considerable overlap between the data from these disturbed sandstones with those from the Forties Sandstone Member in 23/16a-5, suggesting strong links between the two. Some assemblages fall in the zone of overlap between Maureen and Forties. This is regarded as a reflection of the mixing of the two sandstone units through injection.

Garnet geochemical studies were carried out on a subset of samples from the control wells, together with both *in situ* and disturbed intervals from 22/20-A01. The assemblages are all diverse, with a wide range of compositions, grossular ranging to >30%, pyrope to >50% and almandine plus spessartine between 40% and 90%. This reflects a wide diversity in source material. Assemblages in the Forties and Maureen sandstones in the Mungo Field are closely comparable with other Paleocene sandstones in the central North Sea, such as the Forties Sandstone Member of the Forties and Montrose fields (Morton 1987), indicating input from the main northwest/southeast-orientated submarine fans draining through the Moray Firth area. Importantly, none of the sandstones from Mungo field have garnet assemblages that fall outside the range shown by Forties, Lista, Andrew and Maureen sandstones of the central North Sea and Outer Moray Firth. Locally, adjacent to the Jaeren High in the Norwegian sector, garnet assemblages dominated by low grossular, moderate pyrope types have been

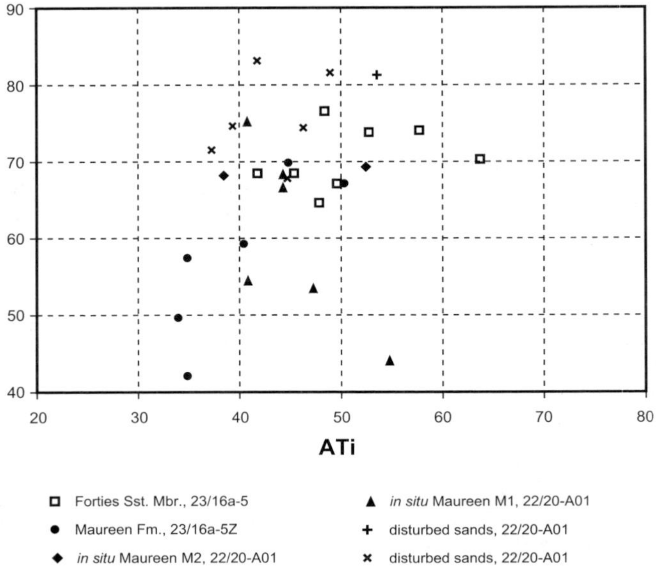

Fig. 5. Crossplots of heavy mineral ratios – garnet:zircon (GZi) vs. apatite:tourmaline (ATi) from control wells and 22/20-A01 well sections to determine parent sand body.

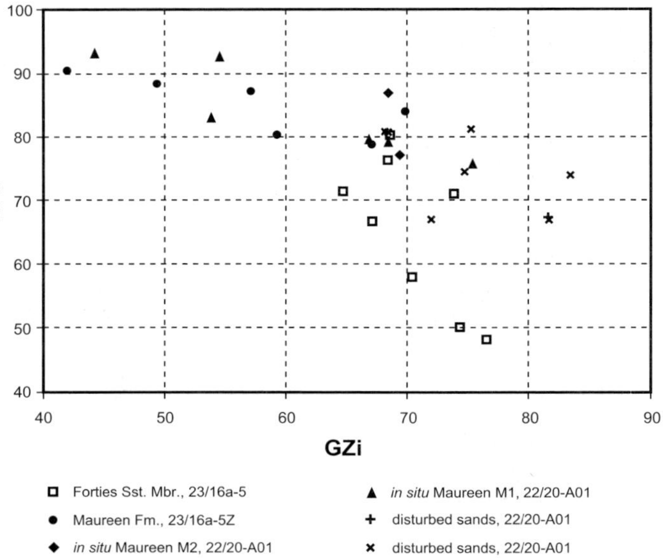

Fig. 6. Crossplots of heavy mineral ratios – garnet:zircon (GZi) vs. zircon:apatite (ZAi) from control wells and 22/20-A01 well sections to determine parent sand body.

found in Paleocene sandstones, and are interpreted as indicating local sourcing. There is no evidence to indicate that sediment from this source influenced the Mungo area.

Further concerns

Despite some emerging clarity in understanding the layering of the Mungo reservoir sands, some concerns still remained. Most significant was the

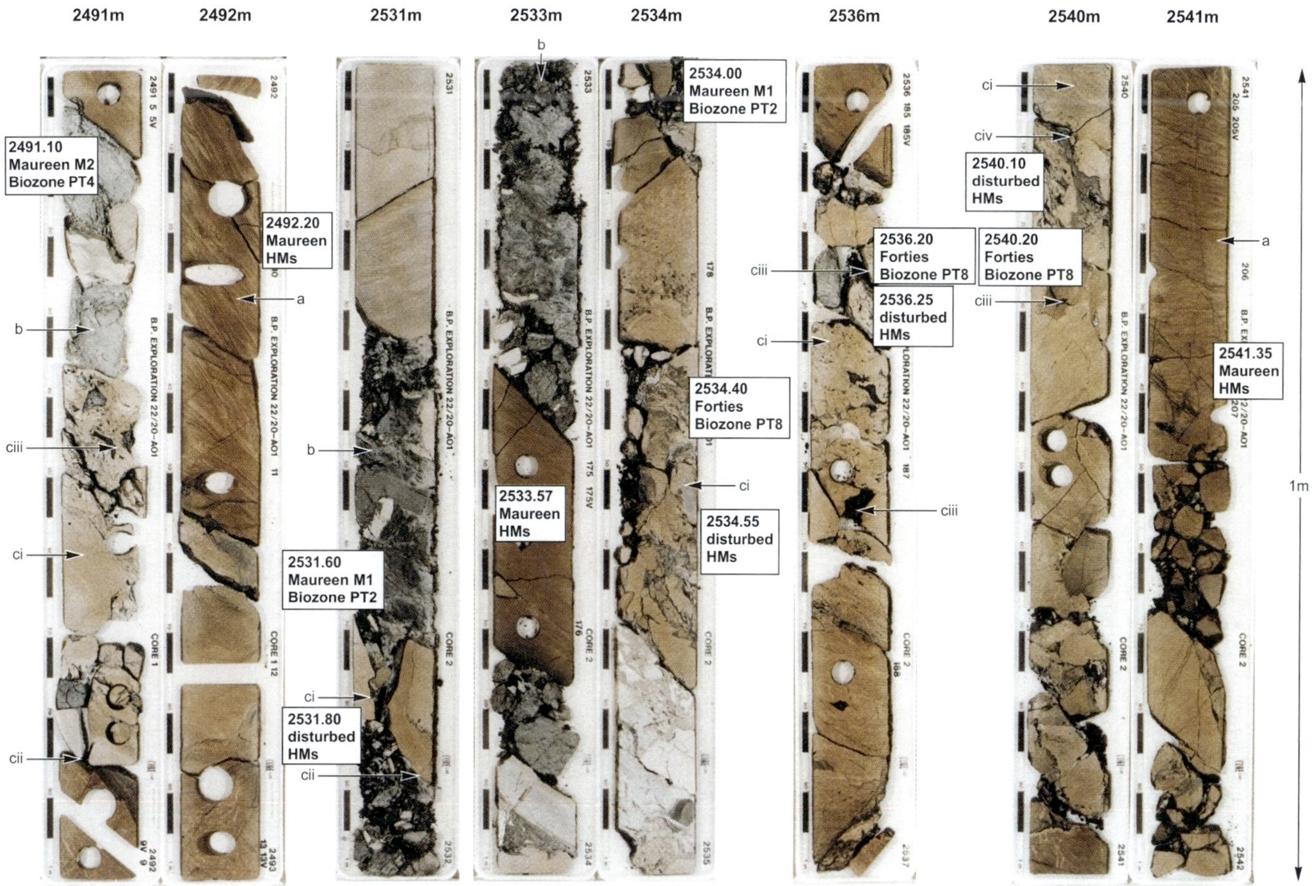

Fig. 7. Interpretation of selected (non-continuous) cores from well 22/20-A01 with data points for biostratigraphic and heavy mineral analyses attached. Note (**a**) the darker, oil stained sands yielding characteristically Maureen Formation heavy mineral suites, showing a concordant relationship with (**b**) greenish-grey bioturbated mudstones biostratigraphically dated as sequential within the Maureen Formation (across biozones PT4 and PT3). Note also (**ci**) tighter (less oil stained) sands with a heavy mineralogy more typical of the Forties Sandstone Member, showing evidence for disturbance, (**cii**) irregular contacts (**ciii**) angular, brecciated, grey-black intraclasts yielding Forties palynomorphs, (**civ**) mixed with muds of Maureen type. Depths in mdbrt.

contention of Thomas (1996) who recorded species of the dinocyst genus *Apectodinium* from core in nearby Everest Field well 22/10a-4. As expected, a large number of *Apectodinium* spp. were recorded from the lower Sele Formation, including the Forties Sandstone Member. More surprisingly *Apectodinium* spp. were also recorded in isolated samples from the Lista and Maureen Formations (both units M1 and M2). If *in situ*, the occurrences would render invalid the widely accepted central North Sea 'golden spike' of the Base *Apectodinium* Datum within the lower part of the Sele Formation. For Mungo Field, the implications of accepting a Maureen Formation inception of the genus *Apectodinium* would obviate the need to explain hitherto anomalous palynomorph assemblages; Maureen and *Apectodinium* assemblages could simply be viewed as being contemporaneous.

Thomas' anomalies appear to have been found in more sandy lithologies and correspond to increases in the relative abundance of terrestrially derived palynomorphs, a reduction in diversity of dinocysts and an increase in amorphous organic matter, all strongly characteristic of the Sele Formation. Thomas suggests that these may not represent *in situ* basinal Lista or Maureen palynofloras, but could be contemporaneous basin margin assemblages, periodically swept into the basin centre.

Another interpretation is that these rare, sporadic and short-lived occurrences of *Apectodinium* spp. in the Montrose Group could represent periodic migration northwards during periods of warmer water conditions (Bujak & Brinkhuis 1998). This could be considered the 'feather edge' of the northward migration of the genus through the Paleocene, prior to the major, more widely recognized migration, the Base *Apectodinium* Datum.

A further explanation for these occurrences is sandstone injection, as in Mungo. However, given the apparent, undisturbed, layer-cake stratigraphy observed over much of Everest Field this possibility is considered unlikely. A more mundane explanation could be sample cross-contamination or mud-cake contamination of what are otherwise likely to have been poorly fossiliferous, sand-prone lithologies.

A further issue challenged the view that Maureen mudstones are never grey-black in colour and laminated as those of the Forties Sandstone Member. In the cored interval of well 22/20-A01, rare, thin (around 5 cm), intervals of apparently undisturbed, black, laminated mudstones grading into characteristically Maureen-type mudstones are seen. On lithological criteria alone, these packages resemble Forties Sandstone Member mudstones. However, where analysed, they have yielded Maureen age plankton, with no benthos, suggesting rare and brief intervals of bottom-water dysaerobia during the deposition of the Maureen Formation. This 'black herring' sounded a note of caution in the reinterpretation of Mungo Field stratigraphy, challenging the assumption that all greyish-black laminated mudstones can be comfortably assigned to the Sele Formation.

Injection model

Extrapolating the injection model beyond well 22/20-A01 involves a substantial shift in approach, from a Forties reservoir with Maureen reworking to a Maureen reservoir with considerable Forties injection. It is a major advance in rationalizing the complexity evident in the biostratigraphical data and understanding the history and layering of the Mungo reservoir. It allows many of the records of seemingly allochthonous Forties palynomorphs to be seen as spurious with respect to reservoir layering. Observations from core of Forties-type mudstone clasts yielding palynomorphs of the same age could no longer be taken as unequivocal evidence of *in situ* Forties Member sediments, albeit preserved as rip-up clasts or vestiges of channel margin collapse. Another explanation that seemed to fit the various datasets better had emerged.

The 'ghosting' of sequential Lista and Maureen taxa can now be postulated as representative of the *in situ* stratigraphy, with the Forties material considered to have been remobilized after deposition. In reality, reworking and slumping is still apparent on Mungo, as well as downhole contamination by caving. In some sections the dataset necessitates carrying equivocal models of reworking, injection and caving, or combinations thereof. Likewise in certain cases, one dataset, perhaps biostratigraphical data from either core, sidewall cores or cuttings, can appear anomalous in comparison with the others. Here a pragmatic decision has to be taken as to which information to credit and which, if any, to discount.

Mechanisms for injection

Once accepted as a potential explanation, a mechanism for the injection of stratigraphically younger sediments into older has to be considered. A prerequisite for clastic intrusions is a sealed, unlithified, sand source body in which an excess pressure has built up (Lonergan *et al.* 2000). The seal on the over-pressured sand body must then be breached suddenly in order to generate the fluid velocities necessary for sediment fluidization and mobilization (Lowe 1975, 1976; Nichols 1995). This overpressuring is most commonly derived from rapid burial of a sealed source body, when the burial rate exceeds the ability of the fluid to escape from within the pore spaces between the sediment grains. High fluid pressures may also have been generated by the migration of hydrocarbon fluids, focused around structures such as salt diapirs. The prime North Sea source rock (and the source of hydrocarbons on Mungo Field), the Kimmeridge Clay Formation, was actively generating hydrocarbons during the Palaeogene (Jolly & Lonergan 2002).

Breaching of the seal around the over-pressured sands may be due to a number of causes, or the interplay thereof. Earthquake-induced liquefaction, tectonic and halokinetic stress associated with faulting (Dixon *et al.* 1995), excessive pore pressures generated by local loading due to renewed deposition (Parize *et al.* 2001), slumping (Hiscott 1979; Strachan 2002) or hydrocarbon migration (Jenkins 1930) are all candidates. Simulation modeling by Parize *et al.* (2001) demonstrates the generation of syn-depositional hydraulic fracturing and injection, both vertically downwards and horizontally, with the rapid deposition of turbidite sands. The impetus from one or more of these factors may provide enough fluid charge to rupture the seal, remobilize and inject sand.

Injection of sandstone dykes and sills is common in deep water turbidites and can be extensive (Truswell 1972; Hiscott 1979; Archer 1984; Huang 1988). Large, seismically resolvable injection on a scale of hundreds of metres has been reported from turbidite oil fields in the North Sea (Jenssen *et al.* 1993; Dixon *et al.* 1995; Lonergan *et al.* 2000). Injections are closely associated with pre-existing fractures and faults. Forceful injection of unconsolidated sand can occur in all directions, but in most sedimentary environments dykes are generally assumed to intrude upwards or laterally, following the hydraulic gradient (Jolly & Lonergan 2002). Nichols (1995) found that in laboratory experiments, downward lobes in the base of coarse sand lenses, presumably comparable to a sand-rich channel, could spread both sideways and downwards when subjected to shaking. Parize & Beaudoin (1998) record downward sand injection from the Numidian flysch, with dykes penetrating up to 150 m below the mother sand. In the Vocontian domain in SE France, O. Parize (pers. comm.) has observed turbidite channel sands with dykes extending up to 300 m downwards into the underlying sediments; the only source for this sand is the channel above. Brodzikowski *et al.* (1987) report gravitational flowage and the collapse and fluidization of unconsolidated Quaternary fluvioglacial sands into tension cracks in Tertiary sediments above extensional faults in Precambrian basement. The fractures are up to 50 m deep and the fill often shows brecciation.

Three possible injection models are suggested to help explain some of the complexity on Mungo Field (Fig. 8):

- Injection as sills or dykes along fractures generated by fluid pressure release upon some trigger, possibly loading by slumps (Fig. 8.1a) or rapid deposition (Fig. 8.1b). With the structural dips generated by diapir growth, sediment injection along discordant fractures could result in material being remobilized structurally upwards or laterally but effectively stratigraphically 'downwards', from younger into older. Forties Sandstone Member sands, with entrained muds, could thereby be injected up or across the structural grain into the Maureen Formation. This model conforms to the 'traditional' concept of injection.
- Injection along faults and fractures generated by gravity, or the extensional fault-driven collapse, of sediments on the diapir flanks, as a response to oversteepening as the diapir grows (Fig. 8.2). Voids open up at the headwall of the slide, perhaps at shallow depth, with an intact cohesive clay layer close to seabed acting as a top seal. These voids are filled by fluidized Forties Sandstone Member sediments driven laterally and downward along these planes into older host sediments. Injection would dominantly be in the form of dykes, which could become sills downwards.
- Injection along tectonic faults propagating upwards from the underlying chalk (Fig. 8.3). The competent chalk may have supported open voids during faulting (observed where cored sections encounter faults), generating a local, temporary, pressure gradient. Fluids may have been more readily mobilized from the overlying unconsolidated clastics, rather than the cohesive lower permeability chalk, allowing the preferential fluidization of the finer-grained Forties material and downwards injection. The emplacement of Forties sediments into the Ekofisk chalk is seen in core in Mungo Field wells 23/16a-4 and 22/20-A13.

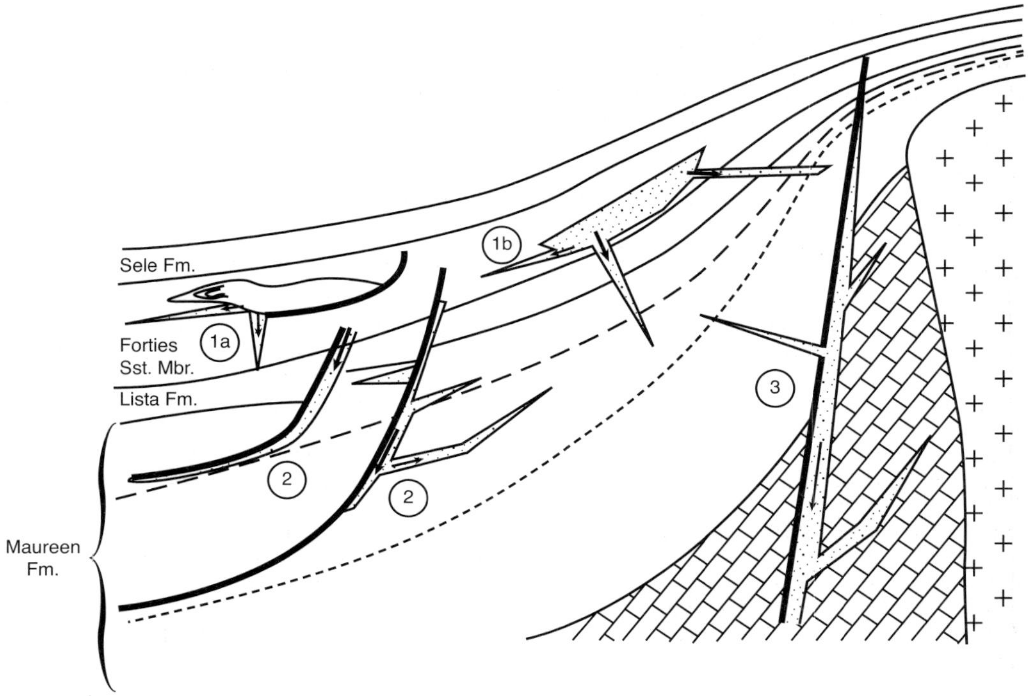

Fig. 8. Schematic cross section of the Mungo diapir, illustrating the three proposed models for sandstone injection of the Forties Member into the Maureen Formation and chalk. 1a, sediment loading due to slumping. 1b, loading by rapid deposition of turbidite flow, with injection of newly deposited sand into the channel margins or base. Shown with post depositional bed rotation due to diapir growth. 2, Injection into headwall voids of gravitational slump faults. 3, Injection into tectonic fault voids propagating upwards from the chalk. Injections may develop complex networks of both dykes and sills with offshoots. Dykes may form both radially or concentric to the diapir, in and out of plane of the figure.

The latter two of these models bear comparison with neptunean dykes, effectively an infill of fractures with a significant downwards element.

Timing of injection

Davidson *et al.* (2000) identified three significant onlapping unconformities typically seen around central North Sea salt diapirs: *intra*-chalk, at top Sele (the nearest seismic pick to the top Forties Sandstone Member) and mid-Miocene. These events are related to enhanced periods of active diapir growth. The resultant tectonic activity may have triggered pressure release by gravitational instability through over-steepening of the diapir flanks and extensional faulting of the diapir carapace.

Cases can be made for either shallow or deep burial overpressure generation and sand injection. The remobilization herein described would appear to relate to the top Forties event, no sand injection being seen above the top Forties Sandstone Member on Mungo Field. Diapir growth after Forties deposition, when the Forties sands were only buried at shallow depths (perhaps tens of metres), would allow the key criteria for injection to be met.

Only Forties sands appear to have been injected, and the Maureen Formation acts as the major host body, with minor evidence for disruption in the Lista and Ekofisk formations. The high net-to-gross Maureen sands possibly underwent early compaction and dewatering before Forties deposition. The sealed low net-to-gross Forties could preserve its undewatered, uncompacted sand fabric and generate over-pressures when stressed. Finer grained, uncompacted Forties sands would be more prone to fluidization than compacted, coarser-grained Maureen sands (Richardson 1971).

Geomechanical contrasts under changing over-pressure conditions in different units could enable different deformation and fracture formation through the stratigraphy. Parize & Beaudoin (1998) note the significance of the varying mechanical properties of strata and differential compaction in controlling the patterns of dykes and sills. Top seal integrity could be preserved in soft ductile shales near the seabed, preventing upwards breaching. The overpressured Forties sands, and more cohesive units of the deeper buried Lista and Maureen formations, could deform in a brittle manner with the formation of open hydraulic fractures related to tectonic activity. These fractures would then be filled by fluidized Forties sand. Rapid hydraulic failure of soft sediments could result in extensive brecciation and entrainment of shale clast breccia into the injected sands. The idea of remobilization during shallow burial is supported by the observation that the shale clasts appear to be formed by brittle brecciation but often show ductile deformation, indicating the shale was competent but still soft.

Once deeply buried, any overpressure is expected to have equalized throughout the whole connected Paleocene sand sequence. Differential pressure drive between the Forties and Maureen would then have become depleted. After Forties deposition the Paleocene does not appear to have achieved significant burial overpressures, pressure probably being released through the regional aquifer or the thin, disturbed, diapir carapace.

A new layering model

The recognition of injection as a process that can help explain key elements of the observed complexity has proven critical in progressing the reservoir description of Mungo Field. Discarding the unlayerable 'Forties mélange' model for a predominantly Maureen reservoir with Forties injection, and the validation of what was hitherto perceived as an anomalous 'ghosting' of an *in situ* stratigraphy, has allowed the Mungo reservoir sands to be realistically layered for the first time.

Five clastic layers or time-slices can be defined on Mungo: Maureen packages M1, M2 and M3, Lista and Forties in their normal stratigraphical order. The layers are defined by gamma maxima bracketing sand-rich packages, individually constrained by bioevents. These form the framework for use in inter-well correlation and reservoir simulation modeling. These are regionally recognized units (except for M3, a local subdivision equating to the upper part of M2 *sensu* Knox & Holloway 1992), but even this recognition of large-scale packages is a major step forward in this complex terrain. Attempts to correlate at a finer resolution have not proved feasible due to the extent of complexity.

The bulk of hydrocarbon production is now seen to be from the Maureen sandstones, the Lista is typically condensed and mud-prone, whilst the Forties Member, formerly the target, is seen to comprise predominantly mudstones and ratty sandstones.

Impact of a revised model

A clearer understanding of the layering of Mungo reservoir allows an increased understanding of the structural evolution of the diapir. Correlating these layers between wells across the field enables changes in the locus of sand deposition to be mapped through time, helping to predict preferred sites of sand accumulation and allowing wells to be more effectively targetted (Fig. 9). A robust layering model gives the option of monitoring the stratigraphical position of the drill-bit in real time through the reservoir, or biosteering.

Stratigraphical evidence for condensed sediment packages suggests that the Mungo diapir was growing by Early Cretaceous times. During the deposition of Late Cretaceous to Early Paleocene chalks the Mungo diapir formed a positive seabed feature, with a thinner, condensed interval accumulated over the top of it. The diapir is likely to have maintained a relatively buoyant position close to the sediment surface, with large thicknesses of sediment accumulating by 'downbuilding' around it (Davidson *et al.* 2000). During the Late Paleocene the diapir appears to have acted as an obstacle to turbidity currents, with sand deposition preferentially focussed in topographic lows on the seabed adjacent to the postulated minor high. Continued loading around these lows would accentuate diapir growth and exacerbate the relative lows as preferred sites for sand accumulation.

The evolving pattern of sand distribution is apparent from net to gross maps generated from Maureen Formation units M1 to M3 (Fig. 10) Turbidite input from the north and NW is deflected around the salt-driven topographic high, with the most extensive M1 sands being deposited on the south and west flanks of the diapir, in addition to local lows to the NE and east. A sand-poor, bypassed area is evident upstream on the NW flank; to the SE a shadow zone, with minimal sand deposition, is also apparent. Loading by M1 clastics leads to growth of the diapir, causing the depositional axis of the ensuing M2 sands to shift a little

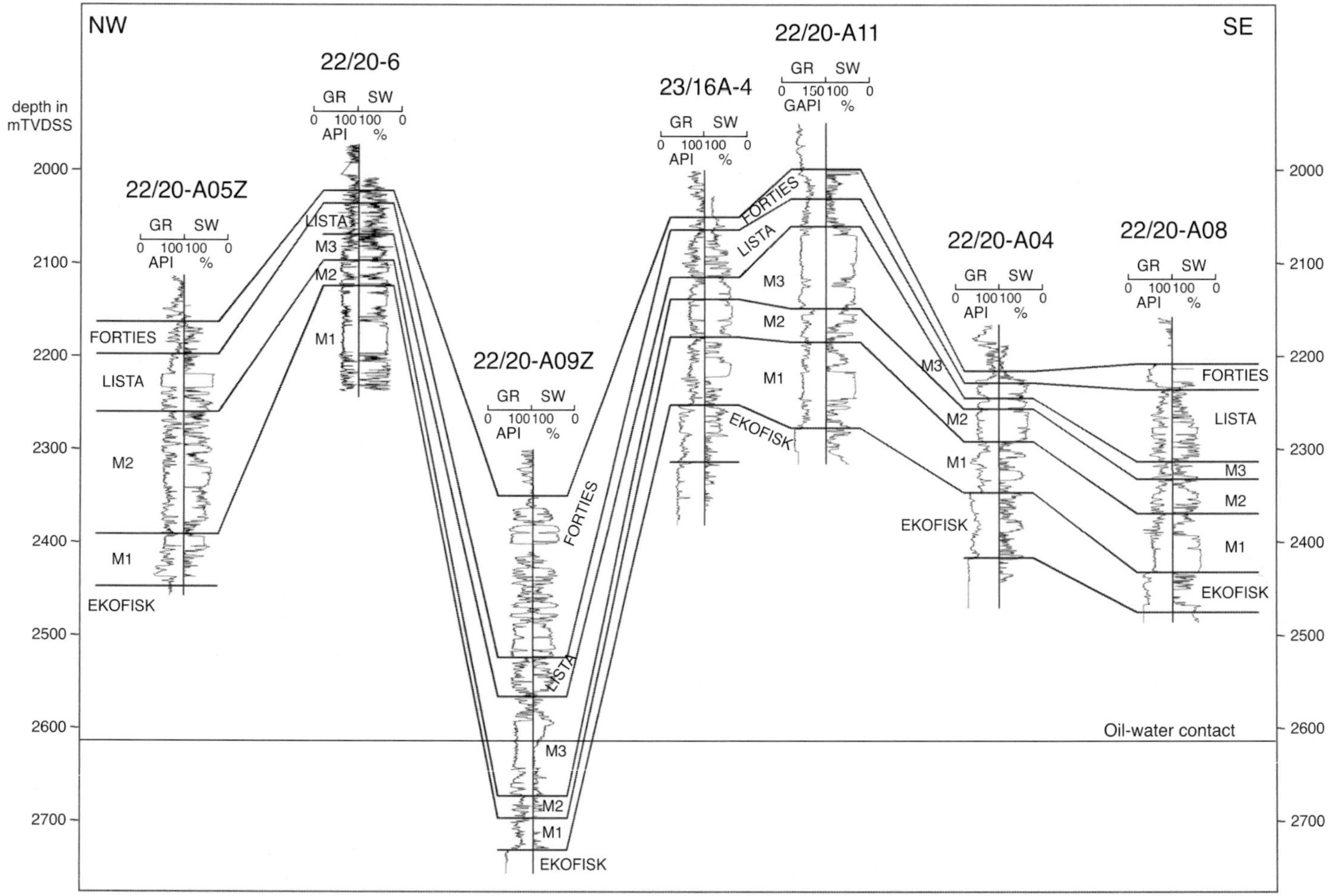

Fig. 9. Inter-well correlation, constrained by biostratigraphic interpretation, across the north east flank of Mungo Field, showing changes in the locus of sand deposition through time and the significance to reservoir development of Maureen Formation units M1 to M3.

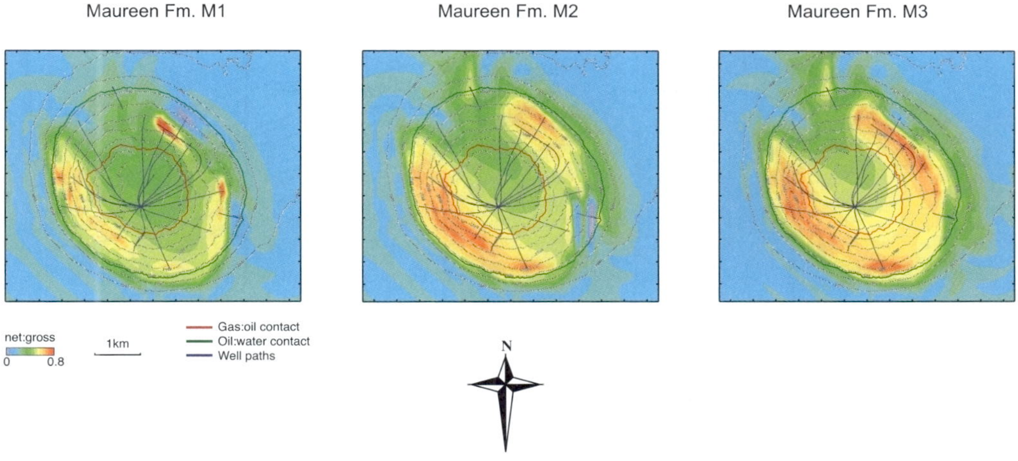

Fig. 10. Maps of net to gross through the units of the Maureen Formation, showing the evolving distribution of sands around Mungo diapir.

further outboard of the diapir, particularly on the NE flank (Fig. 11). M2 sand deposition is more extensive than M1, the main locus to the SW broadens, and the smaller, separated sand packages to the north and east grow and amalgamate. This pattern of two fairways, one to the SW and one to the NE, continues into M3; the southwestern depocentre expands and becomes predominant. By the end of M3 deposition much of the topography around the diapir has been infilled, and the former shadow zone to the SE has received sand input.

The ensuing Lista Formation is interpreted to be primarily an interval of condensed 'background' mudstone deposition. Sandy turbidite deposition returned in the Forties Sandstone Member, though finer grained and with a lower net-to-gross. Both units thin markedly onto structure. Post-production RFT (pressure) data support this view, with the generation of a pressure difference between thin, less well-connected sandstones of the Forties Member and the thicker sandstones of the Maureen Formation.

With this greater understanding of reservoir layering on Mungo, historical anomalies in the well set can be reviewed. For example, in the northwest flank of the field, high-angle development well 22/20-A06 had failed to encounter the predicted Forties sandstone target, instead penetrating a ratty time equivalent package. This was regarded as a local anomaly within the otherwise sand-prone, if unpredictable, 'Forties mélange'. Revision to a Maureen target suggested that drilling deeper would enable reservoir to be intersected. A sidetrack, well 22/20-A06Z, proved this assertion and Maureen sandstones were indeed encountered.

Reservoir sand architecture has been reinterpreted with more emphasis on primary depositional architecture and less on a slumping overprint. Instead, secondary complexity is generated by the injected sands that are expected to increase vertical connectivity. However, textural differences between the injected and host sands may result in lateral capillary flow baffles, reflected by differences in oil staining in the core. Injected sands may also enhance the fluid transmissibilities within the chalk, acting as fluid conduits within the low-porosity/permeability chalk matrix. Hydrocarbon volumes in place, and more significantly produceable reserves, have been re-evaluated on the basis of the total and connected sand volumes.

Conclusions

Mungo remains a very complex field to understand. Unlike many reservoirs, a local biozonation cannot be defined, a result of the disruption and remobilization of the sediments. However, progress has been made from a position in which biostratigraphy struggled to have a role in a mélange model, to one in which reinterpreted biostratigraphy is central to understanding the geology of the field.

Unravelling the stratigraphy of Mungo field stands as a salutory tale when applying biostratigraphy in complex terrains. Extreme

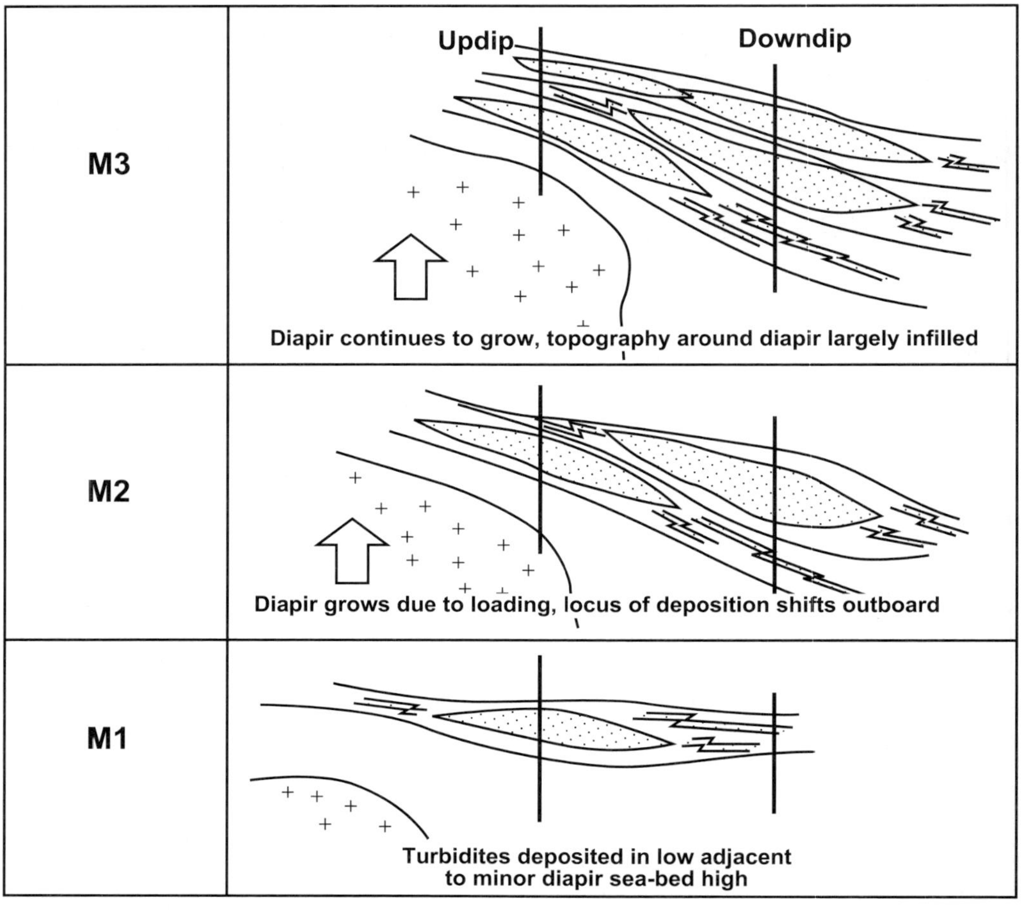

Fig. 11. Schematic illustration of the evolution of sand distribution through a flank section of Mungo diapir in Maureen Formation units M1 to M3.

caution and over-interpretation both beckon the biostratigrapher. The concept of the Base *Apectodinium* 'golden spike' still holds true, but the effects of structuration can greatly complicate its interpretation. Core material was seen as the critical data set, effectively irrefutable, and a geological model was built around it. A more refined interpretation has come from judging which of several, sometimes conflicting, datasets to credit, and which might be spurious.

The detailed integration of biostratigraphical, heavy mineral analysis and lithostratigraphy has driven a fundamental revision of the layering model on Mungo Field. Injection of younger material into older initially ran counter-intuitive, but given due consideration, provides a unifying explanation that ties together the different strands of evidence.

On other North Sea diapir fields, and in other diapir provinces, similar complexity may be anticipated. Lessons learnt on Mungo enable us to 'expect the unexpected' and keep all options open.

The authors wish to acknowledge BP Exploration Ltd for permission to publish. The approval of Mungo Field partners MOC Exploration (UK) Ltd, Murphy Petroleum Ltd and TotalFinaElf Exploration UK plc is gratefully acknowledged. The technical input of BP geologists N. Beveridge, A. Carter, R. Corfield, R. Dixon, A. Hewson, G. Seymour and O. Vaughan, collectively spanning some 15 years, is especially recognized. For batting on through complex microfossil datasets we applaud the biostratigraphic consultancies Biostrat Ltd, Geostrat Ltd, Microstrat Ltd, Millennia Ltd, Robertson Research International Ltd, RPS Paleo Ltd and Varol Research Ltd. The draughting skills of S. Muir and P. Nihell of Schlumberger Evaluation and Production Services UK Ltd (GeoQuest Division) are commended with thanks.

References

AHMADI, Z.M., SAWYERS, M., KENYON-ROBERTS, S., STANWORTH, C.W., KUGLER, K.A., KIRTENSEN, J. & FUGELLI, E.M.G. 2003. Paleocene. *In*: EVANS, D., GRAHAM, C., ARMOUR, A. & BATHURST, D. (eds) *The Millenium Atlas: Petroleum Geology of the Central and Northern North Sea*. Geological Society, London, 235–259.

ANDERTON, R. 1997. Sedimentation and basin evolution in the Paleogene of the northern North Sea. *In*: OAKMAN, C.D. & CORBETT, P.W.M. (eds) *Cores from the Northwest European Hydrocarbon Province: An illustration of geological applications from exploration to development*. Geological Society, London, 39–47.

ARCHER, J.B. 1984. Clastic intrusions in deep-sea fan deposits of the Rosroe Formation, Lower Ordovician, Western Ireland. *Journal of Sedimentary Petrology*, **54**, 1197–1205.

BRINKHUIS, H., ROMEIN, A.J.T., SMIT, J. & ZACHARIASSE, J.-W. 1994. Danian-Selandian dinoflagellate cysts from lower latitudes with special reference to the El Kef section, NW Tunisia. *Geologiska Föreningens Förhandlingar*, Proceedings of the Meeting on the 'Stratigraphy of the Paleocene', **116**, 46–48.

BRODZIKOWSKI, K., GOTOWALA, R., HALUSZCZAK, A., KRZYSZKOWSKI, D. & VAN LOON, A.J. 1987. Soft-sediment deformations from glaciodeltaic, glaciolaucustrine and fluviolacustrine sediments in the Kleszczow Graben (central Poland). *In*: JONES, M.E. & PRESTON, R.M.F. (eds) *Deformation of Sediments and Sedimentary Rocks*. Geological Society, London, Special Publications, **29**, 255–267.

BUJAK, J.P. & BRINKHUIS, H. 1998. Global warming and dinocyst changes across the Paleocene/Eocene epoch boundary. *In*: AUBRY, M.-P., LUCAS, S.G. & BERGGREN, W.A. (eds) *Climatic and Biotic Events in the Marine and Terrestrial Realms*. Columbia University Press, 277–295.

DAVIDSON, I., ALSOP, I., BIRCH, P., ELDERS, C., EVANS, N., NICHOLSON, H., RORISON, P., WADE, D., WOODWARD, J. & YOUNG, M. 2000. Geometry and late-stage structural evolution of Central Graben salt diapirs, North Sea. *Marine and Petroleum Geology*, **17**, 499–522.

DIXON, R.J., SCHOFIELD, K., ANDERTON, R., REYNOLDS, A.D., ALEXANDER, R.W.S., WILLIAMS, M.C. & DAVIES, K.G. 1995. Sandstone diapirism and clastic intrusion in the Tertiary Submarine fans of the Bruce-Beryl Embayment, Quadrant 9, UKCS. *In*: HARTLEY, A.J. & PROSSER, D.J. (eds) *Characterization of Deep Marine Clastic Systems*. Geological Society, London, Special Publications, **94**, 77–94.

DURANTI, D., HURST, A., BELL, C., GROVES, C. & HANSON, R. 2002. Injected and remobilized Eocene sandstone from the Alba Field, UKCS: core and wireline log characteristics. *Petroleum Geoscience*, **8**, 99–107.

HISCOTT, R.N. 1979. Clastic sills and dykes associated with deep-water sandstones, Tourelle Formation, Ordovician, Quebec. *Journal of Sedimentary Petrology*, **49**, 1–10.

HUANG, Q. 1988. Geometry and tectonic significance of Albian sedimentary dykes in the Sisteron area, SE France. *Journal of Structural Geology*, **10**, 453–462.

JENKINS, O.P. 1930. Sandstone dikes as conduits for oil migration through shales. *AAPG Bulletin*, **14**, 411–421.

JENSSEN, A.I., BERGSLIEN, D., RYE-LARSEN, M. & LINDHOLM, R.M. 1993. Origin of complex mound geometry of Palaeocene submarine-fan reservoirs, Balder Field, Norway. *In*: PARKER, J.R. (ed.) *Petroleum Geology of Northwest Europe, Proceedings of the 4th Conference*. Geological Society, London, 123–133.

JOLLY, R.J.H. & LONERGAN, L. 2002. Mechanisms and controls on the formation of sand intrusions. *Journal of the Geological Society*, **159**, 605–617.

KNOX, R.W.O'B. & HOLLOWAY, S. 1992. Paleogene of the Central and Northern North Sea. *In*: KNOX, R.W.O'B. & CORDEY, W.G. (eds) *Lithostratigraphic Nomenclature of the UK North Sea*, Vol. 1. British Geological Survey, Nottingham.

LIU, X. 1995. *Quantitative Analysis of Genetic Sequence Development in the Tertiary North Sea Basin*. PhD thesis, University of Texas at Austin.

LONERGAN, L., LEE, N., JOHNSON, H.D., JOLLY, R.J.H. & CARTWRIGHT, J.A. 2000. Remobilisation and injection in deepwater depositional systems: implications for reservoir architecture and prediction. *In*: WIEMER, P., SLATT, R.M., COLEMAN, J. *et al.* (eds) *Deep-water reservoirs of the world*. Gulf Coast Section of SEPM Foundation 20th Annual Bob F. Perkins Research Conference, 515–532.

LOWE, D.R. 1975. Water escape in coarse grained sediments. *Sedimentology*, **22**, 157–204.

LOWE, D.R. 1976. Subaqueous liquified and fluidized sediment flows and their deposits, *Sedimentology*, **23**, 285–308.

MORTON, A.C. 1987. Influences of provenance and diagenesis on detrital garnet suites in the Forties sandstone, Paleocene, central North Sea. *Journal of Sedimentary Petrology*, **57**, 1027–1032.

MORTON, A.C. & HALLSWORTH, C.R. 1994. Identifying provenance-specific features of detrital heavy mineral assemblages in sandstones. *Sedimentary Geology*, **90**, 241–256.

NICHOLS, R.J. 1995. The liquefaction and remobilization of sandy sediments. *In*: HARTLEY, A.J. & PROSSER, D.J. (eds) *Characterisation of Deep Marine Clastic Systems*. Geological Society, London, Special Publications, **94**, 63–76.

PARIZE, O. & BEAUDOIN, B. 1998. Clastic dykes and sills from Numidian flysch (Sicily and Tunisia): sandy injection related to a high-density turbidity deposit. *AAPG Bulletin*, **72**, 1018.

PARIZE, O., SCHNEIDER, F., ECKERT, S., HADJ HASSEN, F., BEAUDOIN, B. & TROUILLER, A. 2001. La fracturation précoce des massifs argileux: quel role pour la compaction? *In*: BEAUDOIN, B. & TROUILLER, A. (eds) MinAndra Partenariat, Mémoire spécial des sciences de la terre, Ecole des Mines de Paris, 9–12.

POWELL, A.J. 1988. A modified dinoflagellate cyst biozonation for the latest Palaeocene and earliest Eocene sediments from the central North Sea. *Review of Palaeobotany and Palynology*, **56**, 327–344.

REYNOLDS, A.D. 1994. Quantitative analysis of submarine fans in the Tertiary of the North Sea Basin. *Marine and Petroleum Geology*, **11**, 202–207.

RICHARDSON, J.F. 1971. Incipient fluidization and particulate systems. *In*: DAVIDSON, J.F. & HARRISON, D. (eds) *Fluidization*. Academic, London, 25–64.

STRACHAN, L.J. 2002. Slump-initiated and controlled syndepositional sandstone remobilization: an example from the Namurian of County Clare, Ireland. *Sedimentology*, **49**, 25–41.

THOMAS, J.E. 1996. The occurrence of the dinoflagellate cyst *Apectodinium* (Costa & Downie 1976) Lentin & Williams 1977 in the Moray and Montrose Groups (Danian to Thanetian) of the UK central North Sea. *In*: KNOX, R.W.O'B., CORFIELD, R.M. & DUNAY, R.E. (eds) *Correlation of the Early Paleogene in Northwest Europe*. Geological Society, London, Special Publications, **101**, 115–120.

TRUSWELL, J.F. 1972. Sandstone sheets and related intrusions from Coffee Bay, South Africa. *Journal of Sedimentary Petrology*, **42**, 578–583.

The Ravenscar Group: a coeval analogue for the Middle Jurassic reservoirs of the North Sea and offshore Mid-Norway

N. BUTLER[1], M. A. CHARNOCK[2], K. O. HAGER[2,3] & C. A. WATKINS[1]

[1]*Robertson Research International Ltd., Llanrhos, Llandudno LL30 1SA, UK (e-mail nb@robresint.co.uk)*

[2]*Norsk Hydro AS, Research Centre, N-5020, Bergen, Norway*

[3]*Norsk Hydro AS, N-0246, Oslo, Norway*

Abstract: A palynostratigraphical study of the Middle Jurassic Ravenscar Group, Cleveland Basin, northern England involving the integration of miospore and microplankton data with sedimentary facies data has resulted in improvements in the stratigraphical resolution of offshore hydrocarbon-bearing strata in the North Viking Graben and Mid-Norway. The Dogger Formation is of 'earliest' Aalenian age and is correlative with the uppermost Dunlin Group, Drake Formation and Båt Group, Ror Formation. The overlying Aalenian Saltwick and Eller Beck formations correlate with the Brent Group, Broom–Rannoch–Etive–Ness and the Fangst Group, Ile–Not–'lower' Garn genetic packages. The Cloughton Formation is either unrepresented or condensed in the Brent Province and Mid-Norway due to a regional unconformity, which truncates lower Bajocian sediments. The Scarborough Formation is of 'latest' early to 'earliest' late Bajocian age and correlative with the older part of the Tarbert–Heather and the 'upper' Garn–Melke genetic packages. These interpretations contrast markedly with the majority of those published.

Despite their maturity of exploration and development, Middle Jurassic sandstones of the North Sea and offshore Mid-Norway still represent some of the most prolific hydrocarbon reservoirs in northwest Europe. Of these, the Brent and Fangst Groups (Deegan & Scull 1977; Dalland *et al.* 1988) are particularly important and, despite the vast number of wells, the published chronostratigraphic schemes are poorly calibrated with confidently dated, onshore sections. The Ravenscar Group of the Cleveland Basin, Yorkshire (Fig. 1) consists of five formations (Rawson & Wright 1995; Fig. 2). This group comprises similar facies to those encountered in the Brent Province and Mid-Norway with sandstones, siltstones, mudstones and coals of Aalenian to Bathonian age deposited in a coastal/delta plain to shelfal environment. A palynostratigraphical scheme has been developed from quantitative palynological analyses of 114 field samples taken from Ravenscar Group outcrops. This has allowed a chronostratigraphical subdivision of the Ravenscar Group and enabled regional correlation of the constituent formations with coeval sediments of similar facies of the North Viking Graben (Brent Group) and offshore Mid-Norway (Fangst Group).

Methodology and techniques

The analyses were concentrated on three key sections near Whitby, Ravenscar and Scarborough (Fig. 1). Additionally, some of these sections were logged sedimentologically to place the samples in a sedimentary facies context. Samples were selected from mudstones, siltstones and argillaceous sandstones as previous experience has demonstrated these lithologies to be most productive for palynology. Palynological preparations were made using a standard acid maceration and oxidation process with the residue strewn onto a permanent glass slide and analysed under the light microscope. For each sample, a split count technique was undertaken where 200 specimens were counted in the 10 μm fraction. An additional count was made where 100 dinoflagellate cysts were counted, wherever practical, concurrent to the first count, providing a high-resolution, quantitative database. Additionally, each slide was thoroughly scanned to ensure that all minor occurrences were recorded and that the abundance data were representative of the entire slide. A combination of numerical abundance changes in conjunction with evolutionary tops and bases of key palynomorph taxa (mainly dinoflagellate cysts; Williams *et al.* 1998 provide a comprehensive index for all dinoflagellate cyst

From: POWELL, A. J. & RIDING, J. B. (eds) 2005. *Recent Developments in Applied Biostratigraphy*. The Micropalaeontological Society, Special Publications, 43–53. 1747-602X/$15.00

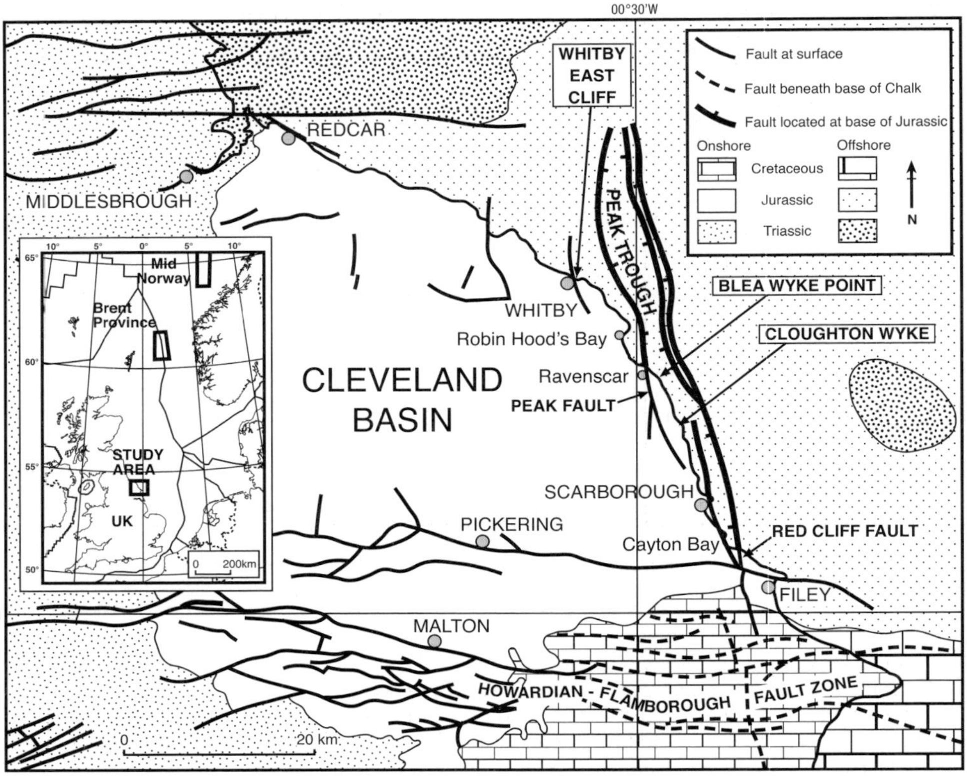

Fig. 1. Location map for the Cleveland Basin and its geographical relationship with the Brent Province and Mid-Norway (after Rawson & Wright 1992).

taxa discussed herein) have been used to constrain the gross stratigraphy. This approach has proven instrumental in many studies particularly of the Lower–Middle Jurassic of the Mid-Norway Heidrun Field (Pedersen *et al.* 1989) and in the Viking Graben to the south (Mitchener *et al.* 1992; Whitaker *et al.* 1992; Bray *et al.* 1995). The abundance events may be manifested as top increases, acmes or base increases of a specific taxon or group of taxa. A number of significant events are established herein and summarized in Figure 3. These are regionally correlative with the Brent and Fangst groups, which has facilitated a refined stratigraphy for these areas.

Previous palynostratigraphical work

Little has been published on the palynology of the Middle Jurassic of Yorkshire. The earliest palynological work was a comprehensive evaluation of miospores from the entire Ravenscar Group concentrating on taxonomy by Couper (1958). However, some stratigraphical data are of interest, particularly regarding the distribution of *Classopollis* spp. that broadly concurs with the distribution recorded herein. Hancock & Fisher (1981) documented the palynofacies of selected parts of the Saltwick, Eller Beck, Cloughton and 'lower' Scarborough formations. Although this information is useful, it provides little foundation for stratigraphical interpretation and correlation. Woollam & Riding (1983) published a palynostratigraphical account on a limited number of field samples. Riding (1984) documented dinoflagellate cyst assemblages from the mid-Toarcian to Aalenian of Blea Wyke Point, Ravenscar. Palynofloral composition compares closely with that recorded herein. The Scarborough Formation was documented by Gowland & Riding (1991) and although a useful sedimentary log was produced (and utilized herein), only eight palynological analyses were undertaken from the entire formation. Hogg (1993) produced a comprehensive account of the largely non-marine Scalby

SYSTEM	STAGE	LITHOSTRATIGRAPHY				
MIDDLE JURASSIC	BATHONIAN (164.4 Ma)	Ravenscar Group	Scalby Formation	60m	Long Nab Member	
	BATHONIAN-LATE BAJOCIAN (? 172.87 Ma)				Moor Grit Member	
	EARLY BAJOCIAN		Scarborough Formation	30m		
			Cloughton Formation	85m	Gristhorpe Member	
					Lebberston Member	9m
	(? 176.5 Ma)				Sycarham Member	
	AALENIAN		Eller Beck Formation	8m		
			Saltwick Formation	57m		
	(180.1 Ma)	Lias Group	Dogger Formation	12m		

Fig. 2. Subdivision of the Middle Jurassic (Aalenian–Bathonian) succession (modified after Rawson & Wright 1992).

Formation. It is apparent that very few palynostratigraphical data have been produced for the more proximal facies of the Saltwick Formation, Sycarham Member and Gristhorpe Member or the marine Eller Beck Formation.

There is considerably more published palynological information regarding the Brent Group. Authors including Graue *et al.* (1987), Mitchener *et al.* (1992), Whitaker *et al.* (1992), Helland-Hansen *et al.* (1992) and Husmo *et al.* (2002) have published stratigraphical schemes, the main conclusion being that the Broom Formation is Aalenian in age, the Rannoch, Etive and Ness formations are Bajocian and the Tarbert Formation is 'latest' Bajocian or Bathonian in age. This interpretation contrasts markedly with the new data presented herein.

Palynostratigraphy of the Ravenscar Group and correlation with the Brent and Fangst Groups

The Lias Group and Dogger Formation were sampled at two locations, namely at Blea Wyke Point where it is 12 m thick and Whitby East Cliff where it is condensed to less than 1 m thick. The Dogger Formation generally comprises argillaceous shelfal sandstones and siltstones that rest unconformably on either the Blea Wyke Sandstone Formation (at Blea Wyke Point) or the older Whitby Mudstone Formation (at Whitby East Cliff). These sediments yield high numbers and a high diversity of dinoflagellate cysts assigned to the Family Phallocystaceae (including *Parvocysta nasuta, Susadinium scrofoides, Susadinium faustum, Ovalicysta hiata, Moesiodinium raileanui, Phallocysta eumekes,* and *P. elongata*). It is possible that these records in the Dogger Formation at Whitby East Cliff may represent reworked occurrences since reworked mud-clasts derived from the Whitby Mudstone Formation are present. However, a diverse and abundant suite of Phallocystaceae (dominated by *Susadinium scrofoides* and *Ovalicysta hiata* with low but consistent numbers of *Phallocysta* and *Parvocysta* spp.) are present in the much thicker Dogger sequence at Blea Wyke Point. The abundance and diversity of this group however decreases sharply in the uppermost part of the formation, a feature that has also been identified in the uppermost Drake Formation (Dunlin Group) in the North Viking Graben (Bray *et al.* 1995) and uppermost Ror Formation (Båt Group) in Mid-Norway (Butler, personal observation). It is possible that these events are a coeval response to a regional regressive phase, which resulted from uplift at the triple junction between the Viking Graben, Moray Firth and Central Graben. However, the top of diverse Phallocystaceae is also identified within the Opalinum ammonite zone in the Bearreraig Sandstone Formation of Skye, Scotland (Riding *et al.* 1991) and consequently

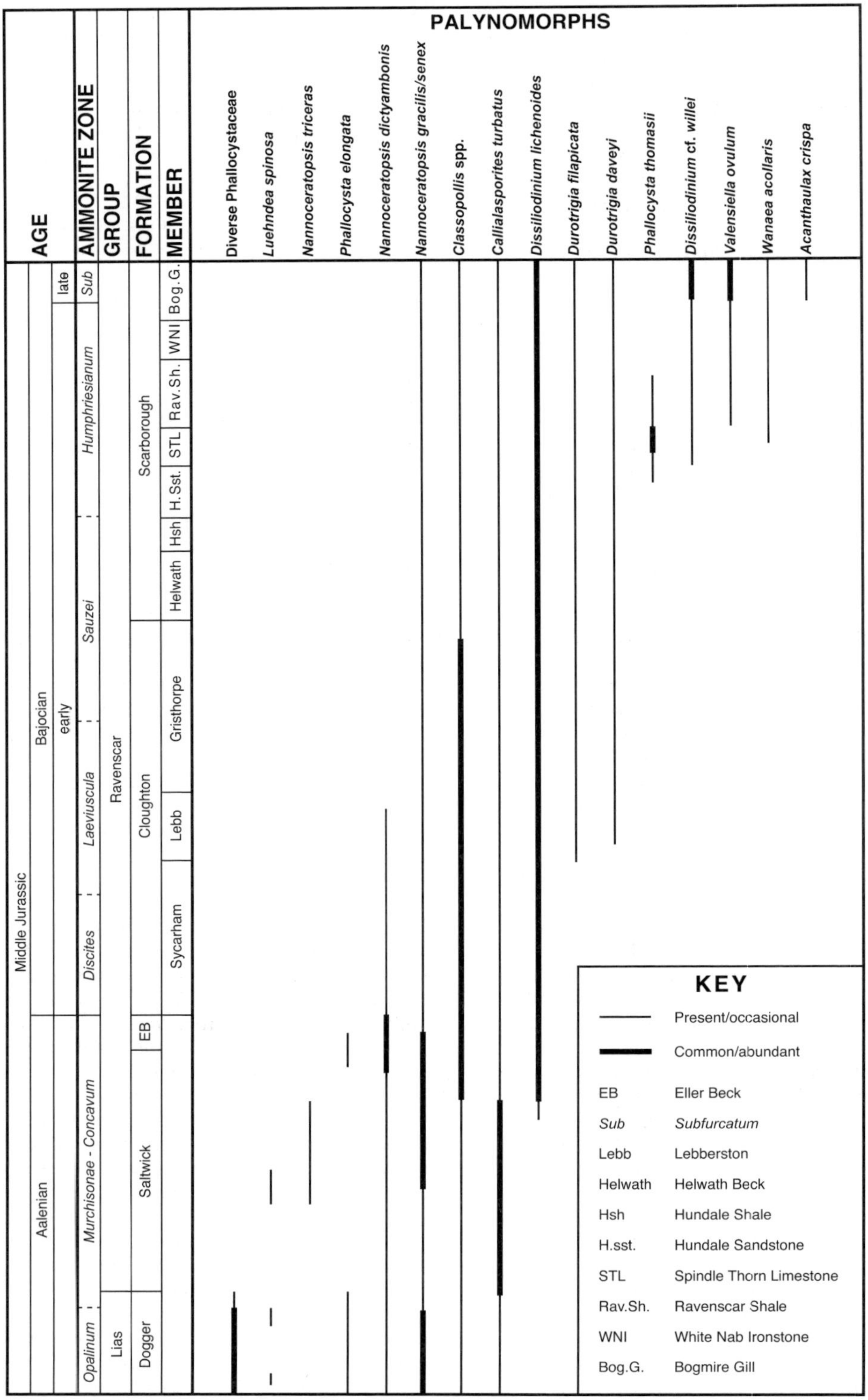

Fig. 3. Ranges of selected taxa in the Ravenscar Group.

its true evolutionary top may also be manifested in the Cleveland Basin. The Dogger Formation is age dated as earliest Aalenian Opalinum ammonite zone on the basis of ammonite faunas recovered from Blea Wyke Point (Cope *et al.* 1980; Riding 1984).

The Saltwick Formation is characterized by sandstones, siltstones, mudstones and coals deposited in a paralic to non-marine environment. This was logged sedimentologically and sampled for biostratigraphy at Whitby East Cliff where it comprises a 33 m thick mudstone-dominated succession (Fig. 4). Although no ammonite faunas have been recovered, the age of the formation is considered to be Aalenian (Murchisonae to Concavum ammonite zones) in age (Cope *et al.* 1980). A major basinward shift in facies is inferred at the base of the formation by the absence of dinoflagellate cysts and the proliferation of pteridophyte spores in conjunction with lacustrine or overbank sedimentary facies. This non-marine aspect persists throughout the lower 10 m of the Saltwick Formation.

The middle to upper interval of the formation is characterised by an increase in dinoflagellate cysts that comprise approximately 1–2% of the total palynoflora; principally *Nannoceratopsis gracilis*, *N. senex*, *N. triceras*, *Mancodinium semitabulatum* and *Mendicodinium morgenrothii*. This assemblage, which broadly substantiates an Aalenian age, has also been identified in the Brent Group, Ness Formation. (The extinction of *Nannoceratopsis triceras* is ammonite constrained as 'earliest' Bajocian, intra-Discites ammonite zone in the UK according to Woollam & Riding 1983.)

Luehndea spinosa has been identified from marine intervals in the 'middle' Saltwick Formation as well as consistently occurring in lower delta plain facies of the Ness Formation. This taxon has an accepted range top (extinction) in the early Toarcian (Woollam & Riding 1983) and its reappearance in these formations is somewhat controversial. In the Brent Province, the event was originally considered to represent reworking of late Pliensbachian to early Toarcian sediments. However, the occurrences of *Luehndea spinosa* in these Aalenian sediments are associated with appropriate ichnofaunas and sedimentary structures that are consistent with a marginal marine, distal environment as opposed to a non-marine, proximate environment. Additionally, increases in other dinoflagellate cyst taxa such as *Mancodinium semitabulatum* and *Nannoceratopsis* spp., in conjunction with gymnosperm pollen associations and lower numbers of pteridophyte spores are noted. If *Luehndea spinosa* was reworked then it would be reasonable to encounter other indigenous late Pliensbachian to early Toarcian taxa such as *Scriniocassis weberi*, *Comparodinium* spp., *Valvaeodinium* spp. and *Maturodinium inornatum* as reworking. This is not the case in either the Saltwick or Ness formations and consequently *Luehndea spinosa* is considered to be *in situ* in these formations. One possible interpretation for the disappearance then reappearance of this taxon could be that the motile stage responsible for *Luehndea spinosa* stopped producing fossilizable cysts at the end of the early Toarcian. Subsequently, during the Aalenian, the motile forms (still present in the water column) had become adapted to life in stressed salinities and had recommenced producing fossilizable cysts. Alternatively, the forms identified as *Luehndea spinosa* in the Saltwick and Ness Formations may be subtly different from these older forms, which became extinct in the early Toarcian. Improved taxonomic work incorporating scanning and transmission electronic microscopy would be necessary to clarify this hypothesis.

Miospores are generally characterized by high numbers of pteridophyte spores throughout. An up-section increase in the xerophytic pollen genus *Classopollis* may be correlative with a similar base increase identified in the Ness Formation (associated with the base of the 'mid-Ness shale'; Bray *et al.* 1995). This increase may be the result of climatic change with an increase in aridity (Vakrameev 1981; Abbink 1998) which commenced in the 'mid' Aalenian. The effects of this increased aridity could be manifested to the north and south of the Mid-North Sea High, affecting both the Cleveland Basin and the North Viking Graben.

The uppermost part of the Saltwick Formation yields a base and up-section increase in *Dissiliodinium lichenoides*. The inception of this dinoflagellate cyst was reported in the uppermost part of the Murchisonae ammonite zone of southwest Germany by Feist-Burkhardt & Monteil (2001). A similar base and base increase is also well documented in the uppermost Ness Formation, where preserved (Butler, personal observation). This event is absent from the Not and 'lower' Garn formations, which are interpreted to predate it.

The Eller Beck Formation was sedimentologically logged and sampled near Cloughton Wyke and is probably latest Aalenian in age based on a lithostratigraphical correlation with the Hydraulic Limestone in South Humberside (Cope *et al.* 1980). It represents a significant marine incursion and is characterized by the

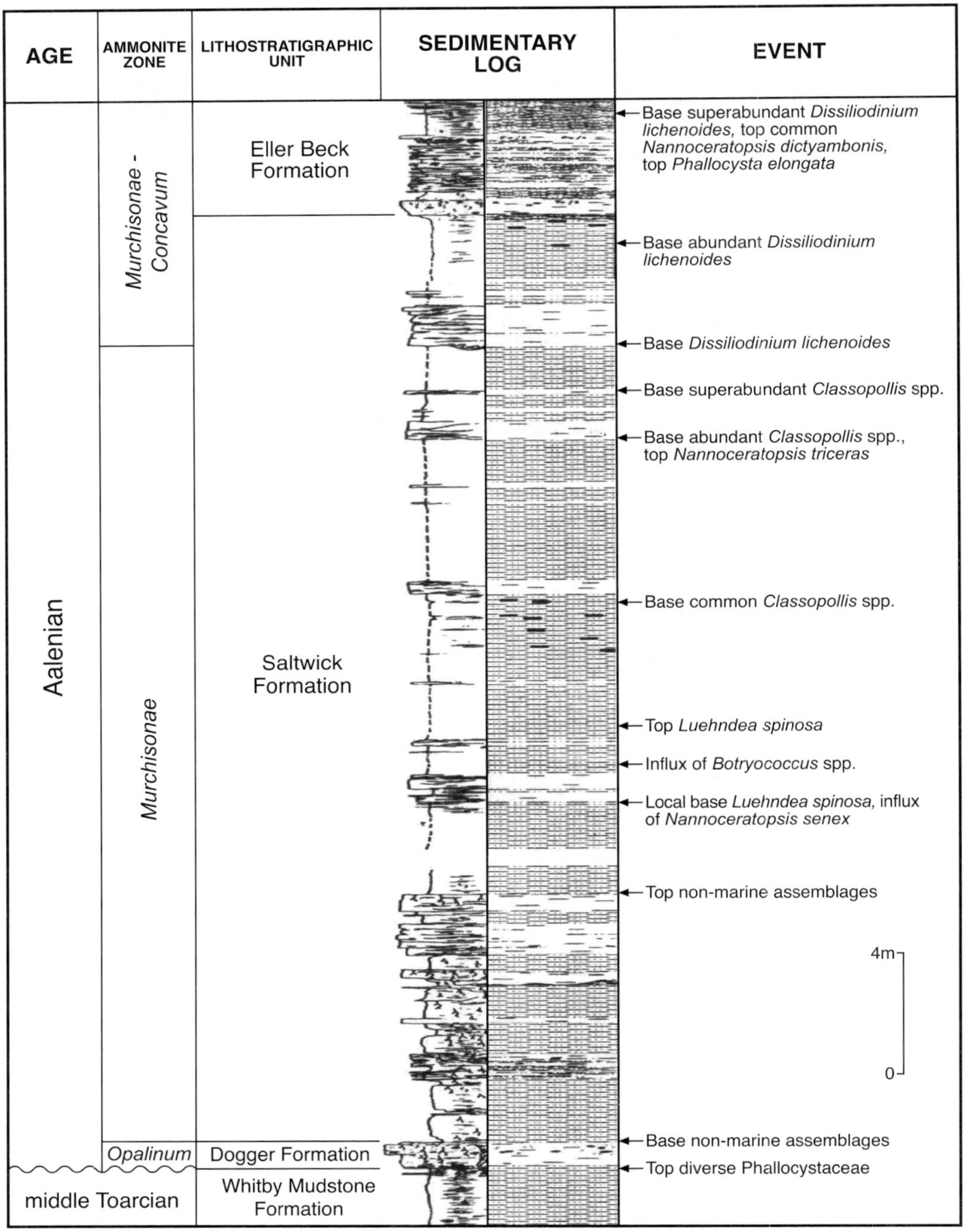

Fig. 4. Stratigraphy of the Saltwick Formation at Whitby East Cliff.

persistence of *Dissiliodinium lichenoides*, in conjunction with *Nannoceratopsis dictyambonis* and *Dissiliodinium psilatum*, all in high relative abundances. Although the formation is considered correlative with the 'uppermost' Ness Formation (where preserved in the North Viking Graben), it post-dates the Not/'lower' Garn Formations which lack these taxa.

The Cloughton Formation was also logged and sampled near Cloughton Wyke and comprises the Sycarham, Lebberston and Gristhorpe members, which are of early Bajocian age (Cope *et al.* 1980). Sediments of this age are significantly truncated or may be condensed in the Brent Province and Mid-Norway.

The inceptions of *Durotrigia filapicata* and *Sentusidinium granulatum* were identified within the Lebberston Member in the Cleveland Basin sections. The true inception is not encountered in the North Viking Graben due to the interpreted stratigraphical break or condensing, with these taxa first appearing in the Tarbert Formation. The co-occurrences of *Durotrigia daveyi* and *Nannoceratopsis dictyambonis*, also within the Lebberston Member (Butler, personal observation), suggests a Laeviuscula ammonite zone age for this unit (Penn *et al.* 1980). These latter two taxa are not encountered in the Brent Province or Mid-Norway.

The top of superabundant *Classopollis* spp. is identified in the uppermost part of the Gristhorpe Member and possibly represents the end of an arid climatic phase during the later part of the early Bajocian. High abundances of this xerophytic pollen genus are also encountered throughout the Aalenian Ness Formation with the true elimination event not identified due to the interpreted truncation at the Ness/Tarbert formational boundary. High numbers of *Classopollis* spp. are generally absent from the Tarbert Formation except in UK Block 211/23 and 211/18 wells where potentially older Tarbert Formation is present (Butler, personal observation).

The Scarborough Formation comprises mudstones, sandstones and silicious carbonates deposited in a progressively deepening environment (Gowland & Riding 1991) and was sampled at Hundale Point to the south of Cloughton Wyke. The distribution of *Phallocysta thomasii* has been carefully documented in the Hundale Point section. Its stratigraphical first and last occurrences have been identified in the middle parts of the Humphriesianum ammonite zone dated Hundale Sandstone and Ravenscar Shale members, respectively, with maximum numbers noted within the upper part of the Spindle Thorn Limestone Member (Fig. 5). This is considered significant for the calibration of similar distributions of *Phallocysta thomasii* within the 'upper' Garn and Melke Formations of the Norwegian Sea and in the Tarbert and Heather formations of the North Viking Graben, the implication being that these formations are, in part, Humphriesianum ammonite zone and therefore 'latest' early Bajocian in age.

The inception of *Dissiliodinium* cf. *willei* has been noted from the uppermost Scarborough Formation, Spindle Thorn Limestone Member, with maximum numbers identified in the Bogmire Gill Member at Hundale Point. This form morphographically resembles *Dissiliodinium willei sensu stricto*, differing in possessing smaller, isolated but hyaline grana, and is perhaps the immediate precursor to *Dissiliodinium willei*. Although the Bogmire Gill Member has been dated in the literature as Humphriesianum ammonite zone by inference (Parsons 1977; Gowland & Riding 1991), marker ammonites are absent. The overall dinoflagellate cyst assemblage is more consistent with the 'earliest' late Bajocian Subfurcatum ammonite zone as described by Feist-Burkhardt & Wille (1992) from Germany (as the Niortense ammonite zone). The absence of *Dissiliodinium willei sensu stricto* from the Hundale Point section is probably due to truncation at the top of the Scarborough Formation with the late Bajocian Garantiana and Parkinsoni zones being absent. The acme of *Dissiliodinium willei sensu stricto* is encountered in Heather and coeval Tarbert formation assemblages in the North Viking Graben, and the Melke Formation in Mid-Norway (Fig. 6). The event is calibrated to the boreal Pompeckji Zone from ammonite-dated core (Regnouf de Vains 1980). This zone is considered to be equivalent to the Parkinsoni and Garantiana ammonite zones in the UK (Callomon 1993).

A stratigraphical break is tentatively identified at the Scarborough Formation, Bogmire Gill Member/Scalby Formation, Moor Grit Member boundary (Leeder & Nami 1979; Riding & Wright 1989) although biostratigraphical control within the Moor Grit Member is lacking. The overlying Long Nab Member yields a similar palynomorph assemblage to the Pentland Group, 'lower' Bryne Formation in the Central Graben which is interpreted as Bathonian in age due to the presence of such miospore taxa as *Verrureticulisporis giganteus*, *Couperisporites jurassicus* and *Quadraeculina anellaeformis* (Butler, personal observation.). The formation is considered age equivalent to the Bathonian Heather and Melke formations of the northern North Sea and Mid-Norway, respectively (post-*Dissiliodinium willei* acme and pre-*Chytroeisphaeridia hyalina* acme).

Conclusions

The correlation of palynoevents identified in the Ravenscar Group with coeval sediments in the North Viking Graben and Mid-Norway is

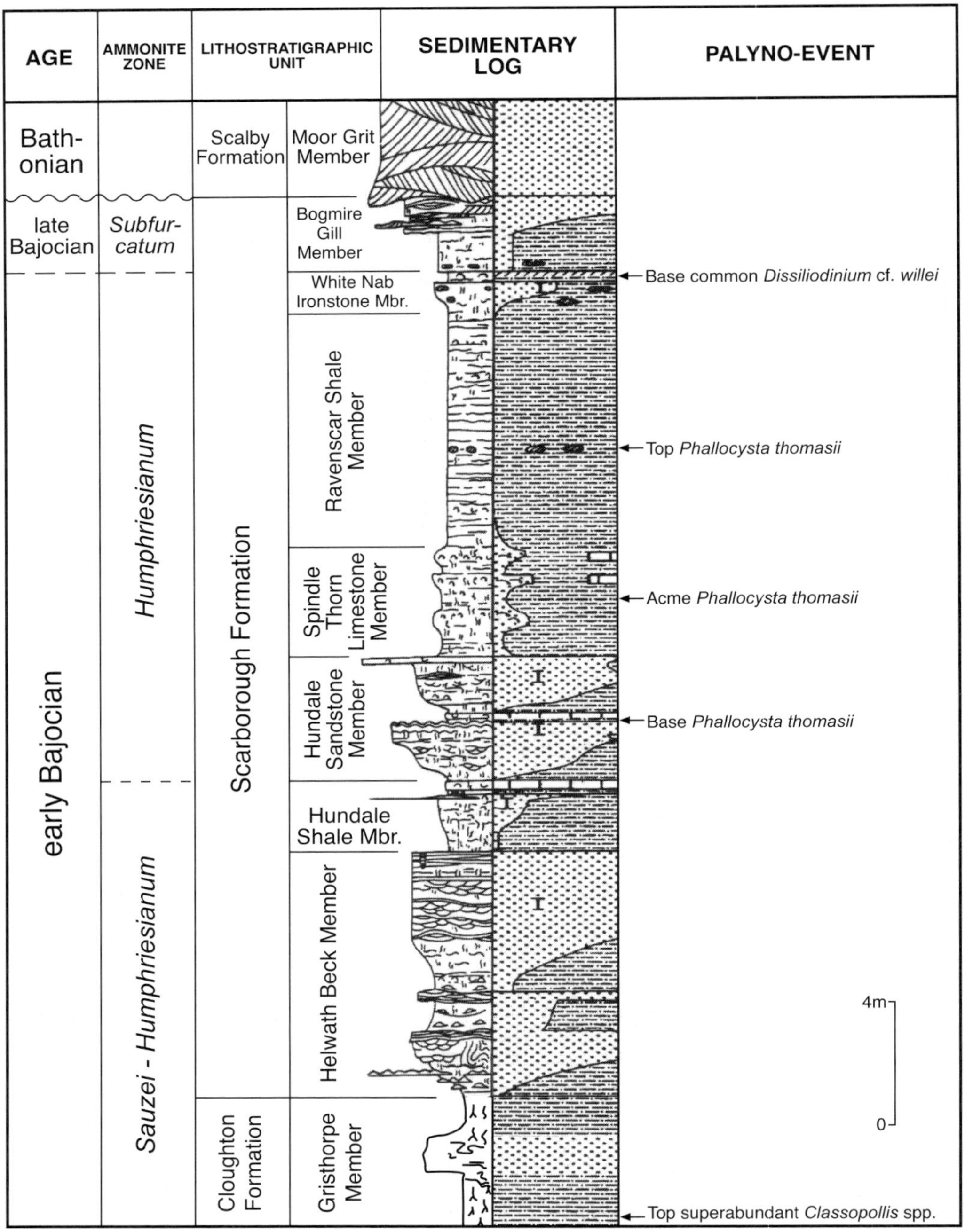

Fig. 5. Stratigraphy of the Scarborough Formation (modified after Gowland & Riding 1991).

summarized in Figure 6 and contrasts markedly with those previously published (e.g. Graue *et al.* 1987; Mitchener *et al.* 1992; Whitaker *et al.* 1992; Helland-Hansen *et al.* 1992; Husmo *et al.* 2002). The shelfal marine Dogger Formation is of 'earliest' Aalenian age and correlative with uppermost Dunlin Group, Drake Formation and Båt Group, Ror Formation. The Aalenian

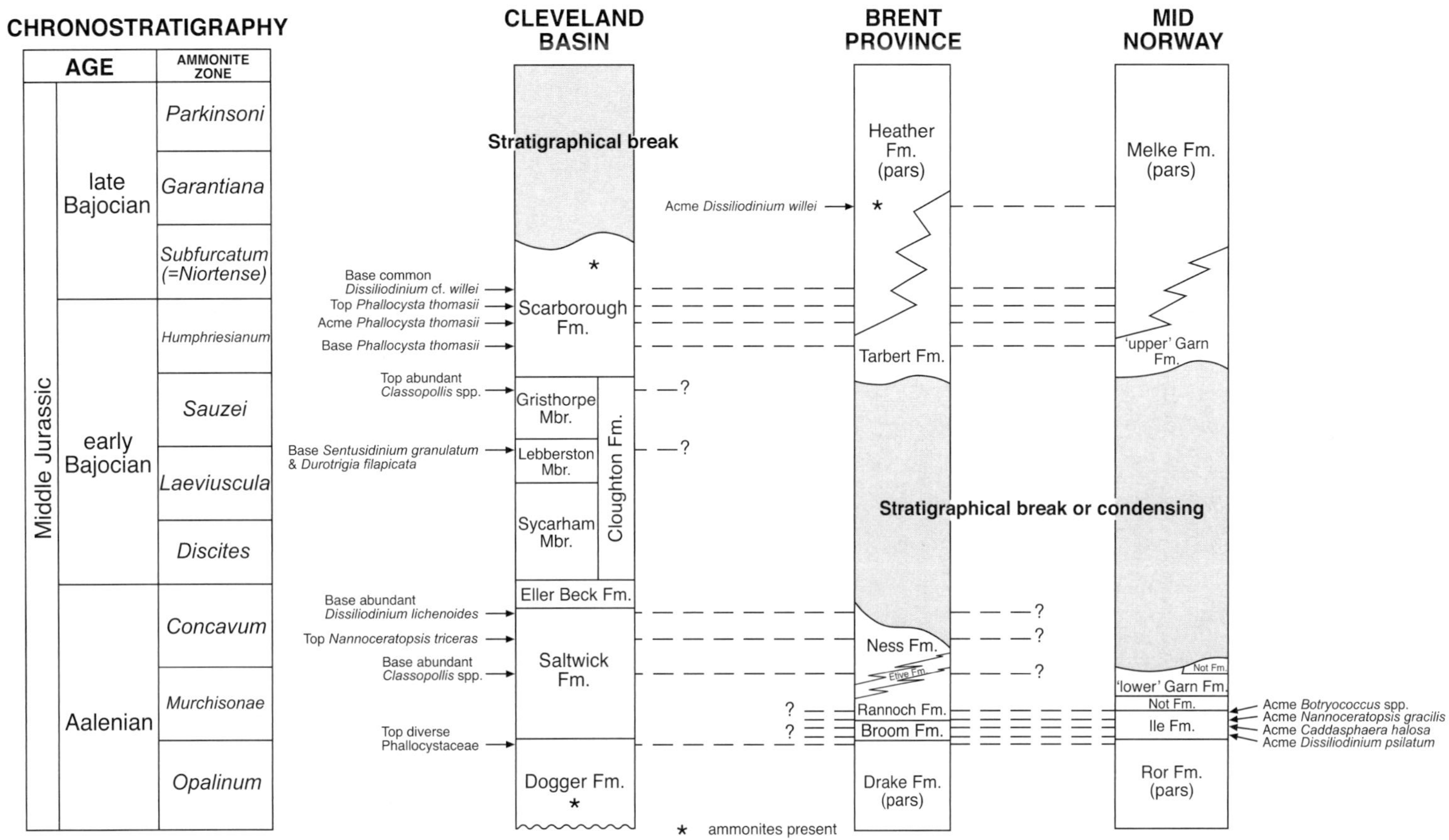

Fig. 6. Correlation of the Cleveland Basin with the Brent province and Mid-Norway.

Saltwick Formation (non-marine/brackish coastal plain/deltaic sediments) and Eller Beck Formation (marine incursion) correlate with the Brent Group, Broom–Rannoch–Etive–Ness and the Fangst Group, Ile–Not–'lower' Garn genetic packages. The variably non-marine to shelfal sediments of the Cloughton Formation are either unrepresented or condensed in the Brent Province and Mid-Norway due to a regional unconformity, which truncates early Bajocian sediments. The Scarborough Formation is a retrogradational package, which was deposited in a shelfal environment and is of 'latest' early to 'earliest' late Bajocian age. This formation is correlative with the older part of the Tarbert–Heather and the 'upper' Garn–Melke genetic packages.

Our thanks are due to the directors of Norsk Hydro a.s. and Robertson Research International Ltd for the use of their technical facilities and permission to publish this paper. We are particularly grateful to J. P. G. Fenton for critically reading the manuscript.

Appendix: list of palynomorph taxa

Dinoflagellate cysts

Chytroeisphaeridia hyalina (Raynaud 1978) Lentin & Williams 1981
Comparodinium spp. Morbey 1975
Dissiliodinium lichenoides Feist-Burkhardt & Monteil 2001
Dissiliodinium psilatum Prauss 1989
Dissiliodinium willei Bailey & Partington 1991
Durotrigia daveyii Bailey 1986
Durotrigia filapicata (Gocht 1970) Riding & Bailey 1991
Luehndea spinosa Morgenroth 1970
Mancodinium semitabulatum Morgenroth 1970
Maturodinium inornatum Morgenroth 1970
Mendicodinium morgenrothii Butler 1995
Moesiodinium raileanui Antonesçu 1974
Nannoceratopsis dictyambonis Riding 1984
Nannoceratopsis gracilis Alberti 1961
Nannoceratopsis senex Van Helden 1977
Nannoceratopsis triceras Drugg 1978
Ovalicysta hiata Bjaerke 1980
Parvocysta nasuta Bjaerke 1980
Phallocysta elongata (Beju 1971) Riding 1994
Phallocysta eumekes Dörhöfer & Davies 1980
Phallocysta thomasii Smelror 1991
Scriniocassis weberi Gocht 1964
Sentusidinium granulatum (Courtinat & Gaillard, 1980) Brenner 1988
Susadinium faustum (Bjaerke 1980) Lentin & Williams, 1985
Susadinium scrofoides Dorhofer & Davies, 1980
Valvaeodinium spp. Morgenroth 1970

Spores

Couperisporites jurassicus Pocock 1970
Verrureticulisporis giganteus Danzé-Corsin & Laveine 1963

Pollen

Classopollis spp. Pflug 1953
Quadraeculina anellaeformis Maljavkina 1949

References

ABBINK, O.A. 1998. Palynological investigations in the Jurassic of the North Sea region. Laboratory of Palaeobotany and Palynology Contributions Series 8, Laboratory of Palaeobotany and Palynology Foundation, Utrecht, 1–191.

BRAY, A.A., BUTLER, N., KIMBER, R.N. & DRAPER, L.F. 1995. Evolution of the Brent delta: An overview of patterns and controls. *In*: OTI, M.N. & POSTMA, G. (eds) *Geology of Deltas*, Balkema, Rotterdam. 97–123.

CALLOMON, J.H. 1993. The ammonite succession in the Middle Jurassic of East Greenland. *Bulletin of the Geological Society, Denmark*, **40**, 83–113.

COPE, J.C.W., DUFF, K.L., PARSONS, C.F., TORRENS, H.S., WIMBLEDON, W.A. & WRIGHT J.K. 1980. *A correlation of Jurassic rocks in the British Isles, Part 2: Middle and Upper Jurassic*. Geological Society, London, Special Publications, **15**, 1–109.

COUPER, R.A. 1958. British Mesozoic microspores and pollen grains, a systematic and stratigraphic study. *Palaeontographica Abt. B*, **103**, 75–179.

DALLAND, A., WORSLEY, D. & OFFSTAD, K. 1988. A lithostratigraphic scheme for the Mesozoic and Cenozoic succession offshore mid and northern Norway. *Norwegian Petroleum Directorate Bulletin*, **4**, 1–65.

DEEGAN, C.E. & SCULL, B.J. 1977. *A Standard Lithostratigraphic Nomenclature for the Central and Northern North Sea*. Institute of Geological Sciences Report, **77/25**, 1–36.

FEIST-BURKHARDT, S. & MONTEIL, E. 2001. Gonyaulacacean dinoflagellate cysts with multiplate precingular archaeopyle. *Neues Jahrbuch für Geologie und Paläontologie, Abhandlungen*, **219**, 33–81.

FEIST-BURKHARDT, S. & WILLE, W. 1992. Jurassic palynology in southwest Germany – state of the art. *Cahiers de Micropaléontologie*, **7**, 1/2, 141–156.

GOWLAND, S. & RIDING, J.B. 1991. Stratigraphy, sedimentology, and palaeontology of the Scarborough Formation (Middle Jurassic) at Hundale Point, North Yorkshire. *Proceedings of the Yorkshire Geological Society*, **48**, 375–392.

GRAUE, E., HELLAND-HANSEN, W., JOHNSON, J., LOMO, L., NOTTVEDT, A., RONNING, K., RYSETH, A. & STEEL, R. 1987. Advance and retreat of the Brent Delta System, Norwegian North Sea. *In*: BROOKS, J. & GLENNIE, K.W. (eds) *Petroleum Geology of Northwest Europe*, **2**. Graham and Trotman, London, 915–937.

HANCOCK, N. & FISHER, M. 1981. Middle Jurassic North Sea deltas with particular reference to

Yorkshire. *In*: ILLING, L.V. & HOBSON, G.D. (eds) *Petroleum Geology of the Continental Shelf of Northwest Europe*. Hayden, London, 186–195.

HELLAND-HANSEN, W., ASHTON, M., LOMO, L. & STEEL, R. 1992. Advance and retreat of the Brent delta: recent contributions to the depositional model. *In*: MORTON, A.C., HASZELDINE, R.S., GILES, M.R. & BROWN, S. (eds) *Geology of the Brent Group*. Geological Society, London, Special Publications, **61**, 109–127.

HOGG, N.M. 1993. *A palynological investigation of the Scalby Formation (Ravenscar Group, Middle Jurassic) and adjacent strata from the Cleveland Basin, North East Yorkshire*. PhD thesis, University of Sheffield.

HUSMO, T., HAMAR, G., HØILAND, O., JOHANNESSEN, E.P., RØMULD, A., SPENCER, A. & TITTERTON, R. 2002. Lower and Middle Jurassic. *In*: EVANS, D., GRAHAM, C., ARMOUR, A. & BATHURST, P. (eds) *The Millennium Atlas: Petroleum Geology of the Central and Northern North Sea*. Geological Society, London, 129–155.

LEEDER, M.R. & NAMI, M. 1979. Sedimentary models for the non-marine Scalby Formation (Middle Jurassic) and evidence for late Bajocian/Bathonian uplift of the Yorkshire Basin. *Proceedings of the Yorkshire Geological Society*, **42**, 461–482.

MITCHENER, B.C., LAWRENCE, D.A., PARTINGTON, M.A., BOWMAN, M.B.J. & GLUYAS, J. 1992. Brent Group: sequence stratigraphy and regional implications. *In*: MORTON, A.C., HASZELDINE, R.S., GILES, M.R. & BROWN, S. (eds) *Geology of the Brent Group*. Geological Society, London, Special Publications, **61**, 45–80.

PARSONS, C.F. 1977. A stratigraphic revision of the Scarborough Formation (Middle Jurassic) of North East Yorkshire. *Proceedings of the Yorkshire Geological Society*, **41**, 203–222.

PEDERSEN, T., HARMS, J.C., HARRIS, N.B., MITCHELL, R.W. & TOOBY, K.M. 1989. The role of correlation in generating the Heidrun Field geological model. *In*: COLLINSON, J.D. (ed.) *Correlation in Hydrocarbon Exploration*. Graham and Trotman, London, 327–338.

PENN, I.E., DINGWALL, R.G. & KNOX, R.W.O'B. 1980. *The Inferior Oolite (Bajocian) sequence from a borehole in Lyme Bay, Dorset*. Institute of Geological Sciences Report, **79/3**, 1–27.

RAWSON, P.F. & WRIGHT, J.K. 1995. Jurassic of the Cleveland Basin, North Yorkshire. *In*: TAYLOR, P.D. (ed.) *Field Geology of British Jurassic*. Geological Society, London, 173–208.

REGNOUF DE VAINS, G. 1980. *Etude de microplankton du Jurassique Moyen (Bathonien–Callovien) dans le Graben de Viking (Mer du Nord)*. PhD thesis, University of Toulouse.

RIDING, J.B. 1984. A palynological investigation of Toarcian to early Aalenian strata from the Blea Wyke area, Ravenscar, North Yorkshire. *Proceedings of the Yorkshire Geological Society*, **45**, 109–122.

RIDING, J.B. & WRIGHT, J.K. 1989. Palynostratigraphy of the Scalby Formation (Middle Jurassic) of the Cleveland Basin, Northeast Yorkshire. *Proceedings of the Yorkshire Geological Society*, **47**, 349–354.

RIDING, J.B., WALTON, W. & SHAW, D. 1991. Toarcian to Bathonian (Jurassic) palynology of the Inner Hebrides, Northwest Scotland. *Palynology*, **6**, 193–202.

VAKRAMEEV, V.A. 1981. Pollen *Classopollis*: indicator of Jurassic and Cretaceous climates. *The Palaeobotanist*, **28–29**, 301–307.

WHITAKER, M.F., GILES, M.R. & CANNON, S.J.C. 1992. Palynological review of the Brent Group, UK Sector, North Sea. *In*: MORTON, A.C., HASELDINE, R.S., GILES, M.R. & BROWN, S. (eds) *Geology of the Brent Group*. Geological Society, London, Special Publications, **61**, 169–202.

WILLIAMS, G.L., LENTIN, J.K. & FENSOME, R.A. 1998. *The Lentin and Williams Index of Fossil Dinoflagellates 1998 Edition*. American Association of Stratigraphic Palynologists Contributions Series, **34**.

WOOLLAM, R. & RIDING, J.B. 1983. *Dinoflagellate Cyst Zonation of the English Jurassic*. Institute of Geological Sciences Report 83/2, 1–41.

Preliminary observations on the micropalaeontological characterization of submarine fan/channel sub-environments, Ainsa System, south-central Pyrenees, Spain

R. W. JONES[1], K. T. PICKERING[2], M. BOUDAGHER-FADEL[2] & S. MATTHEWS[2]

[1]*Area 123G, Building H, BP, Chertsey Road, Sunbury-on-Thames, Middlesex TW16 7LN, UK (e-mail: jonesbob@bp.com)*

[2]*Department of Earth Sciences, University College London, Gower Street, London WC1E 6BT, UK*

Abstract: The submarine fan/channel complex of the Eocene Ainsa System in the south-central Pyrenees, Spain is in the process of being analysed for its microfossil content. The aims of this analysis are to establish the age and palaeoenvironment of the system, and to investigate whether the different types of (channel axis, off-axis and levee/overbank) sub-environments that it contains can be discriminated using micropalaeontological criteria, thus contributing to a hydrocarbon reservoir analogue model. This paper presents some preliminary observations.

The Ainsa System, which crops out in and around the Romanesque medieval town of Ainsa (Fig. 1), comprises three separate submarine fan/channel complexes of Eocene age (Fig. 2). For further details of the sedimentology of the system and the geology of the containing basin the reader is referred to Pickering & Corregidor (2000).

Staff and post-graduate students from the Department of Geological Sciences at University College London (UCL) have been involved in research on the system since 1996, in part funded through the FORCE Deep-Water Clastics Workgroup (a consortium involving the Norwegian Petroleum Directorate and seven oil companies including AMOCO (now BP)). The main aim of this research has been and continues to be to integrate surface outcrop and subsurface borehole and seismic data in order to derive sophisticated descriptive and predictive depositional models for the development of the submarine fan/channel complexes to serve as analogues for petroleum reservoir exploration and exploitation. At the time of writing, the extensive surface outcrop has been mapped and sedimentologically described (Pickering & Corregidor 2000), eight 250 m subsurface boreholes drilled and petrophysically logged, 2000 m of slim-hole core cut and sedimentologically described, and approximately 200 samples biostratigraphically analysed at UCL and at Robertson Research in North Wales (the results of all of which have been evaluated and integrated into our interpretation as appropriate). In addition, some seismic data has been acquired though, unfortunately, this is of somewhat restricted usefulness on account of an unexpectedly poor acoustic impedance contrast between sand bodies and enveloping muds (a function of pervasive carbonate cementation). The surface outcrops have been correlated by means of tracing individual lithostratigraphic and lithofacies units over distances of up to 5 km, both by simply walking them and by use of aerial photographs, while the subsurface boreholes have been correlated by means of sedimentological/petrophysical and, to a lesser extent, biostratigraphic data. The surface outcrops and subsurface boreholes have been tied the one to the other using projection of surface structural dip data into the subsurface. Thus, the spatial relationships of the various units can be regarded as reasonably well established.

Independently, BP's biostratigraphy network has recently been undertaking operational research on the micropalaeontological characterization of submarine fan/channel sub-environments worldwide. The principal objective of this work has been the discrimination of different types of intra-reservoir mudstones with different engineering properties (barriers vs. baffles to fluid flow) and different implications for exploitation strategy and 'biosteering' (see, for instance, Payne *et al.* 1999; Holmes 1999).

This paper presents the results of the first,

From: POWELL, A. J. & RIDING, J. B. (eds) 2005. *Recent Developments in Applied Biostratigraphy*. The Micropalaeontological Society, Special Publications, 55–68. 1747-602X/$15.00

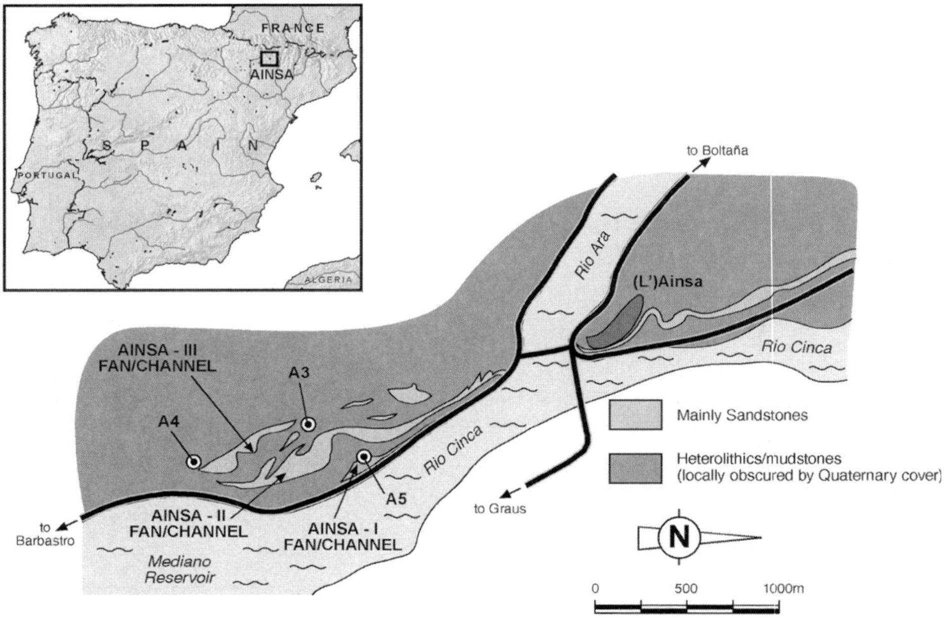

Fig. 1. Location map and field photograph, Ainsa I Fan/Channel Complex. Top, location map showing location of Ainsa and of Ainsa I, III and III fan/channel complexes (inset shows location in context of Iberian Peninsula). Bottom, field photograph, Ainsa I fan/channel complex (from across river valley). Also showing projected positions of studied boreholes. Palaeoflow into plane of photograph and to right (to NW); structural dip 20 degrees into plane of photograph and to left (to SW).

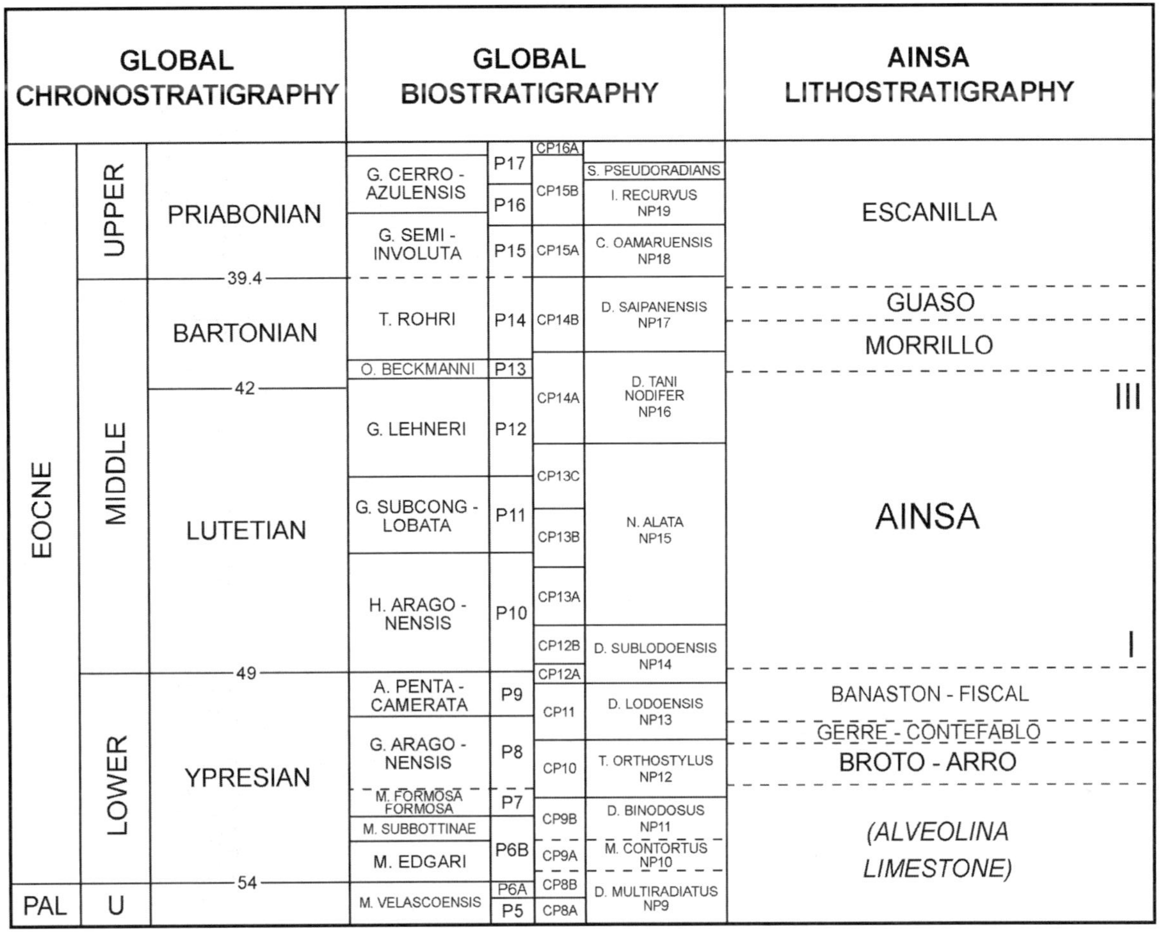

Fig. 2. Stratigraphic column. Global biostratigraphy, standard planktic foraminiferal and calcareous nannoplankton zones (prefixed 'P' and 'NP' respectively). Ainsa lithostratigraphic nomenclature conforms to UCL and BP usage. Calibration best-fit. Note that the Escanilla (?Sobrarbe) Formation, which is conventionally assigned to the Late Eocene, at least locally contains the larger benthic foraminifer *Alveolina*, which ranges no younger than Middle Eocene (RWJ's unpublished field observations).

pilot ('proof of concept') phase of an ongoing collaborative research project between BP and UCL aimed at the micropalaeontological characterization and discrimination of submarine fan/channel sub-environments as represented in selected subsurface boreholes from Ainsa.

A second, follow-up phase is planned partly to test the replicability of the early results presented in this paper, and partly to ensure complete representativeness of sampling, and to address in much more detail issues of vertical variability at, and horizontal variability between, localities (including surface outcrop as well as subsurface borehole localities).

Previous research

As summarized by Jones (1996, pp. 12–13, 17 & 19), there has been a considerable amount of research undertaken over the years on submarine fan or 'flysch-type' and associated faunas. These are also known as '*Rhabdammina*' faunas (Brouwer 1965) or deep-water agglutinated or agglutinating foraminiferal (DWAF) faunas because one of their most characteristic features is the essential absence of a calcareous benthic or planktic component, especially in ancient as opposed to modern submarine fan deposits. This feature could be either primary (depositional) or secondary (diagenetic) or both (i.e. it does not necessarily indicate deposition below the calcite compensation depth). One possible explanation is restricted circulation and hence reduced oxygen concentration at or just below the sediment surface, which factors favour the sustenance and/or preservation of non-calcareous agglutinating foraminifera and the exclusion and/or dissolution of their calc-agglutinating and calcareous counterparts. (This occurs especially at times of lowered sea-level, when the input of organic material and, hence, the release of acid in early diagenetic decomposition reactions is at its highest.) Restricted circulation can in turn, among other alternatives, be attributed to salinity stratification associated with excess fresh-water run-off from a major river system (which, incidentally, would also have the effect of eliminating the normal marine phytoplankton that (as seasonal phytodetritus) is known to constitute an important food source for at least some calcareous benthic species).

Also as noted by Jones (1996):

> "A number of authors have described *vertical variation* in the composition of agglutinating foraminiferal assemblages through interturbiditic mudstones . . .
>
> Many evidently regarded this variation as nothing more than a form of graded bedding (induced by waning flow), in other words that the agglutinate assemblages simply reflected the dynamic properties of their environment . . .
>
> An alternative explanation . . . would envisage the foraminifera behaving as living and discerning organisms rather than as sedimentary particles . . . Thus, Butt (1981) interpreted the differences in vertical variation in the distribution of morphologically . . . primitive and more advanced forms through 20cm interturbiditic mudstones . . . as being due to the greater ability of he former to colonise the barren sea-floor in between turbidity flows . . .
>
> Notably, the most morphologically primitive (tubular unilocular) forms are interpreted by Jones & Charnock (1985) as epifaunal suspension-feeders, which might be expected to favour times or areas characterised by significant amounts of sediment in suspension . . . Also notably (and counter-intuitively), however, uniserial multilocular forms, interpreted . . . as infaunal, proved to most prominent among the pioneer colonisers . . . in experiments conducted by Ellison & Peck (1983) . . . and by Kaminski *et al.* (1988) . . ." [see also Ernst *et al.*, 2000].

and

> "Seiglie *et al.* (1986) described the *horizontal variation* between faunas associated with turbiditic deposits (allochthonous) to those associated with over-bank and basin-plain deposits (quasi-autochthonous" (" . . . the number of indigenous foraminifers increas[ing] with increasing distance from the turbidites"))."

Also in the context of horizontal variation, Bouma *et al.* (1985) and Kohl *et al.* (1985) noted in the case of the modern Mississippi Fan in the Gulf of Mexico that channels are characterized by the non-existent or impoverished development and overbanks by the better development of an autochthonous bathyal benthic fauna. Allochthonous (contemporaneously transported) shelfal, principally calcareous, benthic faunas are present in both settings (as are non-contemporaneously reworked faunas), though more obviously so in sand-prone channels, suggesting that the transportation is associated with density/mass/turbidite flows essentially confined to the channels. Turbiditic/channel deposits of the Monterey Fan in the North-East Pacific, the Baltimore Canyon in the North-West Atlantic and the Amazon Fan in the western equatorial Atlantic are also characterized by transported shallower-water benthic faunas (respectively, Brunner & Ledbetter 1987; Thompson 1992; Vilela & Maslin 1997). Turbiditic deposits from the Monterey Fan are

		168.50m	182.31m	184.04m	203.07m	210.44m	211.43m	217.16m	221.90m
A	*Bathysiphon taurinensis*			9					
	Bathysiphon						5	6	
	Rhabdammina							2	
B	*Recurvoides*						12		
	Glomospira							3	
C	*Dorothia oxycona*		2						
	Ammomarginulina macrospira		4	1					
	Dorothia						1	1	
	Marssonella lodoensis							2	
	Hormosina								2
	Spiroplectammina								1
D	*Cibicidoides*	2			5			5	
	Nonionella	1							
	Nuttallides truempyi		5	6	5			5	
	Cibicidoides eocaenus		6	5	4				
	Lenticulina		6	2		2		1	1
	Cibicidoides dutemplei		4	8	5			10	
	Planulina costata		5	1	2			2	
	Eponides		2					6	
	Anomalinoides			4	2				
	Lenticulina inornata						1		
	Hanzawaia ammophila							2	
	Cibicidoides grimsdalei							4	
	Baggina							2	
	Eponides elevatus							2	
	Baggina dentata							1	
	Cancris							2	
	Cibicides							1	
	Lenticulina limbosa							4	
E	*Uvigerina*	1	13	4			9	4	
	Bulimina		12	12	1			10	
	Fursenkoina		8	3					
	Uvigerina eocaena		14	12				4	
	Bolivina aliformis				1				
	Nodosaria						4		
	Praeglobobulimina pupoides						2		
	Pseudonodosaria inflata						1		
	Virgulinella							1	
	Polymorphina								1
F	*Pseudohastigerina micra*			2					
	Globigerina eocaena				3		3	1	
	Catapsydrax howei						2		
	Acarinina pentacamerata							1	
	Planorotalites pseudomenardii							1	

Fig. 3. Foraminiferal distribution in borehole A3. A, epifaunal suspension-feeding agglutinating foraminifera; B, other epifaunal agglutinating foraminifera; C, infaunal agglutinating foraminifera; D, epifaunal calcareous benthic foraminifera; E, infaunal calcareous benthic foraminifera; F, planktic foraminifera. Location of borehole indicated on Figure 1.

further characterized by size- and shape-sorting of benthic (and planktic) species, while hemipelagic deposits are not (Brunner & Ledbetter 1987).

Material and methods

Twenty-three selected mudstone samples, each consisting of a short length of subsurface borehole core, have each been quantitatively analysed micropalaeontologically and palynologically using standard techniques (the number of samples was restricted by the available budget). These samples are all from the Ainsa I fan/channel system, eight from borehole A5, sited in the axis of the channel and represented essentially by medium-bedded turbidites, eight from A3, off the axis of the channel and represented essentially by thin-bedded turbidites, and seven from A4, on the

		210.00m	214.00m	225.70m	234.00m	237.00m	242.00m	244.00m
A	*Bathysiphon*		5	3	8	5	3	1
B	*Praesphaerammina subgaleata*		6	8	8	8	12	2
	Recurvoides		9	12		10		
	Trochammina		5	2	2	2		
	Adercotryma glomeratum				7			1
	Haplophragmoides				6			
	Cyclammina					6		2
	Recurvoides imperfectus						24	
C	*Caudammina ovula*	1		6				
	Textularia	1						
	Bulbobaculites problematicus			2				
	Ammomarginulina							2
D	*Cibicidoides*				1			
	Cibicidoides dutemplei				7			25
	Lenticulina				2	1		2
	Gyroidina girardana						1	
	Cibicidoides alazanensis							13
	Nuttallides truempyi							2
E	*Angulogerina*	1						
	Bolivina	1						
	Polymorphina	1						
	Quinqueloculina	1						2
	Uvigerina eocaena	2						
	Bulimina			9		4		
	Fissurina orbignyana			1				
	Uvigerina					2		
F	*Globigerina eocaena*	1				1	1	2
	Subbotina linaperta						1	
	Guembelitrioides higginsi							1

Fig. 4. Foraminiferal distribution in borehole A4. Key as per Figure 3. Location of borehole indicated on Figure 1.

laterally equivalent levee/overbank some 500 m distant and represented essentially by hemipelagites (debrites were not sampled). The average vertical spacing over the interval of interest is approximately 5 m. Results are summarized on Figures 3–6.

As noted above, planned future work will address issues of vertical variability at, and horizontal variability between, localities.

Results and discussion

Micropalaeontological recovery varied from non-existent (in the one and only barren sample, namely borehole A5, 13.15 m) to moderate, with total (i.e. planktic plus benthic) abundances ranging from 0–83 specimens and diversities from 0–26 species per sample. The ratio of planktic (free-floating) to benthic (bottom-dwelling) species ranges from 0–0.33 (A5, 45.60 m) in the fossiliferous samples, and averages approximately 0.05 (5%). Benthic foraminiferal abundance ranges from 2–80 specimens, and diversity from 1–23 species, in the fossiliferous samples. Samples yielding fewer than ten specimens have been disregarded for statistical purposes.

Age

Age-diagnostic planktic foraminifera identified in the present study include:

Acarinina broedermanni, A. bullbrooki, A. pentacamerata, and *Globigerina eocaena*, which range in age from late Early–early Middle Eocene (global standard planktic foraminiferal Zones P8–?P12);

Planorotalites pseudoscitula, which ranges no older than Middle Eocene (P10), and, as it co-occurs with *A. broedermanni* in borehole A5, 50.30 m, restricts the age-range of at least this sample to early Middle Eocene (P10–?P12);

Turborotalia pomeroli, which ranges no older than Middle Eocene (P11), and, as it co-occurs with *A. broedermanni* in borehole A5, 45.60 m,

		13.15m	20.50m	32.00m	35.70m	41.50m	45.60m	50.30m	58.00m
A	*Bathysiphon*			4				1	
	Rhabdammina			2					
B	*Cyclammina*		1		2				
	Cyclammina rotundidorsata		1	1					
	Glomospira			2					
	Recurvoides			2	3	9		5	
	Haplophragmoides					12			
	Praesphaerammina subgaleata					18			
C	*Triplasia*		2		1				
	Marssonella crassa			2					
	Dorothia oxycona							4	
	Caudammina ovula							8	1
	Ammomarginulina								1
D	*Cibicidoides eocaenus*			10			3		
	Eponides elevatus			5					
	Hoeglundina elegans			4					5
	Lenticulina arcuatostriata			1					
	Lenticulina			1			4		1
	Nuttallides truempyi			12					8
	Cibicidoides havanensis						1		
	Cribroparrella pteromphalia						1		1
	Cibicidoides mexicanus costatus						1		13
	Cibicidoides							1	
	Cibicidoides dutemplei							1	
	Lenticulina inornata							1	
	Cibicidoides parianus								1
	Cibicidoides eocaenus								14
E	*Bulimina*			1			5	2	
	Globulina						1		
	Bolivina aliformis							1	
	Fissurina							1	
	Nodosaria							2	
	Trifarina								1
	Uvigerina								1
F	*Acarinina brodermanni*						2	4	
	Acarinina bullbrooki						1		
	Globigerina eocaena						3		
	Turborotalia pomeroli						2		
	Guembelitrioides higginis							2	
	Globigerinatheka rubriformis							1	
	Planorotalites pseudoscitula							1	1
	Acarinina primitiva								1

Fig. 5. Foraminiferal distribution in borehole A5. Key as per Figure 3. Location of borehole indicated on Figure 1.

restricts the age-range of at least this sample to early Middle Eocene (P11–?P12).

Interestingly, initial observations on a range of subsurface borehole samples and surface outcrops indicate that contemporaneously transported photic zone (photosymbiont-bearing) larger benthic foraminifera – identifiable with the use of a hand-lens – might provide the basis for establishing an independent stratigraphy, as in the case of the Paleocene Zumaya Flysch, west-central Spanish Pyrenees (RWJ's interpretation of data presented by von Hillebrandt 1965). These include *Assilina maior punctulata*, which ranges from late Early–early Middle Eocene (global standard calcareous nannoplankton zones NP14–NP15), together with several species of *Nummulites*. Such forms are readily reworked on account of their high internal porosities and low pick-up velocities (see, for instance, Jones 1997, and additional references cited therein), and have been observed hydrodynamically sorted into large and small (A and B) forms. Interpreted non-contemporaneously reworked larger benthic foraminifera potentially provide constraints on the timing of structuration (for instance, uplift and erosion of the 'Ilerdian' or latest Paleocene–earliest Eocene *Alveolina* Limestone, now exposed in the Cotiella Nappe).

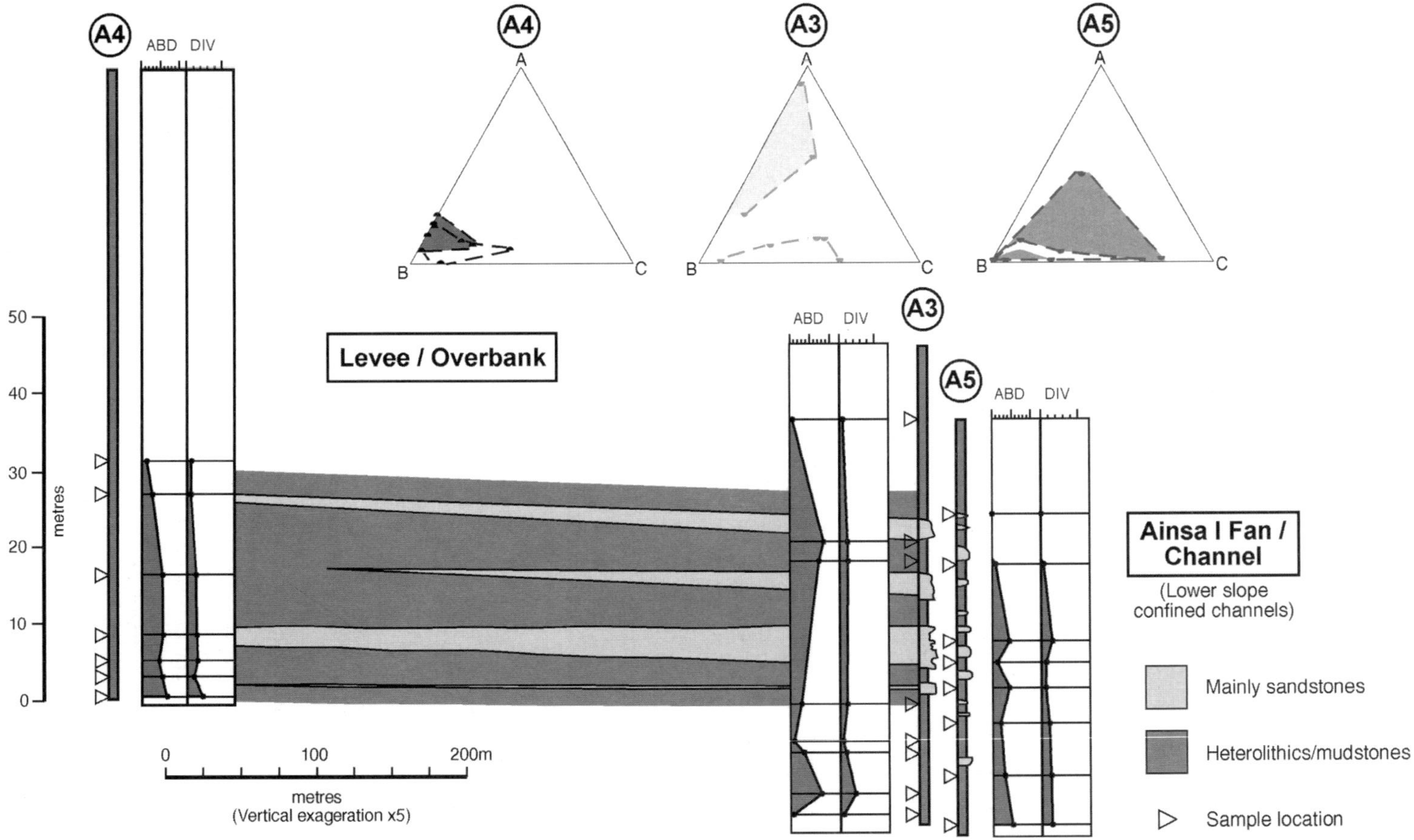

Fig. 6. Correlation between boreholes A3, A4 & A5. Showing foraminiferal abundance and diversity (see also Figure 8). Abundance scale 0–100; diversity scale 0–50. Also showing simplified triangular cross-plots of 'morphogroups' (see also Figures 9–10). Dashed, agglutinating plus calcareous; solid, agglutinating only. Locations of boreholes indicated on Figure 1. Sandstones include debrites as well as turbidites.

	IN	MN	ON	UB	MB	LB	AB
Cibicidoides mexicanus costatus		X	X	X			
Hoeglundina elegans		X	X	X	X	X	(X)
Adercotryma glomeratum		X	X	X	X	X	X
Cibicidoides dutemplei			X	X			
Cibicidoides alazanensis			X	X	X		
Planulina costata			X	X	X	X	
Cibicidoides eocaenus			X	X	X	X	X
Hanzawaia ammophila			X	X	X	X	X
Nuttallides truempyi				X	X	X	X
Cibicidoides havanensis					X	X	X
Cibicidoides grimsdalei						X	X
Cyclammina rotundidorsata						X	X

Fig. 7. Bathymetric range chart of selected species. IN, MN, ON, inner, middle, outer neritic (shelf); UB, MB, LB, upper, middle, lower bathyal (slope); AB, abyssal. Compiled from various sources (see text).

Robertson Research previously recovered early Middle Eocene planktic foraminifera from a condensed section, identified on the basis of a microfaunal abundance maximum, near the top of the Ainsa-I fan/channel complex in borehole A2 that can be tentatively correlated with the 49.0 Ma Maximum Flooding Surface of Haq *et al.* (1987). Staff at UCL recovered early Middle Eocene calcareous nannofossils.

Palaeoenvironmental interpretation

Palaeobathymetry. Three extant species have been recorded whose present-day depth ranges have been reasonably well established and which can therefore serve as proxies for palaeobathymetric interpretation. These are the agglutinating forms *Adercotryma glomeratum* (*A. agterbergi* auctt.), middle shelf to abyss (deeper in lower latitudes) (Charnock & Jones 1990; Jones 1994) and *Cyclammina rotundidorsata*, lower slope to abyss (Jones 1994), and the aragonitic form *Hoeglundina elegans*, middle shelf to abyss, but essentially middle shelf to slope, i.e. above the aragonite compensation depth (Jones 1984). Together, these species indicate deposition in a lower slope or bathyal environment (Fig. 7).

A number of extinct calcareous species have also been recorded whose depth ranges have been established by indirect means (such as backstripping) and which would be more or less consistent with such an interpretation, viz. *Cibicidoides alazanensis, C. dutemplei, C. eocaenus, C. grimsdalei, C. havanensis, C. mexicanus costatus, Hanzawaia ammophila, Nuttallides truempyi* and *Planulina costata* (van Morkhoven *et al.* 1986; Fig. 7). Those with lower depth limits that are inconsistent with the lower slope interpretation, namely *Cibicidoides alazanensis, C. dutemplei* and *C. mexicanus costatus*, are at least arguably allochthonous (transported downslope).

Comparison between channel axis, off-axis and levee/overbank assemblages

Abundance and diversity. Benthic foraminiferal abundance ranges from 2–80 specimens in the fossiliferous samples, averaging 37 in borehole A3 (channel off-axis), 35 in A4 (levee/overbank) and 25 in A5 (channel axis), and diversity from 1–23 species, averaging 9 in A3 (channel off-axis) and 7 in A4 (levee/overbank) and A5 (channel axis) (Figs 3–6). Although there is some overlap, the highest unit diversity (that is, diversity expressed as a function of abundance) is at the channel off-axis site, probably indicating the presence here of both autochthonous and allochthonous, especially allochthonous calcareous, species (Fig. 8). Perhaps significantly, there is more variability in the abundance and diversity values at the channel sites than at the levee/overbank site.

The above observations contrast with the findings of Seiglie *et al.* (1986) (see section on 'previous research' above) and of Jones (1999) (on the Late Paleocene reservoir section in Forties Field in the North Sea) that unit diversity is generally higher in hemipelagites than in interturbidites.

Wall structure groups. The percentage of agglutinating as against calcareous benthic foraminifera ranges from 0–100% overall and averages 73% in A4 (levee/overbank), 37% in

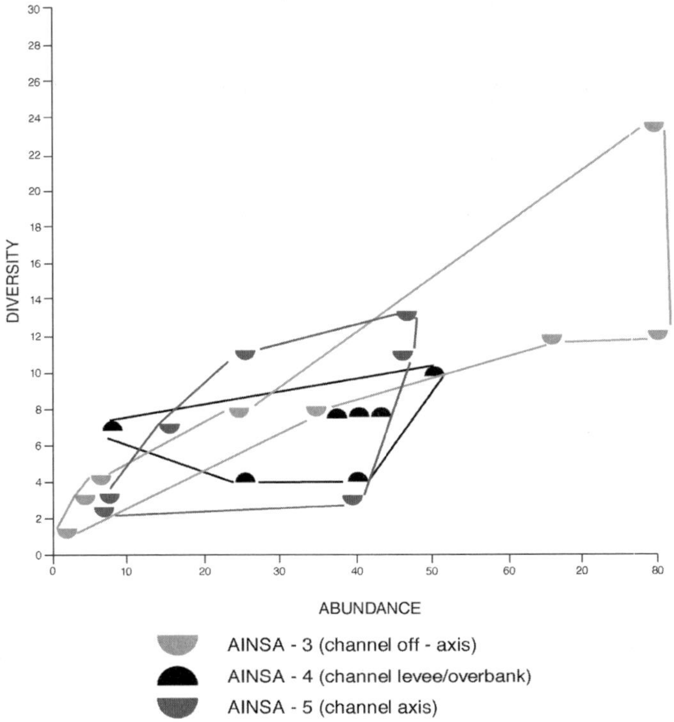

Fig. 8. Cross-plot of abundance vs. diversity ('unit diversity').

A5 (channel axis) and 18% in A3 (channel off-axis). It is important to note that the extent to which the calcareous benthics are autochthonous as against allochthonous (transported downslope), especially in the channel samples, is debatable (see above).

'Morphogroups'. The percentage of epifaunal (surface-dwelling) as against infaunal (sediment-dwelling) morphotypes ranges from 31–100% overall and averages 90% in A4 (levee/overbank), 76% in A5 (channel axis) and 62% in A3 (channel off-axis) (Fig. 9). Again it is important to note that the extent to which the calcareous morphotypes are autochthonous, especially in the channel samples, is debatable.

Disregarding calcareous foraminifera and regarding only agglutinating foraminifera, the percentage of epifaunal morphotypes in still statistically representative samples ranges from 29–100% overall and averages 95% in A4 (levee/overbank), 87% in A3 (channel off-axis) and 71% in A5 (channel axis) (Fig. 10). Incidentally, the range of values from the levee/overbank site is closely comparable with that from the background basin-plain (of the Ainsa-III fan/channel complex) of borehole A6 (RWJ's interpretation of data presented by Farina 2001).

Again, as in the case of abundance and diversity, perhaps significantly, there is more variability in values at the channel sites than at the levee/overbank site, possibly on account of the effects of some form of modification to the life assemblage.

This may have assumed the form of progressive removal through winnowing of hydrodynamically unstable morphotypes proportionately more abundant in epifaunal 'morphogroup' B and resultant concentration of hydrodynamically stable morphotypes proportionately more abundant in infaunal 'morphogroup' C, and disproportionately abundant at the channel axis site (where the large, interpreted infaunal hormosinid *Caudammina ovula* is locally the dominant species).

Note, though, that the disproportionate representation of hydrodynamically intermediate morphotypes of epifaunal suspension-feeding 'morphogroup' A at the off-axis site may reflect

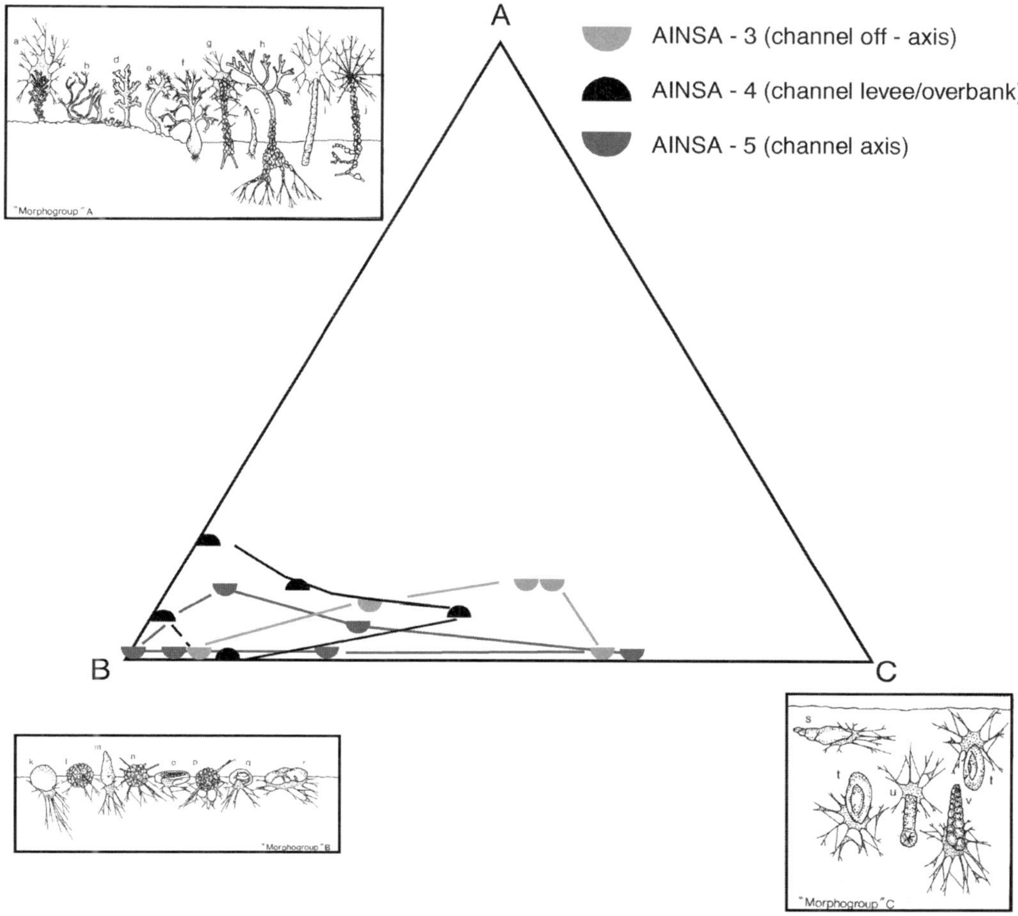

Fig. 9. Cross-plot of 'morphogroups' (agglutinating plus calcareous benthic foraminifera). 'Morphogroups' after Jones & Charnock (1985) and Jones (1996).

an ecological preference for this site, rather than a taphonomic artefact, i.e. the suspension-feeders may have actually lived there rather than been transported there.

The above observations contrast with the findings of Jones (1999) (on the Late Paleocene reservoir section in Forties Field in the North Sea) that epifaunal morphotypes are generally proportionately more abundant in interturbidites than in hemipelagites, and, conversely, that infaunal morphotypes are generally proportionately more abundant in hemipelagites than in interturbidites. The implication is of insufficient time in between density flows to allow recolonization by infaunal morphotypes (leastwise those with sufficient preservation potential not to be lost to the rock record during the fossilization process (Schroder 1986)).

Relationships with ichnofacies. In cores from the Late Paleocene reservoir section in Forties Field in the North Sea (Jones 1999), hemipelagites are characterized not only by high incidences of infaunal foraminifera but also by pervasive *Chondrites*-type burrowing, which is also a conspicuous feature of the outcrops around Ainsa (Hipperson 2001, MS) and of the Zumaya Flysch to the west (Crimes 1973). Such traces have historically conventionally been attributed to the activities of sipunculoid worms, but unbranched forms are at least arguably in part of foraminiferal

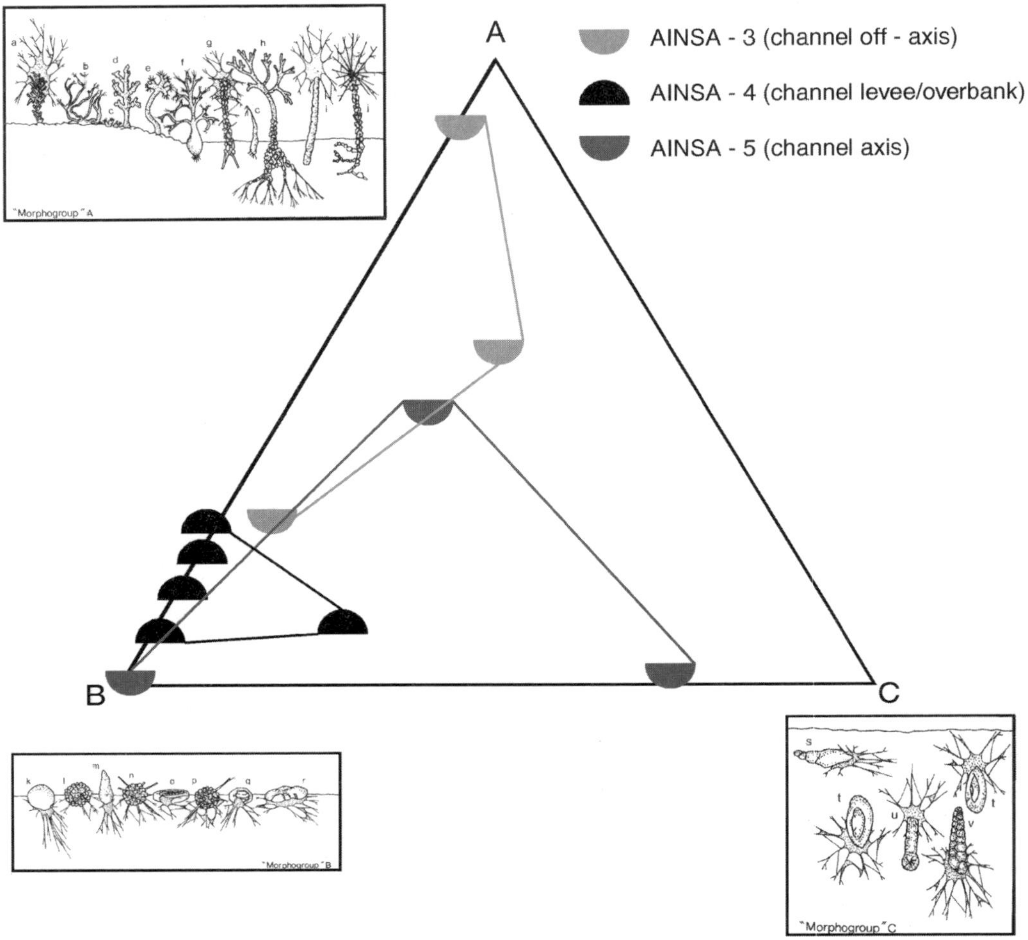

Fig. 10. Cross-plot of 'morphogroups' (agglutinating foraminifera only). 'Morphogroups' after Jones & Charnock (1985).

origin, bearing a strong resemblance to the 'vermicelli' bioturbation of Kaminski *et al.* (1988) (Jones 1999).

Palynology

Palynological assemblages are characterized by varying incidences (at least in part a function of differential transportation or preservation) of:

- terrestrially derived angiosperm (flowering plant) and gymnosperm (conifer) pollen and pteridophyte (fern) spores, indicative of the development of (sub-)tropical highland and lowland vegetation (including mangrove swamps);
- fungal spores and hyphae, indicative of decaying organic matter or soil;
- fresh-water algae;
- marine dinoflagellate cysts and foraminiferal test linings.

There does not appear to be any obvious and systematic difference in palynological assemblage composition between the channel and levee/overbank sites (although interpretation is still ongoing).

Conclusions

(1) The Ainsa I fan/channel complex is of early Middle Eocene age on planktic

foraminiferal and calcareous nannoplankton evidence.

(2) Benthic foraminiferal evidence indicates that it was deposited in lower slope water depths.

(3) Channel axis samples differ from levee/overbank samples in their lower unit diversity of benthic foraminifera. The highest unit diversity is at the channel off-axis site, probably indicating the presence here of both autochthonous and allochthonous species. There is more variability in values at the channel sites than at the levee/overbank site.

(4) Channel samples appear to differ from levee/overbank samples in their generally lower proportion of epifaunal (other than suspension-feeding) and higher proportion of epifaunal suspension-feeding and infaunal morphotypes (although the extent to which the over-representation of at least the infaunal morphotypes might be due to hydrodynamic sorting remains unclear). Again, there is more variability at the channel sites than at the levee/overbank site.

BP is thanked for sponsoring the research on which this paper is based and for providing the word-processing and drafting facilities with which to write up the results. M. Kaminski of the Department of Earth Sciences at UCL is thanked for his assistance with the identification of the agglutinating foraminifera. M. Dykstra, H. Nelson and C. Pirmez are thanked for helpful discussions about the sedimentology of the Ainsa fan/channel system. Two anonymous reviewers are thanked for their constructive comments on the manuscript.

References

BOUMA, A.H., COLEMAN, J.L. & DSDP LEG 96 SHIPBOARD SCIENTISTS 1985. Mississippi Fan: Leg 96 program and principal results. *In*: BOUMA, A.H., NORMARK, W.R. & BARNES, N.E. (eds) *Submarine Fans and Related Turbidite Systems*. Springer, New York, 247–252.

BROUWER, J. 1965. Agglutinated foraminiferal faunas from some turbiditic sequences – Parts I & II. *Proceedings, Koninklijke Nederlandse Akademie van Wetenschappen, Section B*, **68**(5), 309–334.

BRUNNER, C.A. & LEDBETTER, M.T. 1987. Sedimentological and micropalaeontological detection of turbidite muds in hemipelagic sequences: an example from the Late Pleistocene levee of Monterey Fan, Central California continental margin. *Marine Micropaleontology*, **12**, 223–239.

BUTT, A. 1981. Depositional environments of the Upper Cretaceous rocks in the northern part of the Eastern Alps. *Special Publications of the Cushman Foundation for Foraminiferal Research*, **20**, 1–121.

CHARNOCK, M.A. & JONES, R.W. 1990. Agglutinated Foraminifera from the Palaeogene of the North Sea. *In*: HEMLEBEN, C., KAMINSKI, M.A., KUHNT, W. & SCOTT, D.B. (eds) *Paleoecology, Biostratigraphy, Paleoceanography and Taxonomy of Agglutinated Foraminifera*. Kluwer, Dordrecht, 139–244.

CRIMES, T.P. 1973. From limestones to distal turbidites: a facies and trace fossil analysis of the Zumaya Flysch (Paleocene–Eocene, North Spain). *Sedimentology*, **20**, 105–131.

ELLISON, R.L. & PECK, G.E. 1983. Foraminiferal recolonization on the continental shelf. *Journal of Foraminiferal Research*, **13**(4), 231–241.

ERNST, S.R., DUIJNSTEE, I.A.P., JANNINK, N.T. & VAN DER ZWAAN, G.J. 2000. An experimental mesocosm study of microhabitat preferences and mobility in benthic foraminifera: preliminary results. *In*: HART, M.B., KAMINSKI, M.A. & SMART, C.W. (eds) *Proceedings of the Fifth International Workshop on Agglutinated Foraminifera*. Grzybowski Foundation Special Publication, **7**, 101–104.

FARINA, A. 2001. *Micropalaeontological Study of Fine-Grained Turbidites from Eocene Deep-water Sediments, Spanish South-Central Pyrenees*. MSc Thesis, University College London.

HAQ, B.U., HARDENBOL, J. & VAIL, P.R. 1987. Chronology of fluctuating sea levels since the Triassic. *Science*, **235**, 1153–1165.

HILLEBRANDT, A. VON 1965. Foraminiferen-Stratigraphie im Alttertiar von Zumaya (Provinz Guipuzcoa, NW-Spanien) und ein Vergeich mit anderen Tethys-Gebieten. *Bayerische Akademie der Wissenschaften, Mathematisch Klasse, Abhandlungen, Neue Folge*, **123**, 1–62.

HIPPERSON, S. 2001. *Ichnofacies and Deep-Water Clastic Environments: a Core Study, Mid-Eocene Ainsa System, South-Central Pyrenees, Spain*. MSc Thesis, University College London.

HOLMES, N.A. 1999. The Andrew Formation and "biosteering" – different reservoirs, different approaches. *In*: JONES, R.W. & SIMMONS, M.D. (eds) *Biostratigraphy in Production and Development Geology*. Geological Society, London, Special Publications, **152**, 155–166.

JONES, R.W. 1984. *Late Quaternary Benthonic Foraminifera from Deep-Water Sites in the North-East Atlantic and Arctic Oceans*. PhD Thesis, University College of Wales, Aberystwyth.

JONES, R.W. 1994. *The Challenger Foraminifera*. Oxford University Press, Oxford.

JONES, R.W. 1996. *Micropalaeontology in Petroleum Exploration*. Oxford University Press, Oxford.

JONES, R.W. 1997. Aspects of the Cenozoic Stratigraphy of the Northern Sulaiman Ranges, Pakistan. *Journal of Micropalaeontology*, **16**, 51–58.

JONES, R.W. 1999. Forties Field (North Sea) revisited: a demonstration of the value of historical micropalaeontological data. *In*: JONES, R.W. & SIMMONS, M.D. (eds) *Biostratigraphy in Production and Development Geology*. Geological Society, London, Special Publications, **152**, 185–200.

JONES, R.W. & CHARNOCK, M.A. 1985. 'Morphogroups' of agglutinating foraminifera, their life positions and feeding habits and potential applicability in (paleo)ecological studies. *Revue de Micropaleontologie*, **4**, 311–320.

KAMINSKI, M.A., GRASSLE, J.F. & WHITLATCH, R.B. 1988. Life history and recolonization among agglutinated foraminifera in the Panama Basin. *Abhandlungen der Geologischen Bundesanstalt*, **41**, 229–244.

KOHL, B. & DSDP LEG 96 SHIPBOARD SCIENTISTS 1985. Biostratigraphy and sedimentation rates of the Mississippi Fan. *In*: BOUMA, A.H., NORMARK, W.R. & BARNES, N.E. (eds) *Submarine Fans and Related Turbidite Systems*. Springer, New York, 267–273.

PAYNE, S.N.J., EWEN, D.F. & BOWMAN, M.J. 1999. The role and value of 'high-impact biostratigraphy' in reservoir appraisal and development. *In*: JONES, R.W. & SIMMONS, M.D. (eds) *Biostratigraphy in Production and Development Geology*. Geological Society, London, Special Publications, **152**, 5–22.

PICKERING, K.T. & CORREGIDOR, J. 2000. Reservoir-scale study of Eocene confined submarine fans, South-Central Spanish Pyrenees. *In*: WEIMER, P., SLATT, F.M., COLEMAN, J., ROSEN, N.C., NELSON, H., BOUMA, A.H., STYZEN, M.J. & LAWRENCE, D.T. (eds) *Deep Water Reservoirs of the World*. Gulf Coast Section, Society of Economic Paleontologists and Mineralogists, Houston, Texas (Foundation 20th Annual Bob F. Perkins Research Conference).

SCHRODER, C. 1986. Deep-water arenaceous foraminifera from the Northwest Atlantic Ocean. *Canadian Technical Report of Hydrography and Ocean Sciences*, **71**, 1–191.

SEIGLIE, G.A., FLEISHER, R.L. & BAKER, M.B. 1986. Alveovalvulinidae n. fam. and Neogene diversification of agglutinated foraminifers with inner structures. *Micropaleontology*, **32**, 169–181.

THOMPSON, P.R. 1992. Foraminiferal evidence for the sources and timing of mass-flow deposits south of Baltimore Canyon. *In*: ISHIKAZI, K. & SAITO, T. (eds) *Centenary of Japanese Micropaleontology*. Terra Scientific Publishing Company, Tokyo, 97–128.

VAN MORKHOVEN, F.P.C.M., BERGGREN, W.A., EDWARDS, A. *et al.* 1986. *Cenozoic Cosmopolitan Deep-Water Benthic Foraminifera*. Centre des Recherches Exploration-Production Elf-Aquitaine, Pau (Memoire, No. 11).

VILELA, C.G. & MASLIN, M. 1997. Benthic and planktic foraminifers and stable isotope analysis of mass-flow sediments in the Amazon Fan. *Proceedings of the Ocean Drilling Program, Scientific Results*, **155**, 335–351.

Micropalaeontological dissection of the Shu'aiba Reservoir, Saudi Arabia

G. W. HUGHES

Saudi Aramco, Geological Technical Services Division, Box 10646, Dhahran 31311, Saudi Arabia (e-mail: geraint.hughes@aramco.com)

Abstract: Determination of the depositional architecture of carbonate hydrocarbon reservoirs is a necessary prerequisite for pre- and syn-development modelling and simulation. Such reservoirs are typically too thin for internal bedding geometries to be confidently determined from seismic and wireline log evidence, and this aim is best achieved by integrated palaeontological and sedimentological studies. Of these, micropalaeontological analysis and biofacies interpretation of closely spaced core samples can reveal cryptic, high-frequency depositional cycles and three-dimensional palaeoenvironmental information.

In Saudi Arabia, the Shu'aiba Formation is the main reservoir in the Shaybah field, and has been subdivided into a number of biofacies (ShBf-1 to ShBf-10) within four main depositional units, of which the lower unit (Sh1) is regionally extensive, moderately deep marine and of non-reservoir significance. A thicker second unit (Sh2) displays significant lateral and vertical differentiation of a rudist-rimmed shallow carbonate platform in which discrete lagoonal, rudist bank complex and fore-bank biofacies have been determined. Deep platform sediments characterize the uppermost stacked unit (Sh3) of the main buildup, and are of early Aptian age. Flank carbonates onlap the former three units and are assigned a late Aptian age (Sh4).

This fully integrated palaeontological approach to over 50 cored wells from the Shaybah Field has significantly provided a detailed stratigraphic and palaeoenvironmental interpretation that is currently being used to successfully develop this field with 14 billion barrels of crude oil and 25 trillion cubic feet of gas in place. A satisfactory sequence stratigraphic interpretation of the Shu'aiba Formation at Shaybah remains elusive, as neither detailed seismic evidence nor correlatively valuable wireline logs are available.

The Shu'aiba Formation has been studied in great detail in the Emirates and Oman, where it is an important hydrocarbon-bearing reservoir. Key references include Alsharhan & Kendall (1991), Litsey *et al.* (1983), Alsharhan (1985, 1987), Frost *et al.* (1983), Abou-Choucha & Ennadi (1993), Harris *et al.* (1984), Scott (1990), Hamdan & Alsharhan (1991), Azer & Toland (1993), Vahrenkamp & Grotsch (1995), Vahrenkamp (1996), Aktas *et al.* (2000), Immenhauser *et al.* (2000, 2001, 2004), Ziegler (2001), Van Buchem *et al.* (2002*a*, *b*), Russell & Akbar (2002), Pittet *et al.* (2002) and Boote & Mou (2003). Published information on the Shu'aiba Formation in Saudi Arabia is limited (Hughes 1998, 1999, 2000*a*, *b*, 2003*a*, *b*, 2004*a*, *b*; Hughes *et al.* 1999; Salamy *et al.* 2003). In Yemen, the Shu'aiba Formation is termed the Qishn Formation, and forms part of the Mukallah Group, of Barremian to Aptian age (Beydoun *et al.* 1993).

The supergiant Shaybah field lies in the Rub' Al-Khali basin of Saudi Arabia and produces oil and gas from the 400 feet thick reservoir within the Shu'aiba Formation, the top of which is at depth of around 4900 feet (Saudi Aramco 1999). With reserves of over 14 billion barrels of crude oil and 25 trillion cubic feet of gas, it is one of the largest petroleum provinces in the world. The crude is Arabian extra light, a high quality crude grade with a specific gravity of 42 degrees API and a sulphur content of less than 0.7% (Saudi Arabian Embassy press release, March 10, 1999; *www.saudiembassy.net*). Production of 500 000 BOPD commenced in July 1998. Oil is trapped in a SW–NE-trending anticlinal structure roughly 60 km long and 12 km wide (Jenden *et al.* 2003).

The Shu'aiba Formation at Shaybah is mostly of early Aptian age and extends over an area of approximately 700 km^2. It consists entirely of carbonates that were deposited under a variety of water depths and hydraulic energies on an extensive carbonate platform during the Aptian, and in the stratigraphic equivalents in the United Arab Emirates has been considered the most geologically complex reservoir (Russell & Akbar 2002). The formation consists entirely of carbonates that mostly accumulated as a rudist-rimmed shallow marine platform on the southern flank of an intra-shelf basin (Fig. 1) and has an average thickness of approximately 400` (Saudi Aramco 1999; Hughes 1998,

From: POWELL, A. J. & RIDING, J. B. (eds) 2005. *Recent Developments in Applied Biostratigraphy*. The Micropalaeontological Society, Special Publications, 69–90. 1747-602X/$15.00

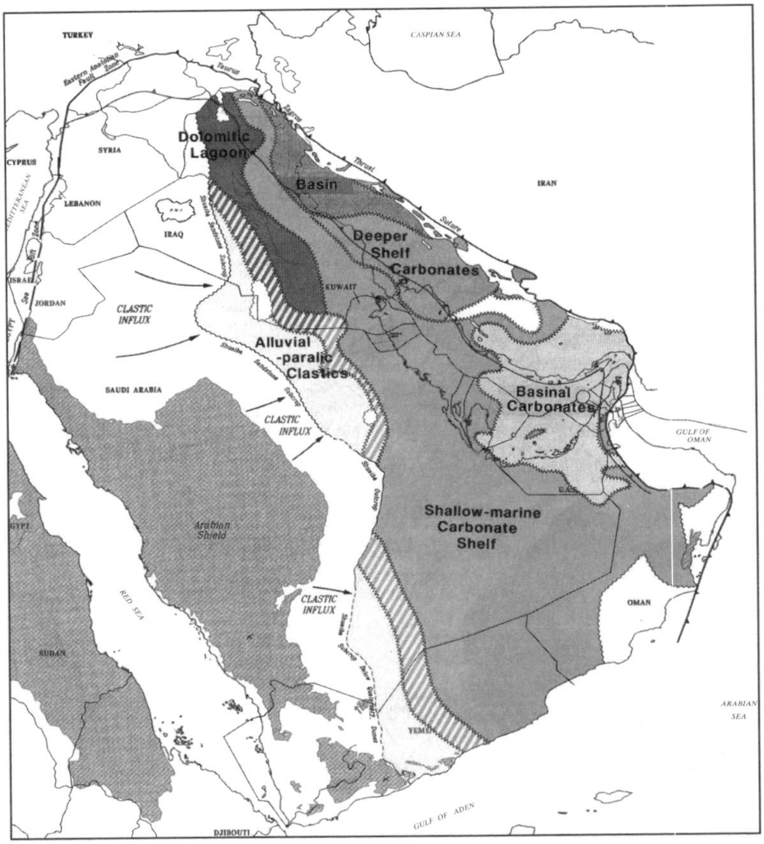

Fig. 1. Palaeogeography of Aptian Arabia showing the location of extensive carbonate platform and the intra-shelf basin (Grabowski & Norton 1995).

figure 3). The basin margin had a slope of only a few minutes but never exceeded 2° (Aldabal & Alsharhan 1989), the rim of which was preferentially colonised by rudists (Grabowski & Norton 1995, fig. 15).

The Shu'aiba Formation is well exposed in Oman, but with the exception of thin carbonates interbedded with clastics of the Biyadh Formation in central Arabia, and attributed to the Shu'aiba Formation by Moshrif & Kelling (1984), the Shu'aiba Formation is not considered to crop out in Saudi Arabia. Study of the Shu'aiba reservoir in the remotely located Shaybah Field prior to hydrocarbon development was forced to concentrate on the wealth of cores recovered during exploration drilling in the 1970s. Absence of good quality seismic data over Shaybah has placed additional importance on core-based information to provide guidance for reservoir facies distribution, unlike the Emirates (Calavan *et al.* 1992; Fischer *et al.* 1994, 1997) where seismic resolution is better owing to the thinner cover of sand dunes, and has assisted structural interpretation of the Shu'aiba rimmed carbonate platform.

Semi-quantitative micropalaeontology was performed on thin sections of core plugs from 17 000 feet of core from vertically cored wells (Fig. 2), with semi-quantitative macropalaeontology based on analysis of all cores. Deep marine, planktonic foraminiferal-bearing sediments testify to episodic transgressions at certain levels, and are persistently present off the east flank of the field. The macropalaeontological and micropalaeontological components, when integrated with petrographical, sedimentological, petrophysical and wireline log information has provided important information for mapping depositional facies and establishing a reservoir model at Shaybah (Aktas *et al.* 2000).

A satisfactory sequence stratigraphic model for the Shu'aiba Formation at Shaybah remains

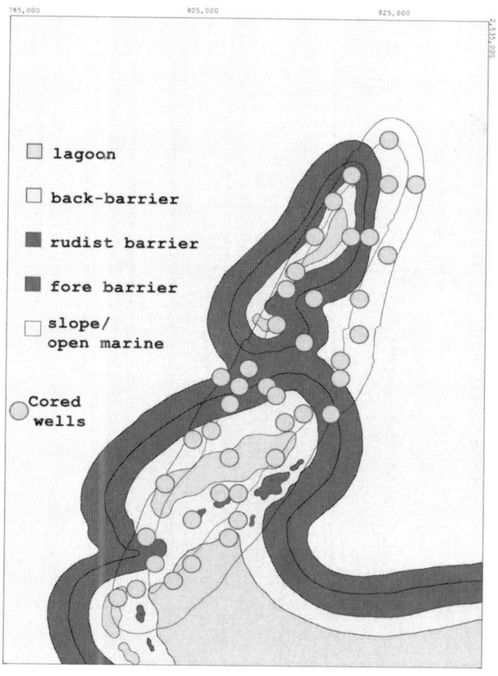

Fig. 2. Distribution of cored wells in Shaybah Field displaying a generalized distribution of environmental variations.

elusive, as the micropalaeontological evidence cannot provide time lines, and the seismic data cannot only provide any intra-formational surface features but the top of the Formation is impossible to image. Wireline logs, especially gamma, have to be scale enhanced to produce character and inter-well log character is difficult to compare with confidence.

The palaeontological data gained for the Shu'aiba Formation has, therefore, provided vertical palaeoenvironmental provinces that are accurate at the well location, but the lateral connectivity of these is based purely on similarity of biofacies character. This approach has, at least, provided an otherwise unattainable depositional model in which associated lithofacies belts can be predicted and used to guide reservoir development. Further consideration will be necessary to solve future development problems that may be associated with intra-reservoir discontinuities. Field-wide unconformities and sequence interpretations are still evolving for the Shu'aiba Formation in this region. The sequence stratigraphic interpretation of the Shu'aiba succession remains to be completely understood, but future advances in this approach will definitely be assisted by the application of biofacies for system tract determination and subsequent spatial distribution of the various orders of intra-Shu'aiba sequences.

Lithostratigraphy

In Saudi Arabia the Shu'aiba Formation overlies the Buwaib Formation with probable disconformity and is unconformably overlain by the Nahr Umr Formation, a fine-grained, deeper marine clastic equivalent of the Khafji Formation of the Wasia Group (Powers *et al.* 1966; Powers 1968; Alsharhan & Nairn 1986, 1997). There is no formal subdivision of the Shu'aiba Formation in Shaybah, but a four-tiered, sequence-bounded scheme has been established using biofacies constraints, here proposed as informal Sh1 to Sh4 members of the Shu'aiba Formation.

The basal transgressive unit consists of dense, dark-grey orbitolinid and planktonic foraminifera bearing carbonates informally termed the Biyadh Formation in Saudi Arabia and equivalent to the Kharaib Upper Dense in the UAE and Oman. The upper part of this unit consists entirely of *Lithocodium* platform sediments. The second layer consists of a variety of biofacies including *Lithocodium aggregatum, Palorbitolina lenticularis* and a variety of back-bank and bank-crest rudists. During the early Aptian a widespread *Lithocodium* platform suffered differential subsidence. Basinal sediments of the Lower Bab Member were deposited in the subsiding part while on the platform a complex of *Lithocodium*, orbitolinid-bearing lagoonal sediments, caprinid and caprotinid rudist bank facies were aggrading and prograding into the basin. The upper unit of the Shu'aiba in the main part of Shaybah Field is also of early Aptian age and is distinguished from the underlying beds by the predominance of agriopleurid rudists within an extensive deep lagoon that was rimmed by a discontinuous and poorly developed rudist bank complex. This episode was terminated towards the end of early Aptian times by a eustatic sea level fall in which sedimentation was confined to the basin (upper Bab Member), while the emergent platform was exposed, karstified and topographically reduced to an almost level surface (Guo *et al.* 2004). The late Aptian coccolith-bearing basinal sediments onlapped the edge of the platform with some progradation into the basin, as described for Bu Hasa by Witt & Gokdag (1994). During the late Aptian, the entire area was transgressed by the Nahr Umr flooding event (Sharland *et al.* 2001). In the UAE and Oman, Van Buchem *et al.*

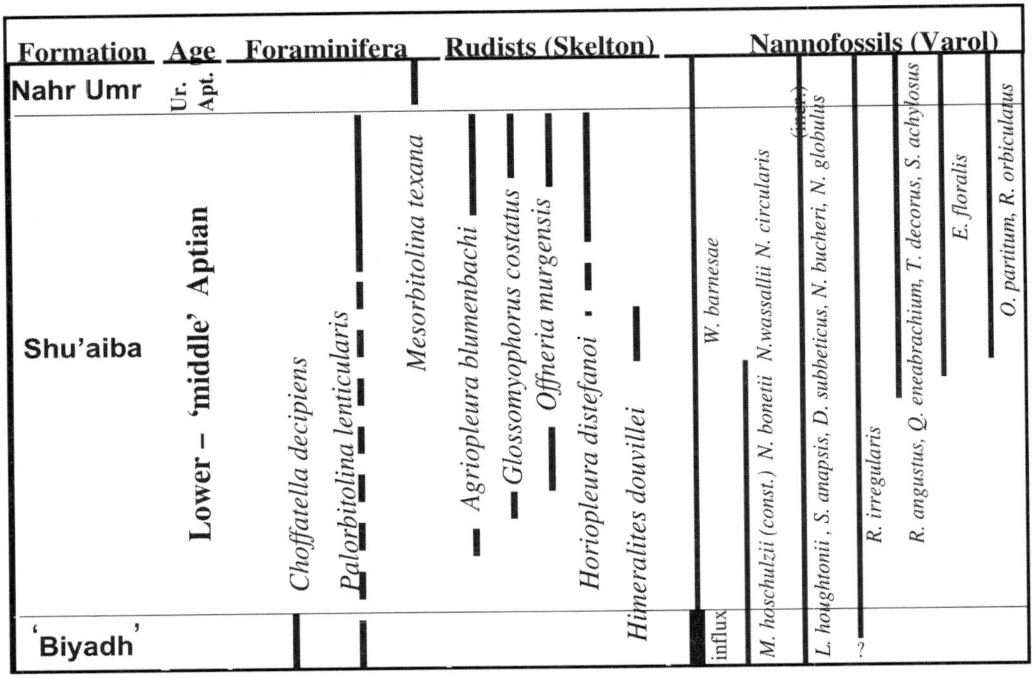

Fig. 3. Biostratigraphy of the Shu'aiba Formation in Shaybah (late Aptian nannofossil evidence is confined to localities off the eastern flank of the Shaybah carbonate platform).

(2002*a*, *b*) consider the Lower Shu'aiba, equivalent to the three lower units of the Shu'aiba Formation of Saudi Arabia as a single sequence, with a separate sequence for the Upper Shu'aiba, equivalent to the off-structure, basinal late Aptian sediments of the Shu'aiba Formation in Saudi Arabia.

Biostratigraphy

Throughout this contribution, the use of the qualifiers 'early' and 'late' follow an abridged version of the International Stratigraphic Guide (Murphy & Salvador 1999, 2000) when applied to the Aptian age. The Aptian has been subdivided, on ammonite stratigraphy, into two subdivisions (Lower and Upper) in the Boreal realm, but a threefold subdivision in the Tethyan realm, following the French division into Bedoulian (*D. deshayesi* to *D. furcata* zones), Gargasian (*E. subnodosocostatum* to *P. melchioris* zones) and Clansayesian (*A. nolani* to *H. jacobi* zones) (Erba 1996). The Lower Aptian is equivalent to the Bedoulian and the Upper Aptian to the Gargasian and Clansayesian. In the absence of ammonites in the Shu'aiba Formation, foraminifera, calcareous nannofossil and rudist biostratigraphy provides calibration with the Bedoulian. The age of the Shu'aiba Formation within the region has been discussed by various authors (Harris *et al.* 1984; Alsharhan & Nairn 1986; Simmons & Hart 1987; Hughes Clarke 1988; Scott 1990; Alsharhan & Kendall 1991; Fischer *et al.* 1994; Witt & Gokdag 1994; Masse 1997; Jones *et al.* 2004).

The Shu'aiba carbonates of the Shaybah field are dated as early Aptian (Fig. 3), based on the presence of the rudist species *Offneria murgensis, Agriopleura* cf. *blumenbachi, Agriopleura* cf. *marticensis* and *Glossomyophorus costatus* (Masse 1985, 1992, 1999; Skelton & Masse 2000; Skelton 2000*a*; Skelton *et al.* 1999) and the absence of *Eoradiolites* spp. (Masse & Gallo Maresca 1997). As the early–late Aptian boundary is conventionally marked by a drastic reduction in rudist species (Masse 1999); the continued presence of these rudist species in the uppermost Shu'aiba Formation platform carbonates at Shaybah confirm an early Aptian age. Schemes for stratigraphically using the larger benthonic foraminifera *Orbitolina* were established by Schroeder (1975) and Moullade *et al.* (1985). The presence of *Palorbitolina lenticularis* in the Shu'aiba Formation of Shaybah confirms an Early Aptian age

(Moullade, pers. comm., Moullade *et al.* 1985, 1998; Husinec 2001; Simmons 1994; Simmons *et al.* 2000). The benthonic foraminiferal species *Choffatella incipiens* within the lowermost Shu'aiba and of evidence Hauterivian to Aptian age provides only supportive age control for the Shu'aiba. Planktonic foraminifera have been assigned to *Hedbergella* species, with possible *H. infracretacea, H. delrioensis* and *H. planispira* determined from randomly oriented thin sections; *Wondersella athersuchi* (Banner & Strank 1987) has not been encountered. Calcareous nannopalaeontology has indicated the presence of a variety of early Aptian species (Fig. 3). The presence of *Salpingoporella dinarica* throughout the section confirms an age no younger than Aptian (Simmons *et al.* 1991).

No faunal or lithological manifestations of the Selli global anoxic event, of latest early Aptian age (Cobianchi *et al.* 1999), have been detected in the Shu'aiba carbonates probably because of the platform environment. In certain deeper marine carbonates of the Zumul and Ramlah regions of NE of Shaybah, calcareous nannofossil evidence also indicates early and late Aptian ages for Bab equivalent sediments that are consistent with the regionally determined ages for the deeper sediments of the Bab equivalent (Varol, written comm.; Aktas *et al.* 2000; Hughes *et al.* 1999). Scott (1990) has concluded an early Aptian age for the Shu'aiba Formation, and a late Aptian age for the Bab Member in Abu Dhabi, based on ammonite faunas (Hassan *et al.* 1975), but no ammonites have been recovered from the Shu'aiba Formation in Shaybah (Kennedy & Simmons 1991). On the northeast flank of Shaybah, nannofossil evidence for a late Aptian age is consistent with the presence of the single rudist species *Horiopleura distefanoi*. No specimens of *Eoradiolites* have ever been identified, but the coexistence of *Horiopleura* with *Eoradiolites* species has been documented from the Early Albian of Oman (Immenhauser *et al.* 2001), thus confirming the post-early Aptian stratigraphic range of *Horiopleura*.

Vahrenkamp (1996) and Folmi *et al.* (1994) have attempted carbon isotope stratigraphy of the Shu'aiba, but this method remains to be proved useful at Shaybah (Hughes *et al.* 1999). The carbon isotope record at the Aptian type section (Moullade *et al.* 1998) displays two marked positive excursions within the early Aptian but these are not evident in the samples analysed. There is, therefore, evidence that the platform carbonate facies of the Shu'aiba Formation at Shaybah is entirely of early Aptian age, and that a considerable unconformity separates the Shu'aiba from the Upper Aptian sediments of the overlying Nahr Umr Formation. The absence of Upper Aptian sediments at this unconformity is also noted in Oman (Immenhauser *et al.* 2001), where an inexplicable absence of the expected strongly negative carbon isotope character related to prolonged exposure has been discussed by Boote & Mou (2003). Strontium isotope analysis of selected cores produced results of limited value because of the apparent lack of detail of the standard Sr seawater curve for the Aptian (Smalley *et al.* 1994; Veizer *et al.* 1997). Limited carbon and oxygen isotope studies of the upper Shu'aiba Formation in Oman are presented by Immenhauser *et al.* (1999).

Biofacies and their palaeonenvironmental interpretation

Biocomponents of the Shu'aiba Formation at Shaybah (Fig. 4) have been described in detail in the various publications by the author. Microfossils include planktonic and benthonic foraminifera, calcareous algae, ostracods, sponge spicules and ostracods, and macrofossils include rudists, chondrodontid oysters, corals, gastropods, echinoid debris and the encrusting calcareous alga *Lithocodium aggregatum*.

Discrete biofacies have been determined from the semi-quantitative species distribution data, with emphasis placed on the stacking arrangement, or tiering, of the various biofacies in order to facilitate palaeoenvironmental interpretations (Figs 5 and 6) and their successive stages (Fig. 7). These determinations have been greatly assisted by reference to the benchmark use of early Cretaceous calcareous algal depth interpretations using submarine light attenuation principles to calibrate benthonic and planktonic foraminifera by Banner & Simmons (1994).

Semi-quantitative micropalaeontological and macropalaeontological analysis of closely spaced samples reveals localized changes in relative species abundance in addition to localized absolute extent within each well. These are readily related to high-frequency depositional cycles, the characteristics of which vary in response to the environmental differences from well to well. Integration of the three-dimensional variations in the macro and microbiofacies has enabled a broad subdivision of the Shu'aiba into three stacked major depositional layers. The lower palaeoenvironmental layer (lower Shu'aiba) is a moderately deep carbonate shelf with a laterally extensive, rather uniform biocomponent composition. It is overlain by a thicker layer (middle Shu'aiba) in which rudist colonies became established,

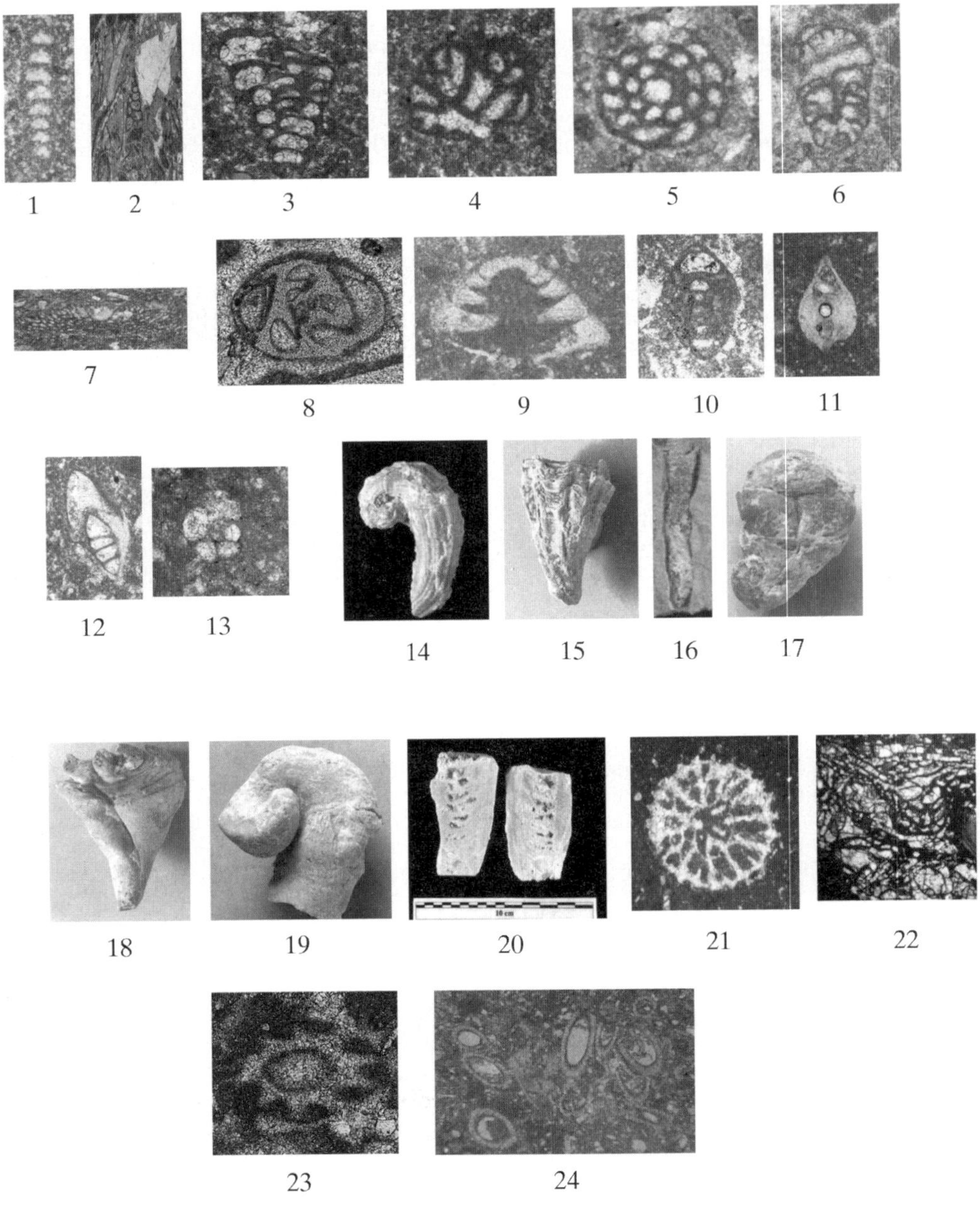

Fig. 4. Important biocomponents of the Shu'aiba Formation at Shaybah (dimension represents length). (1–17 are foraminifera, 14–20 are rudists, 22–24 are calcareous alga) 1, *Reophax* sp. (1.2 mm); 2, (0.3 mm), *Bigenerina* sp. (1.2 mm); 3, *Praechrysalidina infracretacea* (1.2 mm); 4, *Arenobulimina* sp. (0.4 mm); 5, *Debarina hahounerensis* (0.8 mm diameter); 6, *Cribellopsis neoelongata* (0.7 mm); 7, *Palorbitolina lenticularis* (2.5 mm); 8, *Quinqueloculina* sp. (0.4 mm); 9, *Neotrocholina* sp. (0.5 mm diameter); 10, *Nautiloculina oolithica* (0.5 mm diameter); 11, *Lenticulina* sp. (0.4 mm diameter); 12, undifferentiated rotalid (0.4 mm diameter); 13, *Hedbergella delrioensis* (0.3 mm diameter); 14, *Glossomyophorus costatus* (76 mm); 15, *Agriopleura* cf. *marticensis* (52 mm); 16, *Agriopleura cf. blumenbachi* (90 mm); 17, 18, *Horiopleura* cf. *distefanoi* (70 mm; 93 mm); 19, *Oedomyophorus shaybahensis* (37.5 mm width); 20, *Offneria murgensis* (60 mm); 21, solitary coral (5 mm); 22, *Lithocodium aggregatum* (25 mm across); 23, *Coptocampylodon lineolatus* (0.7 mm); 24, *Salpingoporella dinarica* (0.3 mm lumen diameter).

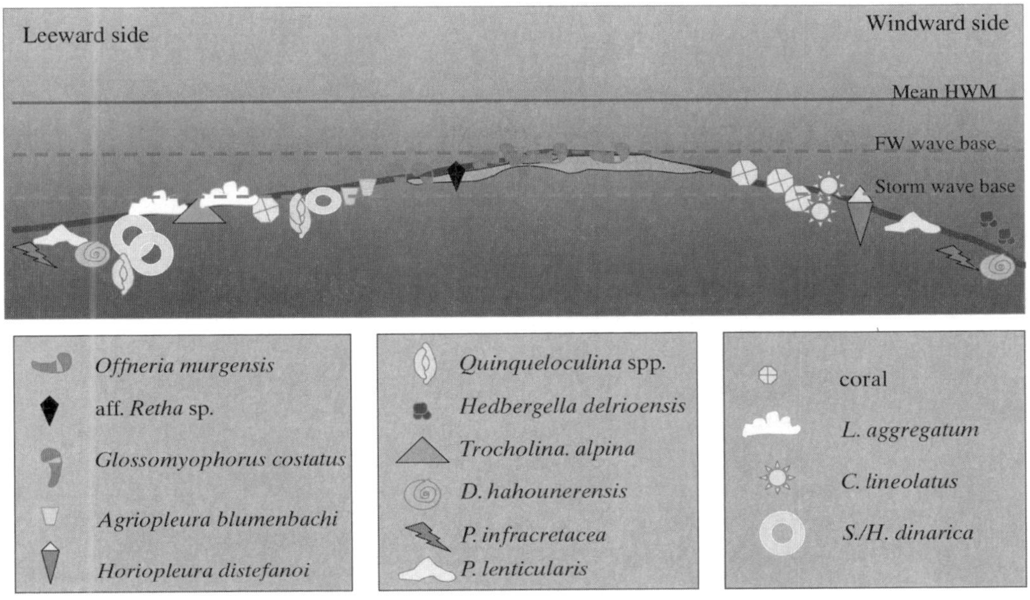

Fig. 5. Interpreted provincial preferences of selected Shu'aiba rudists, foraminifera and calcareous alga.

possibly on the rims of fault-controlled blocks causing differentiation of the platform into shallow lagoon, back-bank, fore-bank and offshore marine provinces (Figs 2, 5, 6 and 7). The biofacies distribution has enabled the rudist bank to be mapped, together with the lagoonal and open marine regimes to be located. It is possible that faults that terminate at the boundary between the lower and middle Shu'aiba layers, interpreted from formation micro scanner (FMI) logs (K. Sadler, oral comm.) caused small submarine topographic highs that were suitable for rudist colonization. A similar structural episode commenced at the Biyadh-Shu'aiba contact in the Al Huwaisah Field of Oman (Boichard *et al.* 1995). Variations in the positions of specific rudist and foraminiferal facies, supported by log correlation, suggest that the rudist bank complex aggraded, retrograded and prograded during the 'middle' Shu'aiba, presumably in response to changes in sea level.

The upper palaeoenvironmental layer (upper Shu'aiba) contains a characteristic micro- and macrofaunal assemblage. A facies dislocation is recognized at the base of this rudist-rimmed lagoon that resulted from a basinward shift of the underlying rudist-rimmed lagoon. The rudist bank complex is narrower than that of the 'middle' Shu'aiba. Many of the foraminiferal species are the same as those that characterize the 'lower' Shu'aiba, and may suggest moderately deep lagoonal conditions. It is possible that the absence of rudists in the 'lower' Shu'aiba resulted from the unrestricted open marine conditions. Within the present study, it has been decided that assignment to informal environmental regimes is probably more accurate than estimates of specific water depths.

The following informal biofacies definitions can be assigned by future micropalaeontologists in the Saudi Arabian Shu'aiba Formation, and also will assist calibration with biofacies in the adjacent successions of Abu Dhabi and Oman. Their distribution within the Shu'aiba Formation at Shaybah is illustrated in Table 1.

Hedbergella Biofacies (ShBf-1)

Principal components: Hedbergella delrioensis, H. planispira.

Accessory microbiocomponents: Lenticulina spp., costate rotalid species.

Accessory macrobiocomponents: Lithocodium aggregatum.

Location in the Shu'aiba Formation: lower Shu'aiba, Sh-1 member.

Palaeoenvironment: This deep, open marine biofacies of the Shu'aiba forms part of the basal Shu'aiba, and is partly equivalent to the

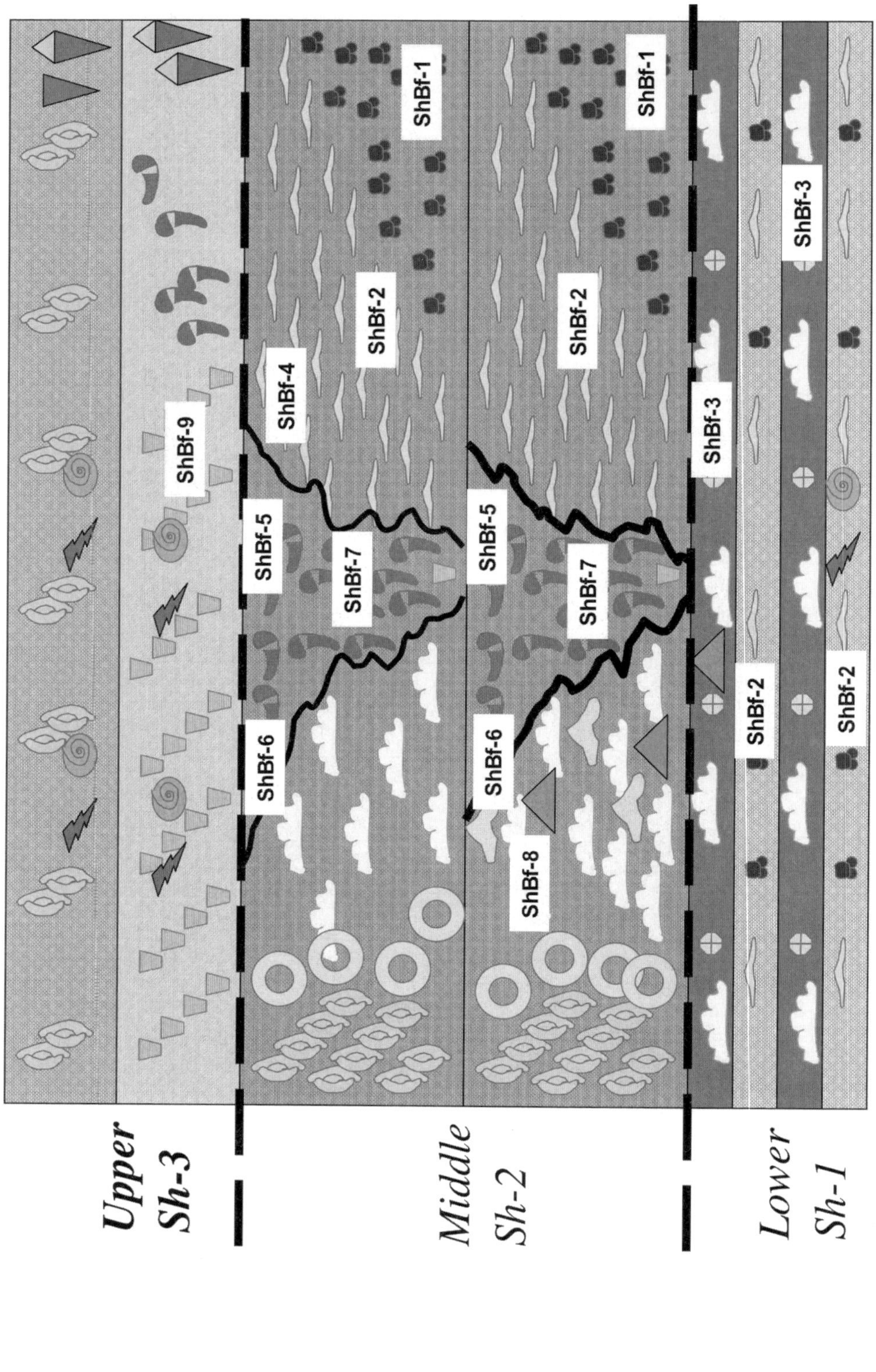

Fig. 6. Distribution of various biocomponents within the lower, middle and upper Shu'aiba during the early Aptian.

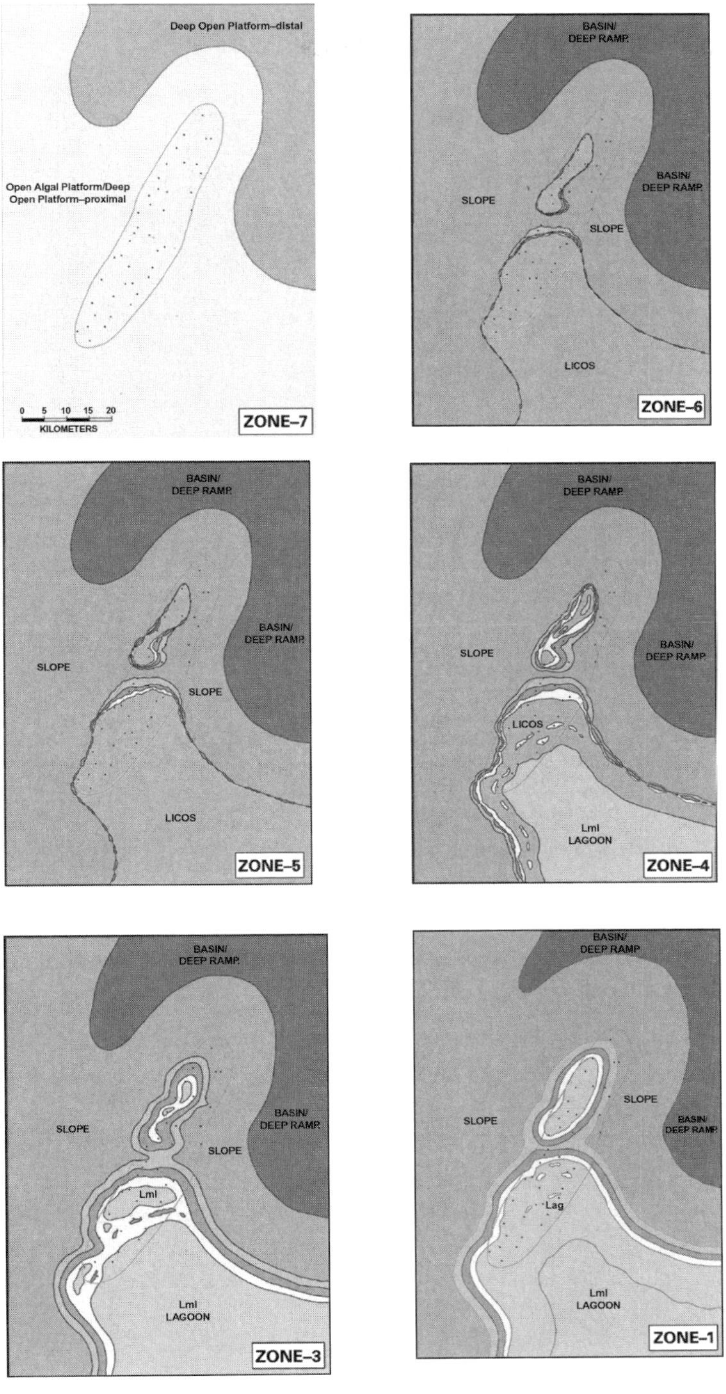

Fig. 7. Evolution of the Shaybah rudist-rimmed platform, illustrated in stages corresponding to biofacies-determined major depositional cycles. See labels for palaeoenvironmental provinces. Zone 7 equates with the Lower Shu'aiba, Zones 6–3 with the Middle Shu'aiba and Zone 7 with the Upper Shu'aiba. The Late Aptian palaeoenvironment is not displayed.

Table 1. *Shu'aiba biofacies, layers, diagnostic biocomponents and palaeoenvironment*

Biofacies	Shu'aiba Layer (see Figs 6 and 7)	Biofacies	Dominant taxa	Accessory taxa	Macropalaeontological biofacies	Subenvironment
ShBf-1	Sh-1	*Hedbergella delrioensis*	*Hedbergella delrioensis; H. planispira*	*Lenticulina* spp., costate and keeled rotalids	*Lithocodium aggregatum*	Deep open marine
ShBf-2	Sh-1, 4	*Hedbergella delrioensis Palorbitolina lenticularis*	*Palorbitolina lenticularis*	*Hedbergella delrioensis, Lenticulina* spp., *Rotalia* spp., *Textularia* spp., miliolids		Moderately deep open marine
ShBf-3	Sh-1, 3	*Praechrysalidina infracretacea*	*Praechrysalidina infracretacea, Vercorsella arenata, Debarina hahounerensis, Salpingoporella dinarica*	miliolids; *Cribellopsis* sp., cf. *Pfenderina* sp.	*Agriopleura blumenbachi; A. marticensis; Lithocodium aggregatum;* coral; *Acroporella* spp.	Open marine shelf
ShBf-4	Sh-2	*Rotalia*	*Praechrysalidina infracretacea; Bolivina* spp.	*Salpingoporella dinarica, Coptocampylodon lineolatus, Cylindroporella arabica*	Echinoid debris	Rudist fore-bank
ShBf-5	Sh-2, 3	*Offneria*			*Offneria murgensis*	Rudist bank crest
ShBf-6	Sh-2	*Offneria-Oedomyophorus*		*Palorbitolina lenticularis; Textularia* spp.	*Offneria murgensis Oedomyophorus shaybaensis*	Rudist proximal back-bank
ShBf-7	Sh-2, 3	*Glossomyophorus*	*Glossomyophorus costatus, Palorbitolina lenticularis; Textularia* spp.	*Neotrocholina* spp., *Praechrysalidina infracretacea*	*Glossomyophorus costatus; Lithocodium aggregatum;* caprinid debris	Rudist distal back-bank
ShBf-8	Sh-1, 2, 3	*Lithocodium*	*Lithocodium aggregatum, Neotrocholina* spp.	*Arenobulimina* sp., *Hedbergella delrioensis, Debarina hahounerensis,* smooth and costate rotalids		Open marine shelf
ShBf-9	Sh-2, 3	*Agriopleura*	*Trocholina alpina, Hedbergella delrioensis, oligosteginids; Debarina hahounerensis*	smooth and costate rotalids	*Lithocodium aggregatum;* coral	Deep lagoon
ShBf-10	Sh-4	*Horiopleura distefanoi*		small, undifferentiated miliolids	coral	open marine reef/ bank slope

carbonate equivalent of the clastic Biyadh Formation. The foraminifera and calcareous algae of this layer are remarkably similar across the entire study area, and indicate a regionally similar palaeoenvironment. Stable and regionally extensive platform carbonate palaeoenvironmental conditions are concluded. Species identification is not always possible owing to the random orientation of the thin sections, but specimens resemble *Hedbergella delrioensis* and *Hedbergella planispira*. The morphologically characteristic planktonic foraminiferal species *Wondersella athersuchi* (Banner & Strank 1987) described from the uppermost Shu'aiba carbonates offshore Abu Dhabi has not been identified in the thin sections examined in the Shaybah region. Calcispheres are present with the planktonic foraminifera within facies considered to be moderately deep (Leckie 1986; Banner & Simmons 1994).

Hedbergella – Palorbitolina Biofacies (ShBf-2)

Principal components: Hedbergella delrioensis, Palorbitolina lenticularis.

Accessory microbiocomponents: Lenticulina spp., costate rotalid species, *Textularia* spp., small undifferentiated miliolids, *Salpingoporella dinarica*.

Accessory macrobiocomponents: Lithocodium aggregatum, coral.

Location in the Shu'aiba Formation: Sh-1 and Sh-4.

Palaeoenvironment: In Sh-1, this deep marine, lowermost, dense part of the Lower Shu'aiba equates with the Biyadh Formation, and consists of dark-grey, orbitolinid packstone with planktonic foraminifera that passes up into cream-coloured wackestone/boundstone characterized by an abundance of encrusting *Lithocodium aggregatum*. Large, low trochoid *Palorbitolina lenticularis* are dominant and well packed. Vilas *et al.* (1995) assign this composition and fabric as 'outer shelf'. Large, low trochoid *Palorbitolina lenticularis*, *Textularia* spp. and rotalid foraminifera are dominant, with subordinate miliolids and rare planktonic foraminifera. Bivalve and echinoid debris with *Salpingoporella dinarica* are well represented, with rare coral fragments. This biofacies is considered to represent a shallower part of the *Hedbergella* Biofacies described above. In the Sh-4 member, this biofacies forms a thick succession located on the outer flanks of the Shu'aiba rudist bank complex, and is considered to represent the shallower equivalent of the intrashelf basinal unit known as the Bab Member in the United Arab Emirates. This environmental regime overlies the 'deep' open marine-upper slope biofacies, and is characterized by the consistent presence of planktonic foraminifera *Hedbergella* spp., and benthonic including rotalids, low trochoid *Palorbitolina lenticularis* and *Textularia* spp., in the absence of rudist debris. Echinoid debris, especially high-ribbed echinoid spines, is common. Thin-walled miliolids and *Lenticulina* spp. are subordinate, together with thin-walled bivalve debris, multi-ribbed small echinoid spines and *Coptocampylodon lineolatus*, pteropods and spicules are also present. Orbitolinid morphology has been used to determine palaeoenvironment (Moullade *et al.* 1985; Vilas *et al.* 1995; Husinec *et al.* 2000; Husinec 2001; Simmons *et al.* 2000) with high conical forms typically of the 'shallow infralittoral zone', or inner shelf, and occur only exceptionally in outer shelf facies. The low conical and larger forms are not so restricted. Other authors (Sartorio & Venturini 1988) place the orbitolinids within the inner shallow platform and also on the outer platform, but absent from the rudist-dominated build-up on the edge of the inner platform. By using the depth ranges of associated algae, Banner & Simmons (1994) have provided depth ranges for *Palorbitolina lenticularis* of 5–60 m, with a preferred range of 10–60 m. The significance of water depth as a sole control on species depth distribution has not been proved, however, and water clarity may have an equally, if not more, important influence.

Praechrysalidina Biofacies (ShBf-3)

Principal components: *Praechrysalidina infracretacea.*

Accessory microbiocomponents: Vercorsella arenata, Debarina hahounerensis, Salpingoporella dinarica.

Accessory macrobiocomponents: Arenobulimina sp., *Pfenderina* sp., small undifferentiated miliolids.

Location in the Shu'aiba Formation: Sh-1 and Sh-3.

Palaeoenvironment: Deep marine open marine platform conditions are concluded for this assemblage, and may represent the shallowest facies of the planktonic foraminiferal-bearing biofacies described above. This biofacies is mostly confined to the Sh-1 member, but a similar biofacies is also developed in Sh-3. Benthonic foraminifera are present throughout most of the Shu'aiba Formation, the most significant being *Palorbitolina lenticularis*, *Praechrysalidina infracretacea*, *Debarina hahounerensis* and *Vercorsella arenata* are well represented within the uppermost and lowermost parts of the Shu'aiba Formation. *Textularia* spp. are present throughout most of the Shu'aiba Formation, but are less well represented within the shallower successions. *Neotrocholina* spp. is rare within the Shu'aiba Formation and is only found within the lower part of the open algal platform, where it is associated with *Lithocodium aggregatum* and rare planktonic foraminifera. Rotalid foraminifera are consistently well represented within the deeper marine facies as well as the lagoonal facies, but are absent from the bank and associated environments. A variety of rotalids has been recorded, and includes costate *Lenticulina* species.

Rotalia Biofacies (ShBf-4)

Principal components: Rotalia spp., *Praechrysalidina infracretacea, Bolivina* spp.

Accessory microbiocomponents: Salpingoporella dinarica, Coptocampylodon lineolatus, Cylindroporella arabica.

Accessory macrobiocomponents: echinoid debris, finely comminuted rudist debris.

Location in the Shu'aiba Formation: Sh-2.

Palaeoenvironment: Sediments that accumulated within this regime are mostly derived from the bank crest, and therefore consist predominantly of coarse and fine-grained debris from the recumbent rudist *Offneria murgensis*. The moderately high sedimentation rates within this regime are not considered to have been suitable for the low-energy, low-sedimentation rates conducive to the establishment of *Lithocodium aggregatum* (Banner *et al.* 1990). Fragments of *Lithocodium aggregatum* and the presence of scattered *Glossomyophorus costatus* are therefore possibly allochthonous and transported through inter-bank channels from the deep back-bank or lagoon. The elevator rudist *Glossomyophorus costatus* mostly grew in protected, low-energy regimes with only occasional episodes of physical disturbance (Gili *et al.* 1995), and it is possible that such an environment was locally available in this regime, and also permitted the dasyclad alga *Coptocampylodon lineolatus*. An allochthonous source for many of the species is suggested by the coarse grain size, which would typically exclude *in situ* benthonic foraminifera such as *Palorbitolina lenticularis*, miliolids and *Textularia* spp.

Offneria Biofacies (ShBf-5)

Principal components: Offneria murgensis.

Accessory microbiocomponents: small undifferentiated miliolids.

Accessory macrobiocomponents: none.

Location in the Shu'aiba Formation: Sh-2; minor development in Sh-3.

Palaeoenvironment: rudist bank crest. Rudist morphology has been characterized by Skelton & Gili (1991) and Gili *et al.* (1995) as being related to an elevator, recumbent or clinger lifestyle, based upon the relative hydraulic energy levels and rates of sedimentation. The spatial distribution of these various growth habits is described by Skelton (2001).This approach has been critical towards the use of the Shu'aiba rudists as palaeoenvironmental indicators, and this study has enabled a refinement of the scheme to individual species preferences. *Offneria murgensis* is the only recumbent form identified although rare elevator forms have been recently recognized (Hughes 2002, 2003*a*; Hughes *et al.* 2003, 2004*a*, *b*; Siddiqui *et al.* 2004); no clinger morphotypes have been recognised. Grainstones consisting of monospecific *Offneria murgensis* debris suggest sites of high-energy shallow crests of rudist biostromes within fair-weather wave-base. The empty shells were easily broken to provide very porous sand that was swept predominantly into the lagoon, but also to the open marine flanks of the bank, possibly through inter-bank channels. Stossel & Bernouli (2000) have described a similar process from the Cenomanian of the Maella platform. Rudist mounds and shoals have been described from the Aptian–Albian carbonates of Arizona, up to 180 m diameter (Hartshorne 1989) and Texas (Kerans *et al.* 1997). Flank beds of storm-transported rudists in coastal Croatian exposures display dips of 12–15° (Hughes 2003*b*), but elsewhere *Offneria* grainstones could theoretically reach dips of 30° (Schlager & Camber 1986; Kenter 1990; Everts & Reijmer 1995).

Offneria – Oedomyophorus Biofacies (ShBf-6)

Principal components: Offneria murgensis, Oedomyophorus shaybahensis.

Accessory microbiocomponents: Palorbitolina lenticularis, Textularia spp., small undifferentiated miliolids.

Accessory macrobiocomponents: rare *Offneria nicolinae* (Skelton 1997; Skelton *et al.* 1999).

Location in the Shu'aiba Formation: Sh-2.

Palaeoenvironment: proximal flank of the rudist bank. The very shallow, high-energy conditions associated with the deposition of this facies is considered responsible for the absence of fine sedimentary grains such as foraminifera. The predominance of the recumbent caprinid rudist *Offneria murgensis* is indicative of unstable, current-swept conditions where sediment by-passing predominated over accumulation (Gili *et al.* 1995; Skelton *et al.* 1995). The new robust caprinulinoidinid elevator rudist *Oedomyophorus shaybahensis* (Skelton *et al.* 1999; Skelton 2004) seems to have adapted to high rates of sedimentation in a moderately high energy environment but sheltered from the high energy bank crest, but in a shallower and higher energy zone than that of *G. costatus*. The dasyclad alga *Coptocampylodon lineolatus* is rare. The sediments of the shallower regimes consist almost entirely of pallial canal fragments derived from *Offneria murgensis* Masse.

Glossomyophorus Biofacies (ShBf-7)

Principal components: Glossomyophorus costatus, Offneria debris, Palorbitolina lenticularis, Textularia spp.

Accessory microbiocomponents: Neotrocholina sp., *Praechrysalidina infracretacea.*

Accessory macrobiocomponents: Lithocodium aggregatum.

Location in the Shu'aiba Formation: Sh-2 and Sh-3.

Palaeoenvironment: distal flank of the rudist bank. The elevator rudist *Glossomyophorus costatus* and abundant bivalve debris dominates this moderately shallow depositional regime. *Offneria* debris, in the form of fragments of pallial canals, is a common component of the sediment, and was derived from the physical destruction of *Offneria murgensis* shells on the biostromal crest and its immediate flanks. The co-occurrence of elevators and recumbent debris is a common observation (Gili *et al.* 1995). Elevator rudists lived partially embedded in bioclastic sands and calcareous sandy muds that accumulated around them, which implies that they occupied low to moderate energy environments (Gili *et al.* 1995). *Glossomyophorus costatus* is particularly well represented within the Shu'aiba Formation. It is here interpreted to have occupied moderately shallow marine conditions either as localized highs within the shelf or lagoon, in the moderately sheltered position leeward, back-bank location and possibly windward of the higher-energy rudist bank dominated by *Offneria murgensis*, based on the spatial relationships with other rudist species in this study. Elevator rudist mud-sticker communities usually grew under restricted carbonate platform conditions and formed tabular biostromes (Skelton *et al.* 1995; Gotz 1999). Palaeobathymetric preferences of recumbent forms are difficult to estimate precisely, but an association of *Glossomyophorus costatus* with *Palorbitolina lenticularis* and the calcareous alga *Lithocodium aggregatum* and *Salpingoporella dinarica* would suggest depths within the range of 5 to 20 m (Banner & Simmons 1994). *Glossomyophorus costatus* is typically found within a fragmented *Offneria* matrix, overlying *Agriopleura* or *Lithocodium-Orbitolina* facies, and mostly overlain by *Offneria* facies. Elevator accumulations have been defined elsewhere (Skelton & Gili 1991) as of metric thickness and essentially lenticular or tabular in shape, showing no topographical relief. They are characterized by displaying low species diversity, and are often monospecific to paucispecific. Miliolid foraminifera are well represented with subordinate *Palorbitolina lenticularis* (high trochoid), and *Trocholina* spp. *Lithocodium aggregatum* is locally present.

Lithocodium Biofacies (ShBf-8)

Principal components: Lithocodium aggregatum, Neotrocholina spp.

Accessory microbiocomponents: Hedbergella delrioensis, Debarina hahounerensis, Arenobulimina spp., smooth and costate rotalids.

Accessory macrobiocomponents: Lithocodium aggregatum, coral.

Location in the Shu'aiba Formation: Sh-1, Sh-2 and Sh-3.

Palaeoenvironment: deep lagoon. Calcareous algae in the Shu'aiba Formation are dominated by the encrusting codiacean *Lithocodium aggregatum* (Banner *et al.* 1990; Schmid & Leinfelder 1996; Dupraz & Strasser 1999; Koch *et al.* 2002; Jones *et al.* 2004). This species forms a number of localized occurrences within the Shu'aiba Formation, especially within the lower half of the section, where it is concentrated in beds that overlie those containing planktonic foraminifera. This association suggests cycles that shoal upwards from moderately deep water. Thick *Lithocodium*-dominated successions are typical within the 'algal platform' facies, where they are either overlain by *G. costatus*-bearing sediments or represent a lateral, deeper marine equivalent within the lagoon. The *Lithocodium*-dominated deep lagoonal sediments are considered to have aggraded to fill the accommodation space that the adjacent, more-rapidly aggrading rudist banks had already filled. Lower *Lithocodium* accumulations are typically encrusting, and are interpreted to display lateral continuity. Upper parts of the *Lithocodium* accumulations, however, display individual pillared growth that may imply shallower conditions. Subspherical bodies, or oncoliths, of *Lithocodium* have also been observed in the upper parts and testify to elevated water energy conditions possibly within wave base. *Hensonella dinarica* is a calcareous dasyclad alga that is also concentrated in the uppermost and lowermost parts of the Shu'aiba Formation, but is also sporadically represented within the deeper back-bank *Glossomyophorus costatus* facies within the lagoon. *Coptocampylodon lineolatus* is a dasyclad alga that is well represented within the deeper marine sediments; preservation of its aragonitic composition may be explained by probable penecontemporaneous transportation from the shallower part of the platform. Corals include solitary, branching, platy and massive forms, and are rare within the Shu'aiba Formation. Microsolenid corals (Rosen *et al.* 2003) are well represented in the lower part, in association with the basal beds of *Lithocodium aggregatum*. Isolate simple corals are locally present. The relationship with *Lithocodium* suggests oligotrophic conditions, favoured by corals, and the microsolenid – *Lithocodium* association would suggest conditions of clear water with minimal sediment input. Elevated gamma values of the *Lithocodium* facies may be attributed to concentrations of finely dispersed potassium from clay particles that may have travelled far from a terrestrial source, although there are no shale beds associated with this biofacies. The initiation of rudists is associated with the presence of muddy carbonates, and it may be that slightly elevated muddy conditions preferentially favoured the filter-feeding rudists over the photosynthesis-reliant corals and *Lithocodium* rather than a competition for space.

This facies gradually shallows to depths where elevator rudists such as *Agriopleura* cf. *marticensis*, *Agriopleura* cf. *blumenbachi* and *Glossomyophorus costatus* became established ('foramol' facies of Carannante *et al.* 1994, 1999*a*, *b*).

Agriopleura Biofacies (ShBf-9)

Principal components: Agriopleura cf. *marticensis*.

Accessory microbiocomponents: Agriopleura cf. *blumenbachi*, *Hedbergella delrioensis*, *Debarina hahounerensis*, *Praechrysalidina infracretacea*, *Vercorsella arenata*, *Salpingoporella dinarica*, *Neotrocholina* spp., undifferentiated small miliolids.

Accessory macrobiocomponents: Lithocodium aggregatum, coral.

Location in the Shu'aiba Formation: Sh-2 and Sh-3.

Palaeoenvironment: deep lagoon. This environment is mostly represented by sediments with high benthonic species diversity and the consistent presence of the elevator rudist *Agriopleura* spp. and the dasyclad alga *Salpingoporella dinarica*. Two morphotypes of *Agriopleura* include the common 'V-type' widely conical or squat *A. marticensis* and the 'U-type' thin elongate *A.* cf. *blumenbachi* (Hughes 2003*a*).

Miliolids (*Quinqueloculina* spp.), *Debarina hahounerensis*, *Praechrysalidina infracretacea* and *Textularia* spp. are consistently present, with subordinate *Everticyclammina hedbergi*, *Orbitolina* spp. and *Vercorsella arenata*. *Archaealveolina* sp. and *Nautiloculina* sp. are rare but confined to this regime. *Agriopleura* cf. *marticensis* and *Agriopleura* cf. *blumenbachi* are considered to have preferred quiet, shallow platform interiors with net sediment accumulation, where extensive tabular bodies were developed. They are often associated with chondrodontid oysters, and thereby form a typical lagoonal assemblage (Cucchi & Pugliesi 1999; Caffau *et al.* 1999). *Agriopleura* species are interpreted to have occupied a range of environments (Hughes 2003*a*) although net positive sedimentation and low energy conditions are suggested by its elevator lifestyle. The tightly fitting, countersunk free (left) valve invites various speculations regarding possible environmental reasons for this peculiar adaptation. *Agriopleura* is typically present above the *Lithocodium–Orbitolina* facies and below the *Glossomyophorus* facies, where it is considered to represent moderately deep, low-energy conditions at the beginning of a new depositional cycle. It is especially well developed, however, within the upper part of the Shu'aiba Formation where moderately deep marine shelf conditions are concluded.

Dark green-grey mudstone beds are typically

present within this unit in the northern part of Shaybah field, and may represent soil surfaces. Many wells additionally display karst features in this layer and therefore support an episode of emergence and dissolution prior to the deposition of the overlying Nahr Umr Formation. The karst features include pyritic mud and breccia-filled cracks that extend at least 30 feet into the top of the Shu'aiba.

Horiopleura Biofacies (ShBf-10)

Principal components: Horiopleura cf. *distefanoi.*

Accessory microbiocomponents: undifferentiated small miliolids.

Accessory macrobiocomponents: coral.

Location in the Shu'aiba Formation: Sh-4.

Palaeoenvironment: Along the northeast flank of the Shaybah platform, corals and the localized presence of the elevator rudist *Horiopleura* cf. *distefanoi* represent a steep, fore-bank environment of late Aptian age.

Biocomponent variation

Various biocomponent associations have been recognized that suggest palaeoenvironmental and palaeobathymetric attributes for species without previous published palaeobathymetry. Figure 6 summarizes the palaeoenvironmental interpretation for benthonic and planktonic foraminifera, 'pteropods', calcispheres, rudists, calcareous algae and echinoid debris of the Shu'aiba Formation. Repeated complete or incomplete vertical stacking cycles of *Agriopleura*, *Glossomyophorus* and *Offneria*, respectively, are considered to have been influenced by small-scale sea level changes typical of the Cretaceous 'greenhouse' conditions (Skelton *et al.* 1997). This biotic feedback in the rudist facies provides important guidelines for depositional layer recognition, and may represent a response to complex palaeobiological and palaeophysical conditions, as described by Gili *et al.* (1995), Gili & Skelton (1999) and Gotz (1999).

A series of seven palaeoenvironmental reconstructions (Fig. 7) have been based on the existing interpretation of depositional layers. The seven time-slices assist the determination of the lateral and vertical reservoir continuity, as described by Payne *et al.* (1999). Each layer has been assigned informally to zones 7 to 1, in ascending order that has been applied to developing a reservoir zonation scheme. Zone 7 equates with the laterally extensive open marine palaeoenvironment assigned to the planktonic foraminiferal-*Orbitolina-Lithocodium* biofacies of the Biyadh, or 'lower' Shu'aiba. Zones 6 to 3 display the initiation, development and expansion of the differentiated rudist-rimmed platform. Combined Zones 1/2 shows a marked restriction of the width of the rudist bank complex and equates with the lagoon-dominated, *Agriopleura*-associated 'upper' Shu'aiba.

Discontinuities in the vertical distribution of various species are marked on the chart and depths of coincident appearance or disappearance up-section are noted and provide events that potential signify depositional cycle boundaries or environmental changes within depositional cycles. In many cases, the localized presence of planktonic foraminifera, *Vercorsella arenata*, *Debarina hahounerensis* and *Praechrysalidina infracretacea* and low profile *Palorbitolina lenticularis*, for example, have been interpreted as the transgressive, deeper part of a depositional cycle of which the succeeding shallower part equivalent to the highstand is represented by the absence of these species and the presence of simple miliolids and biserial agglutinated foraminifera. Rudists have been similarly used following their calibration with certain foraminiferal biofacies. In the lower part of the Shu'aiba Formation, small-scale cycles are represented by alternations of planktonic foraminiferal, low trochoid orbitolinids and deeper benthonic foraminifera with *Lithocodium aggregatum*, of which the latter species is considered to represent the relatively shallower part of the depositional cycle.

Studies of the Shu'aiba carbonate cores from Shaybah have mostly revealed a highly variable distribution of the biocomponents, resulting from the contemporaneously different depositional environments. Palaeoenvironmentally significant biofacies have been determined but it is difficult to establish a refined conventional pan-Shu'aiba biozonation scheme owing to the palaeoenvironmental complexity.

Morphogroups

A novel way of using agglutinated foraminiferal test morphology with feeding behaviour, and then extending this interpretation to enhancing the palaeoenvironment, was proposed by Jones & Charnock (1985). The four main groups include A (suspension feeders), B (surface herbivores), C (infaunal detrital/bacterial scavengers) and D (epifaunal herbivores). The Shu'aiba foraminifera can also be considered in this way (Fig. 8), and an extended scheme is here

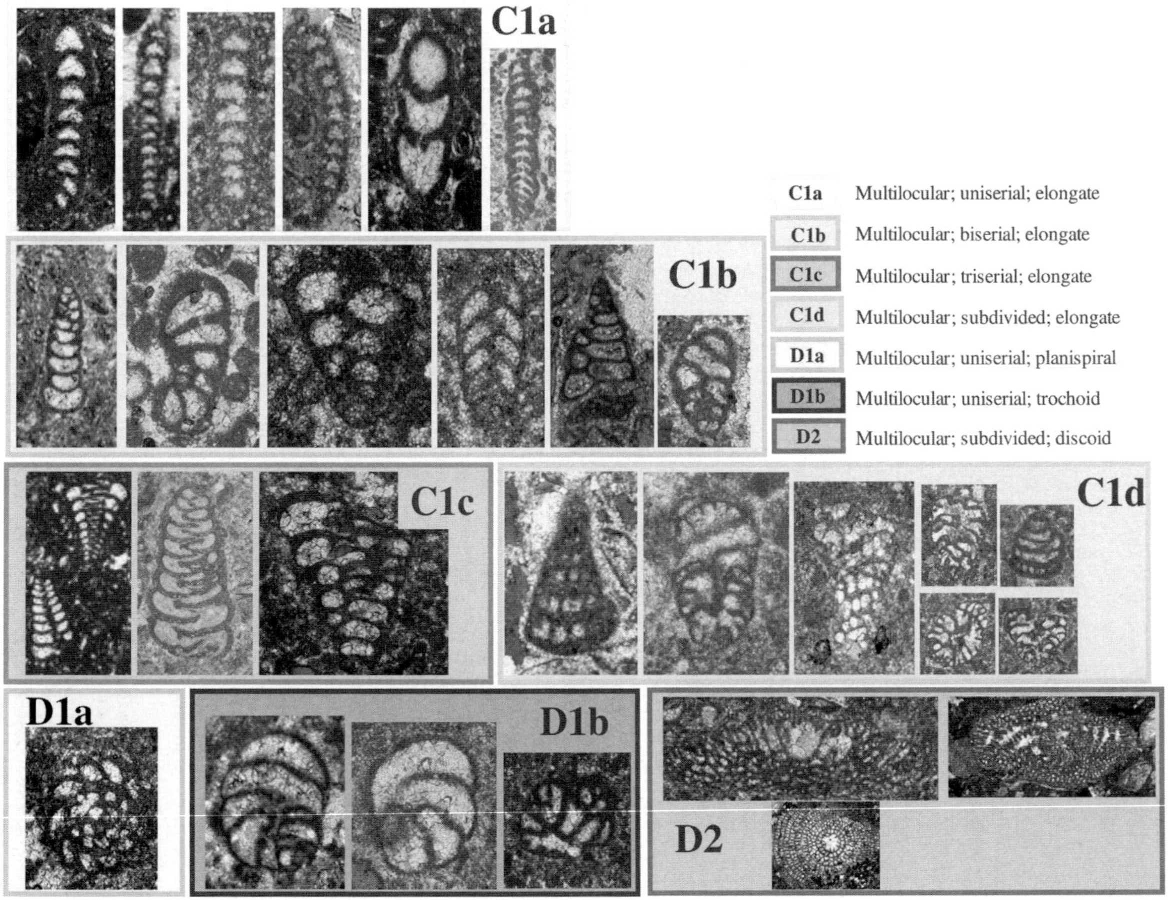

Fig. 8. Suggested modification of agglutinated foraminiferal morphogroup scheme of Jones & Charnock (1985). C1a, various species of *Reophax*. C1b, various species of *Ammobaculites* (two images on left) and *Textularia* (four images on right). C1c, *Praechrysalidina infracretacea*. C1d, *Vercorsella arenata* (image on left) and species of *Cribellopsis* (six images on right). D1a, *Debarina hahounerensis* (one image on left). D1b, species of *Arenobulimina*. D2, *Palorbitolina lenticularis* and *Choffatella decipiens*.

proposed for morphogroups C and D, as there are no representatives of morphogroups A and B. With reference to Figure 8, morphogroup C1 has been subdivided into C1a to d. C1a includes the uniserial forms that typify the deeper lagoon environments in the Shu'aiba Formation, and can display a variety of chamber enlargement and number variations. Morphogroup C1b includes biserial chamber arrangement, with all the subtle variations, and is typical of the deeper lagoon as well as the open platform and deeper rudist back-barrier. Morphogroup C1c includes the triserial forms, here typified by *Praechrysalidina infracretacea*, and found only in the open platform, moderately deep environments. Morphogroup C1d includes the elongate trochospirally coiled forms with chamberlets, including *Vercorsella arenata* and *Arenobulimina* sp. Morphogroup D1a includes the planispirally coiled forms, such as *Debarina hahounerensis*, found in the open platform conditions as well as the deeper parts of the lagoon. It is possible that this morphogroup could be considered as a variant of morphogroup B3 of Jones & Charnock (1985) (Jones, written communication 2003). Morphogroup D1b includes the trochospirally coiled forms, such as cf. *Pfenderina* sp., and is typically present in the open platform conditions as well as the deeper parts of the lagoon. This morphogroup has been suggested, by Jones (written communication 2003), to have possible equivalence to morphogroups B3 or D of Jones & Charnock (1985). Morphogroup D2 includes the complex walled orbitolinid test, typified in the Shu'aiba Formation by *Palorbitolina lenticularis*, and present in a variety of environments that include open platform and deep lagoon (low conical morphotype) and shallow lagoon adjacent to the rudist back bank (elevated conical morphotype). This morphotype is unique, and may be related to special adaptation to support for symbiotic algae.

Summary

Micropalaeontology and macropalaeontology of the Shu'aiba carbonates in the Shaybah reservoir of Saudi Arabia has revealed a number of discrete biofacies. Each biofacies is considered to represent particular palaeoenvironmental conditions that existed during the three major stages of rudist -rimmed platform development during the early Aptian, and the last stage in the late Aptian. The biofacies boundaries have assisted depositional cycle definition and hierarchy. An attempt has been made to define the benthonic foraminiferal test shape to an existing morphotype scheme, but application of this scheme awaits further investigation.

The 'lower' layer of the Shu'aiba Formation, Sh1, includes the orbitolinid-dominated unit known regionally as the Biyadh (Saudi Arabia), 'Kharaib upper dense' (U.A.E.) and Hawar (Oman) and is extensively developed with only slight lateral variations. This polycyclic layer consists of a vertical succession commencing with deep marine beds that contain planktonic foraminiferal (*Hedbergella delrioensis*) and *Palorbitolina lenticularis* (low trochospire) conformably overlain by beds with calcareous algal (*Lithocodium aggregatum*) and platy, microsolenid and solitary coral beds. The lower planktonic-bearing biofacies also includes the benthonic foraminiferal species *Debarina hahounerensis*, *Lenticulina* spp., *Nodosaria* spp., *Praechrysalidina infracretacea* and *Choffatella decipiens*.

The 'middle layer', Sh-2, is characterized by considerable lateral biofacies variations in which rudists are well represented. Possibly fault-controlled submarine relief caused sudden termination of the laterally extensive 'lower layer', although evidence for this event is limited to image log interpretation. Whether a regional tectonic event took place, and was responsible for regional rudist colonization is yet to be confirmed. Colonization of local submarine highs by rudists resulted in differentiation of the basal middle layer into lagoon, rudist bank and slope environments, each of which is characterized by a unique biofacies. The interpreted lagoon environment typically contains a lower biofacies that includes encrusting *Lithocodium aggregatum* and platy corals together with the benthonic foraminifera *Neotrocholina* sp., *Palorbitolina lenticularis* and various unornamented miliolids. In the upper part of this regime, increasingly restricted environmental conditions caused a decrease in biotic diversity with a concentration of miliolids. The rudist assemblages are spatially clustered to form a network of banks, and display an orderly species range from the deeper lagoonal flank of the bank complex with small elevator morphotypes (*Agriopleura* spp. and *Glossomyophorus costatus*). Two distinct *Agriopleura* morphotypes are present, including a rapidly enlarging form assigned to *A.* cf. *marticensis* and a cylindrical form assigned to *A.* cf. *blumenbachi*. Elevated energy conditions approaching the bank crest enabled colonization by robust elevator rudists (*Oedomyophorus shaybahensis*) and rare specimens of the pencil-like caprinid elevator *Offneria* aff. *nicolinae*. Highest energy conditions of the bank crest were responsible for colonization by the recumbent caprinid rudist *Offneria murgensis*. The fore-bank regime is dominated by rudist debris that passes laterally into deeper conditions that are dominated by *Palorbitolina lenticularis* and the planktonic species *Hedbergella delrioensis*. There is evidence for tempestite-triggered allochthonous beds of shallow marine bioclasts within this regime that fine upwards into the autochthonous, planktonic foraminiferal bearing muds.

The marked contrast in biofacies dislocation between the middle (Sh2) and upper layer (Sh3) is related to a basinward shift in the lagoonal biofacies, suggestive of a sequence boundary, although regional evidence for a similar facies is poorly documented. The upper layer is typically characterized by the widespread occurrence of *Agriopleura* cf. *marticensis*, although *Glossomyophorus costatus* and *Offneria murgensis* are also present but with much reduced extent. The elevator rudist *Horiopleura distefanoi* is particularly well developed along the eastern flank of the region. The upper layer is dominated by deep lagoonal conditions in which foraminiferal diversity is typically high. Species include *Praechrysalidina infracretacea*, *Debarina hahounerensis*, *Vercorsella arenata*, textularids and miliolids. Within the lagoon, the dasyclad alga *Hensonella/Salpingoporella dinarica* is well represented. On the outer flanks of the banks, this species is replaced by *Coptocampylodon lineolatus*. The upper layer contains localized layers or pods of grey, non-calcareous, pyritic clay that may be related to karst-fill that took place during the phase of emergence at the top of the Shu'aiba Formation.

This contribution is published with the permission of the Saudi Arabian Oil Company (Saudi Aramco) and the Ministry of Petroleum. The comments provided by the anonymous editors of the manuscript are gratefully acknowledged. Reference to rudist taxonomy and their interpreted palaeoenvironmental preferences is based on numerous discussions with P. Skelton of the Open University, UK. Dr Osman Varol, of Varol Research, provided nannopalaeontological data and numerous discussions.

References

ABOU-CHOUCHA, M.K.A. & ENNADI, I.M. 1993. Possible stratigraphic traps in the Shuaiba Formation onshore Abu Dhabi, *Proceedings of the Society of Petroleum Engineers Middle East Oil Technology Conference*, Bahrain, April, 1993, **393**, 1–14.

AKTAS, G., HUGHES, G.W., ABU-BSHAIT, A.J. & AL-GARNI, S. 2000. Stratigraphic framework of Shu'aiba Formation, Shaybah Field, Saudi Arabia: a basis for reservoir development of an Aptian carbonate ramp complex (abstract). *GeoArabia*, **5**, 11.

ALDABAL, M.A. & ALSHARHAN, A.S. 1989. Geological model and reservoir evaluation of the Lower Cretaceous Bab Member in the Zakum Field, Abu Dhabi, U.A.E. *Proceedings of the Society of Petroleum Engineers Middle East Oil Technology Conference, Bahrain, March, 1989*, 797–810.

ALSHARHAN, A.S. 1985. Depositional environment, reservoir units evolution, and hydrocarbon habitat of Shuaiba Formation, Lower Cretaceous, Abu Dhabi, United Arab Emirates. *AAPG Bulletin*, **69**, 899–912.

ALSHARHAN, A.S. 1987. Geology and reservoir characteristics of carbonate build-up in giant Bu Hasa oil field, Abu Dhabi, United Arab Emirates. *AAPG Bulletin*, **71**, 1304–1318.

ALSHARHAN, A.S. & KENDALL, C.G.ST.C. 1991. Cretaceous chronostratigraphy, unconformities and eustatic sea-level changes in the sediments of Abu Dhabi, United Arab Emirates. *Cretaceous Research*, **12**, 370–401.

ALSHARHAN, A.S. & NAIRN, A.E.M. 1986. A review of the Cretaceous formations in the Arabian Peninsula and Gulf: Part 1. Lower Cretaceous (Thamama Group) stratigraphy and palaeogeography. *Journal of Petroleum Geology*, **9**, 5–60.

ALSHARHAN, A.S. & NAIRN, A.E.M. 1997. *Sedimentary Basins and Petroleum Geology of the Middle East*, Elsevier, Amsterdam.

AZER, S.R. & TOLAND, C. 1993. Sea level changes in the Aptian and Barremian (Upper Thamama) of offshore Abu Dhabi, UAE. *Proceedings of the Society of Petroleum Engineers Middle East Oil Technology Conference, Bahrain, April, 1993*, 141–154.

BANNER, F.T. & SIMMONS, M.D. 1994. Calcareous algae and foraminifera as water-depth indicators: an example from the Early Cretaceous carbonates of northeast Arabia. *In*: SIMMONS, M.D. (ed.) *Micropalaeontology and Hydrocarbon Exploration in the Middle East*. British Micropalaeontological Society Publications Series, Chapman and Hall, London, 243–252.

BANNER, F.T. & STRANK, A.R.E. 1987. On *Wondersella athersuchi*, a new stratigraphically significant hedbergellid foraminiferan from the Cretaceous Shuaiba Formation of the Middle East. *Journal of Micropalaeontology*, **6**, 39–48.

BANNER, F.T., FINCH, E.M. & SIMMONS, M.D. 1990. On *Lithocodium* (Elliott) (calcareous algae); its paleobiological and stratigraphical significance. *Journal of Micropalaeontology*, **9**, 21–36.

BEYDOUN, Z.R., BAMAHMOUD, M.O. & NANI, A.S.O. 1993. The Qishn Formation, Yemen: lithofacies and hydrocarbon habitat. *Marine and Petroleum Geology*, **10**, 364–372.

BOICHARD, R.A.P., AL-SUWAIDI, A.S. & KARAKHANIAN, H. 1995. Sequence boundary types and related porosity evolutions: example of the Upper Thamama Group of Field 'A' offshore Abu Dhabi, United Arab Emirates. *In*: AL-HUSSEINI, M.I. (ed.) GEO *'94, The Middle East Petroleum Geosciences, Gulf PetroLink, Bahrain*, 191–201.

BOOTE, D.R.D. & MOU, D. 2003. Safah Field, Oman: retrospective of a new-concept exploration play, 1980 to 2000. *GeoArabia*, **8**, 367–430.

CAFFAU, M., PUGLIESE, N. & TSAKITIDOU, E. 1999. Evolution of a ramp in the late Cenomanian at Zolla (Trieste Karst, Italy). *In*: HOFLING, R. & STEUBER, T. (eds) *Fifth International Congress on Rudists, Abstracts and Field Guides*, Erlanger Geologische Abhandlungen, Erlangen, **3**, 135–138.

CALAVAN, C.W., HAGERTY, R.M., MITCHELL, J.C. & SCHUTTER, S.R. 1992. Integrated reservoir study and modelling of the Shu'aiba Reservoir, Bu Hasa Field, Abu Dhabi, United Arab Emirates. *ADNOC/SPE Proceedings, Paper SPE 24509*, 299–312,

CARANNANTE, G., CHERCHI, A. & SIMONE, L. 1994. Chlorozoan versus foramol lithofacies in Upper Cretaceous rudist limestones. *Palaeogeography, Palaeoclimatology, Palaeoecology*, **119**, 137–154.

CARANNANTE, G., GRAZIANO, R., PAPPONE, G., RUBERTI, D. & SIMONE, L. 1999*a*. Depositional system and response to sea level oscillations of the Senonian rudist-bearing carbonate shelves. Example from the Central Mediterranean areas. *Facies*, **40**, 1–24.

CARANNANTE, G., GRAZIANO, R., LAVIANO, A., RUBERTI, D., SIMONE, L., SIRNA, G. & TROPEANO, M. 1999*b*. Central-southern Italy Senonian rudist-bearing carbonate shelves (1): depositional model. *In*: HOFLING, R. & STEUBER, T. (eds) *Fifth International Congress on Rudists, Abstracts and Field Guides*, Erlanger Geologische Abhandlungen, Sonderband, **3**, 10–12.

COBIANCHI, M., LUCIANI, V. & MENEGATTI, A. 1999. The Selli level of the Gargano Promontory, Apulia, southern Italy: foraminiferal and calcareous nannofossil data. *Cretaceous Research*, **20**, 255–269.

CUCCHI, F. & PUGLIESI, N. 1999. The carbonate sequence of the Karst at Trieste (Italy). *In*: HOFLING, R. & STEUBER, T. (eds) *Fifth International Congress on Rudists, Abstracts and Field Guides*, Erlanger Geologische Abhandlungen, Sonderband, **3**, 131–133.

DUPRAZ, C. & STRASSER, A. 1999. Microbialites and micro-encrusters in shallow coral bioherms (Middle to Late Oxfordian, Swiss Jura Mountains). *Facies*, **40**, 101–130.

ERBA, E. 1996. The Aptian Stage. *Bulletin de l'Institut Royal des Sciences Naturelles de Belgique, Sciences de la Terre*, **66**, 31–43.

EVERTS, A.J.W. & REIJMER, J.J.G. 1995. Clinoform composition and margin geometries of a Lower Cretaceous carbonate platform (Vercors, SE France). *Palaeogeography, Palaeoclimatology, Palaeoecology*, **119**, 19–35.

FISCHER, K.C., MOELLER, U. & MARSCHALL, R. 1994. Development of an exploration concept for the Shu'aiba Formation (Thamama Group) using seismic sequence and facies analysis in combination with forward modelling. *In*: AL-HUSSEINI, M.I. (ed.) *GEO '94, The Middle East Petroleum Geosciences, Gulf PetroLink, Bahrain*, 377–386.

FISCHER, K.C., MOELLER, U. & MARSCHALL, R. 1997. Development of an exploration concept for the Shu'aiba Formation (Thamama Group) using seismic sequence and facies analysis in combination with forward modelling. *In*: PALAZ, I. & MARFURT, K.J. (eds) *Carbonate Seismology*. Geophysical Developments Series 6, Society of Exploration Geophysicists, 407–416.

FOLMI, K.B., WEISSERT, BEISPRING, H.M. & FUNK, H. 1994. Phosphogenesis, carbon-isotope stratigraphy and carbonate-platform evolution along the Lower Cretaceous northern Tethyan margin. *Bulletin Geological Society of America*, **106**, 726–746.

FROST, S.H., BLIEFNICK, D.M. & HARRIS, P.M. 1983. Deposition and porosity evolution of a lower Cretaceous rudist buildup, Shuaiba Formation of Eastern Arabian Peninsula. *In*: HARRIS, P.M. (ed.) *Carbonate Buildups*. Society of Economic Paleontologists and Mineralogists, Core Workshop, **4**, 381–410.

GILI, E. & SKELTON, P.W. 1999. Organism-environment feedbacks in elevator rudist congregations. *In*: HOFLING, R. & STEUBER, T. (eds) *Fifth International Congress on Rudists, Abstracts and Field Guides*, Erlanger Geologische Abhandlungen, Sonderband, **3**, 22.

GILI, E., MASSE, J.P. & SKELTON, P.W. 1995. Rudists as gregarious sediment dwellers, not reef-builders on Cretaceous carbonate platforms. *Palaeogeography, Palaeoclimatology, Palaeoecology*, **118**, 245–267.

GOTZ, S. 1999. Elevator rudist mudsticker communities: a self-regulating sedimentary buffer-system? *In*: HOFLING, R. & STEUBER, T. (eds) *Fifth International Congress on Rudists, Abstracts and Field Guides*. Erlanger Geologische Abhandlungen, Sonderband, **3**, 23–24.

GRABOWSKI, G.J. JR. & NORTON, I.O. 1995. Tectonic controls on the stratigraphic architecture and hydrocarbon systems of the Arabian Plate. *In*: AL-HUSSEINI, M.I. (ed.) *GEO '94, The Middle East Petroleum Geosciences*, Gulf PetroLink, Bahrain, 413–430.

GUO, L., SIMMONS, M.D. & BLANC, J.-P. 2004. Palaeokarst and porosity development in Cretaceous carbonate reservoirs, central Oman. *GeoArabia*, **9**, 75.

HAMDAN, A.R.A. & ALSHARHAN, A.S. 1991. Palaeoenvironments and palaeoecology of the rudists in the Shuaiba Formation (Aptian) United Arab Emirates. *Journal of African Earth Sciences*, **12**, 569–581.

HARRIS, P.M., FROST, S.H., SEIGLIE, G.A. & SCHNEIDERMANN, N. 1984. Regional unconformities and depositional cycles, Cretaceous of the Arabian Peninsula. *In*: SCHLEE, J.S. (ed.) *Interregional Unconformities and Hydrocarbon Accumulation*. American Association of Petroleum Geologists, Memoirs, **36**, 67–80.

HARTSHORNE, M. 1989. Facies architecture of a Lower Cretaceous coral-rudist patch reef, Arizona. *Cretaceous Research*, **10**, 311–336.

HASSAN, T.H., MUDD, G.C. & TWOMBLEY, B.N. 1975.

The stratigraphy and sedimentation of the Thamama Group (Lower Cretaceous) of Abu Dhabi. *Proceedings of the 9th Middle East Arab Petroleum Conference*, 1–11.

HUGHES, G.W. 1998. Middle East Aptian rudist-foraminiferal-algal associations and their possible Arabian Gulf analogue. *In*: MASSE, J.-P. & SKELTON, P.W. (eds) *Quatrième Congrés International sur les Rudistes*. Geobios, Memoir Special, **22**, 147–158.

HUGHES, G.W. 1999. Bioecostratigraphy of the Aptian Shu'aiba Formation, Saudi Arabia. *In*: HOFLING, R. & STEUBER, T. (eds) *Fifth International Congress on Rudists, Abstracts and Field Guides*. Erlanger Geologische Abhandlungen, Sonderband, **3**, 25–27.

HUGHES, G.W. 2000*a*. Bioecostratigraphy of the Shu'aiba Formation of Saudi Arabia. *GeoArabia*, **5**, 545–578.

HUGHES, G.W. 2000*b*. Saudi Arabian Late Jurassic and Early Cretaceous agglutinated foraminiferal associations and their application for age, palaeoenvironmental determination, sequence stratigraphy and carbonate reservoir architecture. *In*: HART, M.B., KAMINSKI, M.A. & SMART, C.W. (eds) *Proceedings of the Fifth International Workshop on Agglutinated Foraminifera, Plymouth, September, 1997*. Grzybowski Foundation Special Publications, **7**, 149–165.

HUGHES, G.W. 2003*a*. *Agriopleura* morphotypes of the Lower Aptian Shu'aiba Formation of Saudi Arabia. *Geologia Croatica*, **56**, 133–138.

HUGHES, G.W. 2003*b*. Review: 6th International Rudist Congress in Rovinj, Croatia. *GeoArabia*, **8**, 151–156.

HUGHES, G.W. 2004*a*. Palaeoenvironments of Early Aptian agglutinated foraminifera of Saudi Arabia. *In*: BUBIK, M. & KAMINSKI, M.A. (eds) 2004. *Proceedings of the Sixth International Workshop on Agglutinated Foraminifera*. Grzybowski Foundation Special Publications, **8**, 195–207.

HUGHES, G.W. 2004*b*. Bioecostratigraphy of the Aptian Shu'aiba Formation, Saudi Arabia. *In*: HOFLING, R. (ed.) Fifth International Congress on Rudists. *Courier Forschungsinstitut Senckenberg*, **247**, 233–245.

HUGHES, G.W., AKTAS, G., G., VAROL, O. & SKELTON, P.W. 1999. New evidence and implications for age determination of the rudist-bearing Shu'aiba Formation, Saudi Arabia. *In*: HOFLING, R. & STEUBER, T. (eds) *Fifth International Congress on Rudists, Abstracts and Field Guides*. Erlanger Geologische Abhandlungen, Sonderband, **3**, 27–28.

HUGHES, G.W., SIDDIQUI, S. & SADLER, R.K. 2003. Shu'aiba rudist taphonomy using computerized tomography and image logs, Shaybah Field, Saudi Arabia. *GeoArabia*, **8**, 585–596.

HUGHES, G.W., SIDDIQUI, S. & SADLER, R.K. 2004. Computerised tomography reveals Aptian rudist species and taphonomy. *Geologia Croatica*, **57**, 67–71.

HUGHES CLARKE, M.W. 1988. Stratigraphy and rock unit nomenclature in the oil-producing area of interior Oman. *Journal of Petroleum Geology*, **11**, 5–60.

HUSINEC, A. 2001. *Palorbitolina lenticularis* from the northern Adriatic region: paleogeographical and evolutionary implications. *Journal of Foraminiferal Research*, **31**, 287–293.

HUSINEC, A., VELIC, I., FUCEK, L., VLAHOVIC, I., MATICEC, D., OSTRIC, N. & KORBAR, T. 2000. Mid Cretaceous orbitolinid (Foraminiferida) record from the islands of Cres and Losinj (Croatia) and its regional stratigraphic correlation. *Cretaceous Research*, **21**, 155–171.

IMMENHAUSER, A., CREUSEN, ESTEBAN, A.M. & VONHOF, H.B. 2000. Recognition and interpretation of polygenic discontinuity surfaces in the Middle Cretaceous Shu'aiba, Nahr Umr and Natih Formations of Northern Oman. *GeoArabia*, **5**, 299–322.

IMMENHAUSER, A., VAN DER KOOIJ, B., VAN VLIET, A., SCHLAGER, W. & SCOTT, R.W. 2001. An ocean-facing Aptian-Albian carbonate margin, Oman. *Sedimentology*, **48**, 1187–1207.

IMMENHAUSER, A., HILLGARTNER, H. *et al.* 2004. Barremian–lower Aptian Qishn Formation, Haushi-Huqf area, Oman – a new outcrop analogue for the Kharaib/Shu'aiba reservoir. *GeoArabia*, **9**, 153–194.

JENDEN, P.D., AL-AMOUDI, M.A., AL-EID, M.I., SADLER, R.K. & SALAMY, S.P. 2003. Geochemical Fingerprinting of Crude Oils, Shaybah Field, Saudi Arabia: Evidence of Reservoir Charging History (abstract). *American Association of Petroleum Geologists, Annual Meeting, Salt Lake City, May, 2003*, 268.

JONES, R.W. & CHARNOCK, M.A. 1985. 'Morphogroups' of agglutinating foraminifera: their life positions and feeding habits and potential applicability in paleoecological studies. *Revue de Paléobiologie*, **4**, 311–320.

JONES, R.W., SIMMONS, M. & WHITTAKER, J. 2004. Chronostratigraphic and palaeoenvironmental significance of agglutinated and associated larger benthonic foraminifera from the Lower to 'Middle' Cretaceous of the Middle East. *In*: BUBIK, M. & KAMINSKI, M.A. (eds) *Proceedings of the Sixth International Workshop on Agglutinated Foraminifera. Grzybowski Foundation Special Publication*, **8**, 229–235.

KENNEDY, W.J. & SIMMONS, M.D. 1991. Mid-Cretaceous ammonites and associated microfossils from the Central Oman Mountains. *Newsletters on Stratigraphy*, **25**, 127–134.

KENTER, J.A.M. 1990. Carbonate platform flanks: slope angle and sediment fabric. *Sedimentology*, **37**, 777–794.

KERANS, C., FITCHEN, W., ZAHM, L., WARD, B., OSLEGER, D., BARNABY, R. & JENNINGS, J. 1997. Carbonate reservoir heterogeneity styles within a sequence stratigraphic framework: Albian (Cretaceous), Pecos River Canyon: The University of Texas at Austin, Bureau of Economic Geology, RCRL Laboratory Field Trip Guidebook, 130 pp.

KOCH, R., MOUSSAVIAN, E., OGORLEC, D., SKABERNE, D. & BUCUR, I.I. 2002. Development of a *Lithocodium* (syn. *Bacinella irregularis*) – reef – mound – A patch reef within Middle Aptian lagoonal limestone sequence near Nova Gonca (Sabotin Mountain, W. Slovenia). *Geologija*, **45**, 71–90.

LECKIE, R.M. 1986. Paleoecology of mid-Cretaceous planktonic foraminifera: a comparison of open ocean and Epicontinental Sea assemblages. *Micropaleontology*, **33**, 164–176.

LITSEY, L.R., MACBRIDE, W.L., AL-HINAI, K.M. & DISMUKES, N.B. 1983. Shuaiba Reservoir geological study, Yibal Field, Oman. *Proceedings of the Society of Petroleum Engineers Middle East Oil Technology Conference, Bahrain, March, 1983*, SPE 11454, 131–138.

MASSE, J.-P. 1985. Paleobiogeographie des Rudistes du domaine peri mediterraneen a l'Aptian inferieur. *Bulletin Société Géologique de France*, **8**, 715–721.

MASSE, J.-P. 1992. Les rudistes de l'Aptien Inferieur d'Italie Continentale: aspects systematiques, stratigraphiques et paleobiogeographiques. *Geologica Roma*, **28**, 243–260.

MASSE, J.-P. 1997. Stratigraphy and tectonosedimentary evolution of a late Aptian-Albian carbonate margin: the northeastern Jebel Akhdar (Sultanate of Oman). *Sedimentary Geology*, **113**, 269–289.

MASSE, J.-P. 1999. Biological crisis and environmental changes in rudist bivalves: the early Cretaceous record. *In*: HOFLING, R. & STEUBER, T. (eds) *Fifth International Congress on Rudists, Abstracts and Field Guides*. Erlanger Geologische Abhandlungen, Erlangen, Sonderband **3**, 33.

MASSE, J.-P. & GALLO MARESCA, M. 1997. Late Aptian Radiolitidae (rudist bivalves) from the Mediterranean and Southeast Asiatic regions: taxonomic, biostratigraphic and palaeobiogeographic aspects. *Palaeogeography, Palaeoclimatology, Palaeoecology*, **128**, 101–110.

MOSHRIF, M.A. & KELLING, G. 1984. Stratigraphy and sedimentary history of Upper–Lower and Middle Cretaceous rocks, Central Saudi Arabia. *Mineral Resources Bulletin* **28**, Deputy Ministry for Mineral Resources, Jiddah, Saudi Arabia.

MOULLADE, M., PEYBERNES, B., REY, J. & SAINT-MARC, P. 1985. Biostratigraphic interest and paleobiogeographic distribution of early and mid-Cretaceous Mesogean orbitolinids (Foraminiferida). *Journal of Foraminiferal Research*, **15**, 149–158.

MOULLADE, M., KUHNT, W., BERGEN, J.A., MASSE, J.-P. & TRONCHETTI, G. 1998. Correlation of biostratigraphic and stable isotope events in the Aptian historical stratotype of La Bedoule (southeast France). *Compte Rendu de l'Académie des Sciences, Paris*, **327**, 693–698.

MURPHY, A.M. & SALVADOR, A. 1999. International Stratigraphic Guide – an abridged version. *Episodes*, **22**(4), 255–271.

MURPHY, A.M. & SALVADOR, A. 2000. International Subcommission on Stratigraphic Clasification of IUGS International Commission on Stratigraphy. International Stratigraphic Guide – an abridged version. *GeoArabia*, **5**, 231–266.

PAYNE, S.N.J., EWEN, D.F. & BOWMAN, M.J. 1999. The role and value of "high impact biostratigraphy" in reservoir appraisal and development. *In*: JONES, R.W. & SIMMONS, M.D. (eds) *Biostratigraphy in Production and Development Geology*. Geological Society London, Special Publications, **152**, 5–22.

PITTET, B., VAN BUCHEM, F.S.P., HILLGARTNER, H., RAZIN, P., GROTSCH, J. & DROSTE, H. 2002. Ecological succession, palaeoenvironmental change and depositional sequences of Barremian–Aptian shallow-water carbonates in northern Oman. *Sedimentology*, **49**, 555–581.

POWERS, R.W. 1968. *Saudi Arabia*. Lexique Stratigraphique International, 3, Centre National de la Recherche Scientifique, Paris.

POWERS, R.W., RAMIREZ, L.F., REDMOND, C.D. & ELBERG, E.L. 1966. Geology of the Arabian Peninsula, *U.S. Geological Survey Professional Paper* 560-D, Washington.

ROSEN, B.R., AILLUD, G.S., BOSELLINI, F., CLACK, N.J., INSALACO, E., VALLDEPERAS, F.X. & WILSON, M.E. 2003. Platy coral assemblages: 200 million years of functional stability in response to the limiting effects of light and turbidity. *In*: KASIM MOOSA, M.K., SOEMADIARDJO, S., NONTJI, A., SOEGIARTO, A. & ROMIMOHTARTO, K. (eds) *Proceedings of the 9th International Coral Reef Symposium, Bali, Indonesia*, Indonesian Institute of Sciences and State Ministry for Environment, Republic of Indonesia, Jakarta, Indonesia, 255–265.

RUSSELL, S.D. & AKBAR, M. 2002. Rock types and permeability prediction from dipmeter and image logs: Shu'aiba reservoir (Aptian), Abu Dhabi. *AAPG Bulletin*, **86**, 1709–1732.

SALAMY, S.P., AL-MUBARAK, H.K., AL-MALKI, S.S. & AL-AJMI, A.M. 2003. Impact of a comprehensive horizontal well production logging and testing program in Shaybah Field, Saudi Arabia. *Saudi Aramco Journal of Technology*, Winter, 2003, 2–7.

SARTORIO, D. & VENTURINI, S. 1988. Southern Tethys Biofacies. Agip S.p.A., S. Donato Milanese, Italy, 235 pp.

SAUDI ARAMCO 1999. *Shaybah Field, the challenge, the achievement*. 13, The Saudi Aramco Oil Company Public Relations Dept., Dhahran, Saudi Arabia.

SCHLAGER, W. & CAMBER, O. 1986. Submarine slope angles, drowning unconformities and shelf erosion on limestone escarpments. *Geology*, **14**, 762–765.

SCHMID, D.U. & LEINFELDER, R.R. 1996. The Jurassic *Lithocodium aggregatum–Troglotella incrustans* foraminiferal consortium. *Palaeontology*, **39**, 21–52.

SCHROEDER, R. 1975. General evolutionary trends in orbitolinas. *Revista Española de Micropaleontología*, Numero Especial, 117–128.

SCOTT, R.W. 1990. Chronostratigraphy of the Cretaceous carbonate shelf, south-eastern Arabia. *In*: ROBERTSON, A.H.F., SEARLE, M.P. & RIES, A.C. (eds) *The Geology and Tectonics of the Oman*

Region. Geological Society, London, Special Publications, **49**, 89–108.

SHARLAND, P.R., ARCHER, R., CASEY, D.M., DAVIES, R.B., HALL, S.B., HEWARD, A.P., HORBURY, A.D. & SIMMONS, M.D. 2001. Arabian Plate Sequence Stratigraphy. *GeoArabia Special Publication*, **2**.

SIDDIQUI, S., HUGHES, G.W. & SADLER, R.K. 2004. Density-based imaging to supplement FMI images for sedimentary facies modeling. *Saudi Aramco Journal of Technology*, Spring, 60–68.

SIMMONS, M.D. 1994. Micropalaeontological zonation of the Kahmah Group (Early Cretaceous), Central Oman Mountains. *In*: SIMMONS, M.D. (ed.) *Micropalaeontology and Hydrocarbon Exploration of the Middle East*. Chapman & Hall, London, 177–219.

SIMMONS, M.D. & HART, M.B. 1987. The biostratigraphy and microfacies of the Early to mid-Cretaceous carbonates of Wadi Mi'aidin, Central Oman Mountains. *In*: HART, M.B. (ed.) *Micropalaeontology of Carbonate Environments*. Ellis Horwood, Chichester, 176–207.

SIMMONS, M.D., EMERY, D. & PICKARD, A.H. 1991. *Hensonella dinarica*, an originally calcitic Early Cretaceous dasycladacean alga. *Palaeontology*, **34**, 955–961.

SIMMONS, M.D., WHITTAKER, J.E. & JONES, R.W. 2000. Orbitolinids from Cretaceous sediments of the Middle East – a revision of the F.R.S. Henson and Associates Collection. *In*: HART, M.B., KAMINSKI, M.A. & SMART, C.W. (eds) *Proceedings of the Fifth International Workshop on Agglutinated Foraminifera,* Grzybowski Foundation Special Publication, **7**, 411–437.

SKELTON, P.W. 2000. Rudists and carbonate platforms – growing together, dying together. *Accademia di Scienze Lettere e Arti di Modena, Collana di Studi*, **21**, 231–235.

SKELTON, P.W. 2001. The geometry of rudist lithosomes. *Géologie Mediterranéenne*, **28**, 169–171.

SKELTON, P.W. 2004. *Oedomyophorus shaybahensis*, a new genus and species of caprinid rudist from the Lower Aptian Shu'aiba Formation of Eastern Saudi Arabia. *In*: HOFLING, R. & STEUBER, T. (eds) *Fifth International Congress on Rudists*, E. Schweizerbartsche Verlagsbuchandlung (Nagle u. Obermiller), Stuttgart, **247**, 35–47.

SKELTON, P.W. & GILI, E. 1991. Palaeoecological classification of rudist morphotypes. *Proceedings of the 1st Conference on Rudists, Beograd, 1988*, Serbian Geological Society, Special Publication, **2**, 265–287, Beograd.

SKELTON, P.W. & MASSE, J.-P. 2000. Synoptic guide to the Lower Cretaceous rudist bivalves of Arabia. *In*: ALSHARHAN, A. & SCOTT, R.W. (eds) *Middle East Models of Jurassic and Cretaceous Carbonate Systems*. Society of Economic Mineralogists and Paleontologists, Special Publications, **69**, 89–99.

SKELTON, P.W., GILI, E., VICENS, E. & OBRADOR, A. 1995. The growth fabric of gregarious rudist elevators (hippuritids) in a Santonian carbonate platform in the southern Central Pyrenees. *Palaeogeography, Palaeoclimatology, Palaeoecology,* **119**, 107–126.

SKELTON, P.W., SPICER, W. & REES, A. 1997. *Evolving Life and Earth*. Open University S269 Earth and Life, Open University, Milton Keynes.

SKELTON, P.W., HUGHES, G.W. & AKTAS, G. 1999. Rudists of the Shaybah Field, eastern Saudi Arabia. *In*: HOFLING, R. & STEUBER, T. (eds) *Fifth International Congress on Rudists, Abstracts and Field Guides*. Erlanger Geologische Abhandlungen, Sonderband **3**, 65–66.

SMALLEY, P.C., HIGGINS, A.C., HOWARTH, R.J., NICHOLSON, H. & JONES, C.E. 1994. Seawater Sr isotope variations through time: a procedure for constructing a reference curve to date and correlate marine sedimentary rocks. *Geology*, **22**, 431–434.

STOSSEL, I. & BERNOULI, D. 2000. Rudist lithosome development on the Maella carbonate platform margin. *In*: INSALACO, E., SKELTON, W.P. & PALMER, T.J. (eds) *Carbonate Platform Systems: Components and Interactions*. Geological Society, London, Special Publications, **178**, 177–190.

VAHRENKAMP, V.C. 1996. Carbon isotope stratigraphy of the Upper Kharaib and Shuaiba Formations: implications for the Early Cretaceous evolution of the Arabian Gulf region. *AAPG Bulletin*, **80**, 647–662.

VAHRENKAMP, V.C. & GROTSCH, J. 1995. 3-D architecture model of a Lower Cretaceous carbonate reservoir, Al Huwaisah Field, Sultanate of Oman. *In*: AL-HUSSEINI, M.I. (ed.) *GEO '94, The Middle East Petroleum Geosciences*, Gulf PetroLink, Bahrain, 901–915.

VAN BUCHEM, F.S.P., PITTET, B. *et al.* 2002*a*. High-resolution sequence stratigraphic architecture of Barremian/Aptian carbonate systems in northern Oman and the United Arab Emirates (Kharaib and Shu'aiba Formations). *GeoArabia*, **7**, 461–500.

VAN BUCHEM, F.S.P., RAZIN, P., HOMEWOOD, P.W., OTERDOOM, W.H. & PHILIP, J. 2002*b*. Stratigraphic organization of carbonate ramps and organic-rich intrashelf basins: Natih Formation (middle Cretaceous) of northern Oman. *AAPG Bulletin*, **86**, 21–53.

VEIZER, J., BUHL, D. *et al.* 1997. Strontium isotope stratigraphy: potential resolution and event correlation. *Palaeogeography, Palaeoclimatology, Palaeoecology*, **132**, 65–77.

VILAS, L., MASSE, J.-P. & ARIAS, C. 1995. *Orbitolina* episodes in carbonate platform evolution: the early Aptian model from SE Spain. *Palaeogeography, Palaeoclimatology, Palaeoecology*, **119**, 35–45.

WITT, W. & GOKDAG, H. 1994. Orbitolinid biostratigraphy of the Shuaiba Formation (Aptian), Oman – implications for reservoir development: *In*: SIMMONS, M.D. (ed.) *Micropalaeontology and Hydrocarbon Exploration in the Middle East*. British Micropalaeontological Society Publications Series. Chapman & Hall, London, 221–241.

ZEIGLER, M.A. 2001. Late Permian to Holocene Paleofacies Evolution of the Arabian Plate and its Hydrocarbon Occurrences. *GeoArabia*, **6**, 445–504.

Saudi Arabian Permo-Triassic biostratigraphy, micropalaeontology and palaeoenvironment

G. W. HUGHES

Geological Technical Services Division, Saudi Aramco, Box 10646, Dhahran 31311, Saudi Arabia (e-mail: geraint.hughes@aramco.com)

Abstract: The Khuff Formation of Saudi Arabia consists of a succession of shallow marine carbonates, mudstones and evaporites. It is exposed along the basement-flanking outcrop belt of central Saudi Arabia, and also forms important hydrocarbon reservoirs at depth in the east. Micropalaeontological evidence suggests that the Khuff Formation in Saudi Arabia is of latest Permian to early Triassic age. Of the four members, the Permo-Triassic boundary is located at, or slightly above, the boundary between the uppermost Midhnab and Khartam Members in the outcrop belt and within the lower part of the Khuff B carbonate reservoir in the subsurface. Accurate correlation between the exposed succession and the subsurface lithologies remains a tantalizing problem.

Micropalaeontological biocomponents are sufficiently distinct within the late Permian carbonate succession to enable determination of a series of palaeoenvironmentally significant biofacies for which no previous interpretation has been suggested. Tiered biofacies reveal significant deepening and shoaling trends that assist in determining maximum flooding events and depositional cycle boundaries. The subsurface reservoir carbonates display subtle palaeoenvironmental variations of an extensive, generally shallow marine, carbonate platform that ranged from intertidal to depths within intra-shelf basins that probably rarely exceeded storm wave base. The problems of synonomy of the foraminifera have been resolved by a pragmatic approach justified by the analytical time constraints demanded by industrial micropalaeontology.

During the late Permian and early Triassic, the Arabian Peninsula was located on the south-western margin of the Palaeotethys Ocean, and was the site of an extensive shallow marine platform (Murris 1980; Le Nindre *et al.* 1990; Stampfli & Borel 2002). The western margin of this platform is not clearly defined owing to removal of the coastal facies by erosion, but it is considered to have been located some distance off the present-day eastern margin of the Arabian Shield. Scattered across the platform were localized deeper areas considered to be intra-shelf basins, and localized highs that formed the sites of oolitic and bryozoan shoals.

Stratigraphical subdivision within Saudi Arabian carbonate reservoirs, using conventional micropalaeontological methods, is mostly beyond chronostratigraphical resolution. Current demands for biostratigraphical control focus on methods of improving intra-reservoir architecture as well as variations in the palaeoenvironment to improve the distribution and prediction of the best reservoir facies. Carbonate reservoirs of the Khuff Formation in Saudi Arabia, of late Permian to early Triassic age, provide such an example where this demand has been assisted by biofacies characterization and tiering recognition.

Sequence stratigraphical studies rely on the detection of significant stratigraphical surfaces, locating maximum flooding events and developing a hierarchy of vertical shoaling and deepening-upwards trends. Semi-quantitative micropalaeontological analysis of closely spaced thin sections from the Khuff reservoirs of Saudi Arabia, spanning the late Permian to early Triassic, has achieved significant progress towards this goal. Recording of all biocomponents is essential, and includes the entire spectrum from calcareous algae, ostracods, foraminifera, brachiopods and bryozoa.

Saudi Arabian carbonate hydrocarbon reservoirs were examined for their microfauna during early investigations but these studies were primarily for stratigraphy (Steineke *et al.* 1958; Powers *et al.* 1966; Powers 1968). Sequence stratigraphical interpretations reveal that platform carbonates, previously considered to have accumulated in an aggradational fashion, can display considerable complexity resulting from sea-level induced progradational and retrogradational trends. Such complexities now require a search for additional information to identify and constrain the palaeoenvironmental variations and determine the high-resolution depositional cyclicity of the reservoir carbonates. The environmental sensitivity of

From: POWELL, A. J. & RIDING, J. B. (eds) 2005. *Recent Developments in Applied Biostratigraphy*. The Micropalaeontological Society, Special Publications, 91–108. 1747-602X/$15.00

benthic foraminifera, combined with associated microfossils and macrofossils, provides a potentially valuable technique for determining subtle variations in the depositional environment and also provides a proxy for sea-level fluctuations. These variations do mean, however, that local stratigraphical ranges of shallow marine benthonic foraminifera should be expected to display diachronous variations that, unfortunately, decrease their value as intra-reservoir chronostratigraphical markers. Biofacies evidence for deepening acmes, however, provide valuable chronostratigraphical control and assist recognition of depositional facies tracts.

The palaeoenvironmental significance of the Middle East late Palaeozoic benthic foraminifera and associated biocomponents has not been documented except by Hughes (2002, 2003), and is based on an interpretation of the semi-quantitative analysis of thousands of thin sections. The thin sections were prepared from core plugs of subsurface samples, collected 6 inches (~ 15 cm) apart from exploratory and development wells in the Ghawar Field of eastern Saudi Arabia. In addition, samples were analysed from each bed in measured sections from the outcrop belt of the Tuwaiq Mountain escarpment of central Saudi Arabia.

Six palaeoenvironmental regimes have been defined, including the open marine/intra-shelf basin, transitional basin–shoal, shoal complex, deep lagoon, shallow lagoon and shore–intertidal. These categories can be used for all formations here discussed to define the distribution of the palaeoenvironmentally significant biofacies.

For each reservoir, predominantly foraminiferal biofacies have been determined that display palaeoenvironmental variations that have been valuable in terms of improving reservoir characterization. Such biofacies-based palaeoenvironmental information is typically applied in multiwell, field-wide studies to assist three-dimensional mapping of palaeoenvironmental variations. These maps provide templates for distributing lithofacies and reservoir attributes to assist in planning vertical and horizontal development wells.

Lithostratigraphy

Within the region, a variety of formation names have been assigned by different countries that hinder regional correlations and palaeoenvironmental interpretations. Regional equivalents are discussed and clarified by Beydoun (1988, 1991), Alsharhan & Nairn (1994, 1997) and Sharland *et al.* (2001). The Khuff Formation name is retained across the Middle East, although the members are given different names. The Khuff Formation in Saudi Arabia is best described by Powers *et al.* (1968). Aspects of the depositional environment, diagenesis and reservoir characteristics have been investigated by Al-Jallal (1995). A major field-based study of the Khuff by the Bureau de Recherches Géologique et Minères (BRGM) examined the lithostratigraphy, sedimentology and biostratigraphy (Le Nindre *et al.* 1990). Alsharhan & Nairn (1997) provide a review of the Permian succession in Saudi Arabia. The Khuff Formation has been assigned to the basal part of the Arabian Plate megasequence AP6 of Sharland *et al.* (2001), with maximum flooding surfaces designated P20 to P40. A maximum flooding-based stratigraphy, in the style of Galloway (1989) has been proposed by Sharland *et al.* (2000), and implies the presence of a number of sequences within this succession. A major unconformity is present at the Permo-Triassic boundary within the Khuff.

The Khuff Formation is considered as one major, but complex, marine depositional cycle that transgressed over central Arabia unconformably over the Unayzah land surface during the late Permian. Shales of the Sudair Formation, of early Triassic age, unconformably overlie the Khuff carbonates in outcrop. Four members, including siliciclastic, carbonate and evaporite lithologies, have been determined for the outcrop (Le Nindre *et al.* 1990) and include, in ascending stratigraphical order, the Huqayl, Midhnab, Duhaysan and Khartam (Fig. 1). In the subsurface at Ghawar a different four-fold reservoir scheme has been established that includes, also in ascending stratigraphical order, the Khuff D, C, B and A units. This latter scheme is based on four major repeated carbonate–evaporite cycles of which the evaporite forms the seal for the underlying carbonate reservoir. There is no direct relationship between the four units in the outcrop with the four carbonates developed in the subsurface. The Khuff Formation, of late Permian to early Triassic age, includes Khuff reservoirs within the informal B and C members. Equivalence with the formal Khuff lithostratigraphical subdivisions based on outcrop studies (Le Nindre *et al.* 1990) is now closer to being understood.

Various attempts have been made to correlate the Ghawar subsurface succession with the outcrop belt where carbonates are exposed as a result of road-building activities. Although poorly exposed pink anhydrites of the Huqayl Member are exposed, evaporites equivalent to all of the subsurface units are not exposed. The

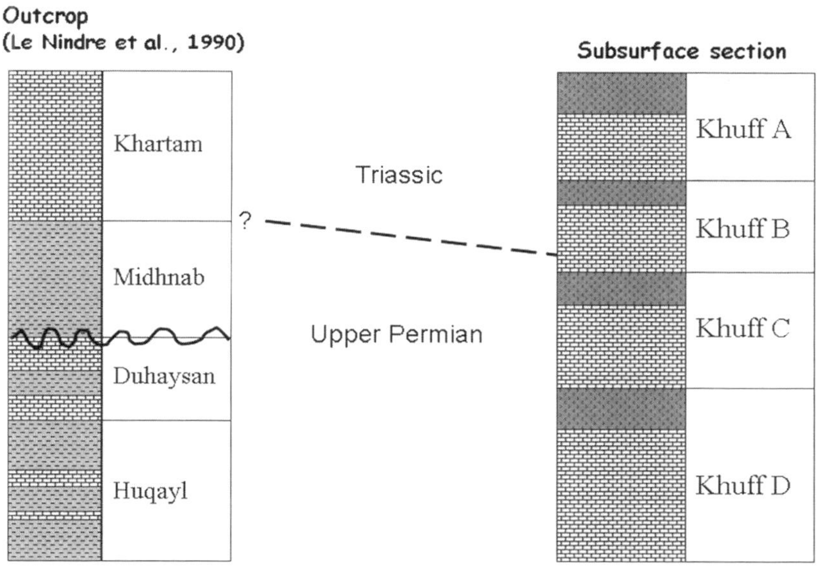

Fig. 1. Members of the Khuff Formation and their possible equivalence in the Khuff subsurface.

well-exposed members of the Khuff Formation, once related to their subsurface reservoir equivalents, offer potentially significant insights into the details of depositional layering of the subsurface reservoirs. A significant advance in understanding the surface to subsurface lithostratigraphical relationships was made during a joint Saudi Aramco-BRGM field trip in November 2002, and the results will be documented elsewhere (R. Kamal and A. Al-Tawil, pers. comm.). In outcrop and subsurface, the basal carbonates are underlain by a genetically associated succession that consists mainly of siliciclastics with locally developed thin carbonate lenses. This basal portion has been called the basal Khuff clastics, or Khuff E (Al-Jallal 1995) or Ash Shuqqah Member (Senalp & Duwaiji 1994) and is considered to represent the transitional, possibly regionally diachronous, terrestrial–shallow marine succession of the early Khuff transgression over the Arabian Platform.

Age

The age of the carbonates of the Khuff Formation across the Arabian Plate spans the middle Permian (Wordian) to early Triassic (Scythian) (Fig. 1). In Saudi Arabia, the Khuff was originally dated (Steineke *et al.* 1958) as probably late Permian, based on poorly preserved brachiopod and nautiloid evidence. The Permo-Triassic boundary is known to be an event when there was mass extinction of many Permian faunas and floras (Steiner *et al.* 2003). The presence of the serpulid worm *Spirorbis phylactaena* (Plate 1) near the uppermost Khuff carbonates in SHD-1 (Le Nindre *et al.* 1990; Vaslet *et al.* 2004) confirms a Scythian (early Triassic) age. Ammonites of Scythian age have also been recovered from the base of the Khartam Member in the Saudi Arabian outcrops (D. Vaslet, oral communication 2002).

In Oman, the oldest carbonates equivalent to the Khuff Formation have been dated as Wordian (Kazanian), based on a combination of brachiopod, foraminiferal and conodont evidence (Angiolini *et al.* 1998, fig. 9). A middle Permian (Midian) age has recently been assigned to reworked limestone blocks from the Aseelah Unit of Oman (Vachard *et al.* 2002). This age difference between Oman, Iran and Saudi Arabia suggests that the basal Khuff is diachronous, with the marine transgression commencing earlier in Oman than in Saudi Arabia (Vachard *et al.* 2002).

In the comprehensive discussion regarding the age of the Khuff Formation by Sharland *et al.* (2001), a late Permian age is assigned to the Huqayl, Duhaysan and Midhnab members, and the Permo-Triassic boundary is placed at the base of the Khartam Member. The most recently published age (Vaslet *et al.* 2004) provides a Wuchiapingian (Dzulfian) age for the Huqayl Member, a Changhsinian age for the

Midhnab Member and an undifferentiated late Permian age for the lower part of the Khartam Member, based on ostracods (Table 1). The age of the upper Khartam Member is confirmed as early Triassic (Scythian) on the abundance of *Spirorbis phlyctaena*. This would place the Permo-Triassic boundary within the Khartam Member, even though the authors state the presence of late Permian foraminifera within the lower Khartam. All samples collected by the author from the Khartam Member carbonates exposure in a roadcut on the Buraydah–Riyadh highway have proved to be barren of foraminifera.

Micropalaeontological and palynological evidence (Le Nindre *et al.* 1990) provide a late Permian age for most of the Khuff, with an earliest Triassic age suggested for the upper part. Micropalaeontological analysis of field samples and published data from water well SHD-1 (Le Nindre *et al.* 1990) has provided a good assemblage of calcareous algae and foraminifera that have been used to date the four members of the Khuff Formation, as recognized in the outcrop belt. The base of the lowest Huqayl Member is placed at the boundary between the Upper Murghabian and the Djulfian, within the late Permian. The entire Huqayl, Duhaysan, Midhnab and basal Khartam members are considered as being of Djulfian (Wuchiapingian) to early Dorashamien (Changhsingian) age. This age is based on the recognition of the *Colaniella parva* biozone of Lys *et al.* (1980) and Lys (1983, 1988), defined by the presence of *Robuloides* cf. *gibbus* at the base of the Huqayl Member, and of *Paradagmarita flabelliformis* at the base of the Midhnab Member. The Permo-Triassic boundary is placed within the lower part of the Khartam Member, based on the presence of Spirorbis *phylactaena*, and the absence of Permian foraminifera above this event. Recent palynological evidence also supports a late Permian age for the lowermost, pre-carbonate clastics (Stephenson *et al.* 2003). Subsurface micropalaeontology (Hughes 2002, 2003; Hughes *et al.* 2004) has revealed the presence of the Permo-Triassic boundary within carbonates of the B member. Recent age assignment to selected benthonic foraminifera by Pronina-Nestell & Nestell (2001) and from written communication with Pronina-Nestell, the Khuff C carbonates are probably of latest Permian, Changsinian age. The Triassic section is barren of foraminifera, but includes the spirorbid tubes of *Spirorbis phlyctaena*.

Benthonic foraminiferal species from the Khuff C reservoir provide a late Permian age, equivalent to the Wuchiapingian and Changsingian stages. The absence of Permian microfauna above the lower part of the Khuff B carbonates has been used as supportive evidence to that provided by sedimentology and carbon isotope studies, to determine the position of the Permo-Triassic boundary.

In addition to contributing to age determination, this study was aimed at investigating, through integration with sedimentology, the palaeoenvironmental control on deposition. This has been particularly useful in discriminating between carbonate mud deposition in a shallow hypersaline setting from those deposited in a deeper marine normal salinity environment. The biostratigraphical event distribution and their characteristics have been used to determine regional palaeoenvironmental variations within depositional sequences, and also provide an independent data source for recognition and confirmation of depositional cycle boundaries.

In this study, biostratigraphy has been confined to semi-quantitative analysis of thin sections and selected plug-ends. A total of 3200 thin sections were analysed. In general, the Khuff reservoirs B and C carbonates contain sparse biocomponents, as there is a predominance of biologically barren grainstones, mudstones and dolomite. At certain levels, however, microfossils are common, and locally abundant, and consist mostly of benthonic foraminifera and calcareous red algae with rare calcareous green algae. Foraminifera are mostly non-fusulinid, but certain parts of the succession contain concentrations of the simple fusulinids *Staffella* spp. and *Nankinella* spp. Brachiopods, ostracods and bryozoa are also present.

The application of non-fusulinid foraminifera to stratigraphically define the age of the Permian remains in its early stages, and there are only a few publications that consider the stratigraphical ranges of such smaller benthonic foraminifera. Within the present study, the identified foraminiferal species are consistent with a latest Permian, Changsingian, age. Understanding of the palaeoenvironmental preferences of the non-fusulinid Permian foraminifera, and Late Palaeozoic species in general, also remains a topic of negligible research, with the best reference (Gallagher 1998) discussing only Lower Carboniferous species.

With reference to the stratigraphical ranges of Lys (1988), the presence of the foraminiferal species *Colaniella parva*, *Dagmarita chanakchiensis*, *Agathammina pusilla, Ichthyolaria latilimbata, Hemigordiopsis* spp., *Hemigordius reicheli*, *Paraglobivalvulina mira,*

Table 1. *Summary of Middle to Upper Permian chronostratigraphic schemes (Jin* et al. *1997, 2000)*

	EPOCH	Proposed Global Stages	W. Tethys Stages
Upper Permian	(Lopingian)	Changhsingian	Dorashamian
Upper Permian	(Lopingian)	Wuchiapingian	Dzhulfian
'Middle' Permian	Guadalupian	Capitanian	Midian
'Middle' Permian	Guadalupian	Wordian	
'Middle' Permian	Guadalupian	Roadian	

Globivalvulina vonderschmitti, Nankinella orbicularis and *Reichelina* spp. indicate a late Permian (Wuchiapingian) age for the Khuff C carbonates of Saudi Arabia. This assemblage compares well with that recorded from the late Permian Dalan Formation in Iran (Kalantari 1982).

The absence of these forms within the upper part of the Khuff B carbonates provides evidence, of a supportive nature, that the reservoir spans the Permo-Triassic boundary. More specifically, the faunal extinction event occurs at a readily recognized unconformity in the Khuff B section. The precise position of the boundary is at the top of a distinctive and widely correlative brecciated palaeosol.

The presence of a coniferous-dominated plant assemblage from the Midhnab Member (Hill & El-Khayal 1983) is used to conclude a late Permian age. Other palaeobotanical references relevant to the Saudi Arabian and Oman region include El-Khayel *et al.* (1980), Lemoigne (1981*a*, *b*), El-Khayel & Wagner (1985) and Broutin *et al.* (1995). The palynofloras from well SHD-1 (Le Nindre *et al.* 1990) are consistent with a late Permian (Djulfian to lower Dorashamien; Wuchiapingian to Changhsingian) age. Recent palynological studies by Stephenson *et al.* (2003) are consistent with this earlier determination, for which Palynological Zone OSPZ6 has been assigned.

Recent micropalaeontological work on the Khuff outcrops (Vaslet *et al.*, written communication; 7 July 2002) confirmed that the Permo-Triassic boundary coincides with the Midhnab-Khartam boundary, although Vaslet *et al.* (2004) placed the boundary within the lowermost Khartam Member. The maximum flooding surface P45 (of Sharland *et al.* 2001) is now considered not to be represented in the outcrop. In the lower Midhnab, maximum flooding surface P40 has been proposed, and dated as late Dorashamien, based on the presence of *Paradagmarita* sp. In conclusion, the carbonates in outcrop are not considered to be older than Lopingian. Foraminifera within the basal Khartam have been considered to be reworked from the Midhnab by Vaslet. The maximum flooding surface within the early Triassic TR10 (of Sharland *et al.* 2001) is interpreted to lie within the lower Khartam.

Of the five major global catastrophic events, the Permo-Triassic boundary is recognized as the most significant, with a loss of almost 95% of Permian species. Despite continued debate regarding the cause of this extinction, its coincidence with a major sea level fall, as in Iran would suggest that the faunal extinction resulted primarily from a destruction of shallow marine ecological niches. Other suggested causes include a major plate reorganization and the resulting climate change, followed by a reduction of stratospeheric ozone and the ensuing mutagenic ultraviolet radiation effects represented in the the palynological assemblages (Holser & Magaritz 1985; Krassilov & Afonin 2000). There is good evidence for a major marine regression at the end of the Permian (Changhsingian) (Wu & Fan 2003), and this would have extensively exposed the shallow marine Khuff carbonate platform. The location of this significant exposure surface within the Saudi Arabian Khuff subsurface has been placed within the lower part of the Khuff B reservior carbonate (Meyer & Hughes 2000, 2002).

Carbon isotope data collected by Chevron from the Khuff reservoirs in Ghawar (Al-Jallal 1989) has recently been reappraised (Swart & Mackenzie 2001; Hughes *et al.* 2004). This has revealed that grainstones of the upper Khuff C reservoir display consistently heavy C isotope values, whereas those from the upper Khuff B reservoir are light. Although this change could result from diagenetic influence (Al-Jallal 1989), the degree of change is considered to be great enough to compare well with the established shift at the Permo-Triassic boundary. Carbon isotope studies of the Permo-Triassic boundary (Holser & Magaritz 1987; Magaritz 1989; Magaritz *et al.* 1992; Wei *et al.* 2003) have revealed a significant trend towards negative, or lighter isotope values above the boundary, in Iran and Italy. Similar trends are continually being documented from various parts of the world, including Kashmir (Baud *et al.* 1989; Brookfield *et al.* 2003), China (Baud 2003), Greenland (Looy *et al.* 2003) Australia and Antarctica (Retallack 1999). In China, Baud (2003) has recently noted that the carbon isotope shift occurs centimetres below the actual boundary, within the late Changhsingian. Although the characteristic negative shift is documented in reefal carbonates of Oman (Krystyn *et al.* 2003), it is of interest to note that it is not present within deep marine Permo-Triassic sediments (Richoz *et al.* 2003). In Iran (Kashfi 2000) the Khuff Formation has been subdivided across the Permo-Triassic unconformity into a lower Dalan Formation and an upper Kangan Formation.

The stable isotopic composition of bulk carbonate samples from the Saudi Arabian Khuff Formation provides an excellent marker for the Permian-Triassic boundary. The boundary is characterized by a change from relatively negative carbon and oxygen isotopic compositions, within the upper part of the Khuff B carnbonates, to more positive values in the lower Khuff B carbonates and typical of the Khuff-C dolomites. Although typically the rocks overlying this boundary are limestones and the rocks below dolomites, this change occurs regardless of whether the mineralogy is only dolomite. The initiation of the change coincides with sedimentological evidence of exposure and the extent of the change varies between 10 feet to over 50 feet in the north Ghawar subsurface (Meyer & Hughes 2002). The variation in thickness is probably a result or the extent of the freshwater lens and the concomitant mixing-zone associated with the exposure.

Stratigraphical studies in the Ghawar Field of Saudi Arabia reveal sedimentological and sequence stratigraphical evidence that support the location of the Permo-Triassic boundary in the lower part of the Khuff B Reservoir. This work identifies the presence of a major unconformity, with a thick interval of rooted, desiccated, bleached and disrupted carbonates below the unconformity, suggestive of a prolonged period of subaerial exposure. Progressively thinning depositional cycles parallel an increase in the frequency of exposure evident at the cycle tops and thereby record the loss of accommodation space due to falling sea level in the carbonates as they approach this better developed unconformity. Above this boundary, metre-scale grainstone cycles formed during the ensuing transgression. Although the sedimentological features themselves do not establish this unconformity is the Permo-Triassic boundary, the loss of Permian faunas noted above coincides with this major unconformity. Micropalaeontological analysis confirms the absence of foraminifera above the lower part of the Khuff B Reservoir, which could simply be explained by an intra-Permian environmental change, and the onset of biologically adverse conditions. The presence, however, of characteristic Triassic palynomorphs and an annelid worm tube attributed to the Triassic form *Spirorbis phlyctaena* in the Khartam Member and upper part of the Khuff B Reservoir strongly suggest that the Permo-Triassic boundary lies within both the lithostratigraphic and subsurface reservoir units. Similar biostratigraphic evidence within the upper Khuff Formation offshore Qatar has also confirmed the Permo-Triassic boundary (Virgone *et al.* 2000).

Biostratigraphy (Table 2)

Foraminifera (Plates 2–4)

The most comprehensive previous work on the micropalaeontology of the Khuff Formation of Saudi Arabia is that by Le Nindre *et al.* (1990) in which the stratigraphical distribution of all identified species is listed with reference to the four members. The Huqayl Member contains a few indeterminate foraminifera, with an algal form assigned to *Pseudovermiporella japonica*. Echinoid fragments are present within the dolostones; this term is preferable to dolomite to avoid ambiguity between the mineral and rock names. Calcareous algae are particularly well represented within the gypsiferous sands and include *Gymnocodium belerophontis* and *Permocalculus fragilis*. Rare bryozoa are present. Foraminifera include *Baisalina pulchra, Globivalvulina cyprica, G. vonderschmitti, Ichthyolaria* cf. *latilimbata, Pachyphloia cukurkoyi, P.* sp., *Robuloides* cf. *gibbus*, *Geinitzina* sp., *Palaeoglomospira* sp., *Hemigordius* cf. *ovatus* and *H. longus.* Grainstones of dolomitic limestone of the Duhaysan Member contain algal debris, foraminifera and fish teeth, while the gypsiferous sands contain gastropods, ostracods and indeterminate algae. Foraminifera include *Geinitzina postcarbonica, Hemigordius permicus* and *Pachyphloia schwageri*. At the base of the Midhnab Member, grainstones include the foraminifera *Globivalvulina vonderschmitti, Hemigordius ovatus, Pachyphloia iranica, Paradagmarita flabelliformis, Protonodosaria* sp. and *Glomospira* sp. The Khartam Member typically contains bivalves and gastropods, a few *Glomospira* sp. foraminifera and the calcareous tube of the polychaete *Spirorbis phylactaena*.

Table 2. *Late Permian and early Triassic microbiofacies of the Khuff Formation and their major biocomponents*

Microbiofacies	Biocomponents
Spirorbis-ostracod-gastropod (Triassic only)	*Spirorbis phlyctaena*, smooth ostracods, smooth gastropods, smooth, small brachiopods
Gastropod-brachiopod-ostracod	smooth ostracods, smooth gastropods, smooth small brachiopods
Ostracod-*Mizzia*-*Gymnocodium*	smooth ostracods, *Mizzia velebitana, Gymnocodium belerophontis* *Aeolisaccus koteri, Anatolipora* sp., *Rivularia* sp.
Agathammina-Hemigordius	*Agathammina pusilla, Agathammina* spp., *Palaeoglomospira* spp., *Baisalina pulchra, Hemigordius* spp., *Hemigordius schlumbergeri*
Agathammina-Hemigordius-Globivalvulina	*Agathammina pusilla, Agathammina* spp., *Palaeoglomospira* spp., *Baisalina pulchra, Hemigordius irregularis, Hemigordius schlumbergeri, Hemigordiopsis* sp., *Globivalvulina vonderschmitti, Paraglobivalvulina globulosa*
Nankinella-Stafella	*Nankinella orbicularis, Stafella hupehensis, Neoendothyra pulchra*
Bryozoa	*Dyscritella* cf. *tenuimuralis*
Pachyphloia-Protonodosaria	*Pachyphloia iranica, P. cukurkoyi, P. iranica colcheni, Paradagmarita* sp., *Dagmarita chanakchiensis, Reichelina pulchra, Cribrogenerina sumatrana, Lingulina semilavata, Protonodosaria rauserae, Nodosaria ochotica, N. farcimeniformios, N. elabugae, N. suchonensis, Frondina permica, Polarisella sagitta*

As the Khuff carbonates in the subsurface and surface are mostly hard and dolomitized, foraminiferal taxonomy is based on randomly oriented thin sections rather than the external appearance of extracted individuals. To gain the maximum taxonomic benefit from such sections, the author has examined illustrations by numerous micropalaeontologists who have published on late Permian foraminifera from the wider region, especially the work of Lys (1988), but also including Glintzboeckel & Rabate (1964), Bozorgnia (1967, 1973), Johnson (1981), Kotlyar *et al.* (1983*a*, *b*, 1999), Lys (1984, 1988), Okimura (1988), Pronina (1988, 1989, 1995, 1996, 1999), Fontaine & Nguyen (1989), Ross & Haman (1989), Vachard & Ferriere (1991), Sartorio & Rozza (1991), Dawson & Racey (1993), Vachard *et al.* (1994), Vachard & Fourcade (1997), Pronina & Nestell (1997), Groves & Wahlman (1997), Kobayashi (1997), Salaj *et al.* (1998), Groves & Boardman (1999), Pronina-Nestell & Nestell (2001), Vachard *et al.* (2002) and Groves *et al.* (2003). For age determination, the species ranges of Groves & Boardman (1999) and Pronina-Nestell & Nestell (2001) have been used, as the stratigraphical ranges of Middle East Permian benthic foraminifera are relatively poorly established.

In the absence of any published information on the palaeoenvironmental preferences of late Permian foraminifera, their morphometric similarity with extant benthonic foraminifera and assemblage stacking patterns permit provisional palaeoenvironmental interpretation, especially when such interpretations have been compared with lithofacies stacking patterns visible in the core. Micropalaeontology has also assisted in establishing potentially correlative palaeoenvironmental events that may be related to depositional sequences. Integration of the biostratigraphic events with core based depositional cycle and sequence boundaries is a critical procedure to identify hierarchical boundaries of depositional cycles.

The Khuff carbonates display a moderately high diversity of benthonic species, with agglutinated, microgranular calcareous and calcareous test walls. The distribution of benthonic foraminifera in the four members of the Khuff Formation as defined in the outcrop and well SHD-1 have been described by Le Nindre *et al.* (1990). The present study of the Khuff in subsurface cores reveals that certain beds within the Khuff D carbonates contain low diversity foraminifera, whereas certain beds within the Khuff C carbonates and the lower part of the Khuff B carbonates contain rich, high diversity and abundance of foraminifera. Agglutinated foraminifera include *Pseudoglomospira* sp., *Palaeotextularia* sp., but most of the foraminifera are calcareous microgranular forms that include *Lasiodiscus* spp., *Nodosinella* spp., *Nodosinelloides ronda, Nodosinelloides longissima camerata, Nodosinelloides pinardae, Cribogenerina sumatrana, Lunucammina* spp., *Pachyphloia* cf. *iranica colcheni, Pachyphloia* cf. *cukurkoyi, Pachyphloia* cf. *pygmobesa, Pachyphloia* cf. *iranica, Pachyphloia pedicula, Pachyphloia cukurkoyi, Robustopachyphloia* spp., *Colaniella media, Verveillina bradyi, Nodosaria saggita, Protonodosaria globifrondina, Protonodosaria rauserae, Frondinonodosaria pyrula, Climacammina* spp., *Palaeotextularia* spp., *Globivalvulina vonderschmitti, Paraglobivalvulina globulosa, Paraglobivalvulina mira, Paraglobivalvulinoides* spp., *Dagmarita chanakchiensis, Cryptoseptida* spp., *Frondina permica, Pseudolangella conica, Langella perforata, Climacammina major* and species of *Ichthyolaria, Calvezina* and *Paradagmarita.* Fusulinid calcareous microgranular foraminifera include *Reichelina cribroseptida, Reichelina pulchra, Neoendothyra pulchra, Nanlingella* spp., *Nankinella orbicularis, Stafella changlinensis* and *Stafella hupehensis.* Miliolid foraminifera include *Agathammina pusilla, Baisalina pulchra, Midiella zaninettiae, Midiella karinae, Pseudomidiella labensis, Hemigordiopsis* spp., *Hemigordius reicheli, Hemigordius schlumbergeri, Graecodiscus kotlyarae, Raphconilia modificata* and *Brunsiella concava.* Dasyclad algae include *Gymnocodium belerophontis* and *Mizzia* spp.. Bryozoan and brachiopod fragments are locally common. The Triassic section contains only thin walled ostracods, brachiopods and *Spirorbis phlyctaena.*

Calcareous algae

Useful references for the Permian calcareous algae are scarce, but include Elliott (1955), Flugel (1977), Delofre (1988, 1992), Roux (1991), Razghalleh & Vachard (1991), Granier & Deloffre (1994), Vachard & Fourcade (1997) and Hughes (2005). In the present study of the Khuff carbonates in surface and subsurface sections, the calcareous algae are dominated by *Gymnocodium belerophontis* (Gymnocodiaceae, red algae) but *Permocalculus plumosus* (Gymnocodiaceae, red algae) and *Mizzia velebitana* (Dasycladaceae, green alga) are also present. *Ortonella* sp. is locally present, but generally rare. The preferred environments of red and green algae have been discussed by Wray (1977) and Banner & Simmons (1994), and while it is not intended to directly extrapolate the depth preferences of extant forms to those of the Permian, it is of interest to note the relatively deeper preference of the red algae. Present day Dasycladaceae (green algae) and *Permocalculus* spp. (red algae) preferentially occupy the uppermost red photic zone; occupy low energy environments, below intensive wave action, or in protected areas, with maximum abundance being between 5 m and 10–15 m water depth. Mixed dasyclad–rhodophyte assemblages are found between 10–15 m and 20–30 m. Present-day Rhodophyta (red algae) occupy the slightly deeper blue-green photic zone, and are typically below 20–30 m.

Simple rectilinear or gently curved, conical forms are locally common within the Khuff C reservoir. They are characterized by a dark, granular composition, and may have a botanical association. They have recently been identified as the cyanobacterial envelope *Aeolisaccus kotori* Radojic (Golubic *et al.* 2003).

Bryozoa

Bryozoa are confined to selected levels within the Khuff C Reservoir, where they are typically rare, but locally common. Bryozoan and crinoidal packstones to grainstones are considered to have formed in an open marine subtidal environment, where they were continually reworked by currents, storm action and bioturbation at, or below, fairweather wave base (Madi *et al.* 1996; Gallagher 1998). At least two forms have been recognized, based on Ernst (2000, 2001) and include the branched form *Dyscritella* cf. *tenuimuralis,* and the encrusting form cf. *Rectifenestella* sp.

Brachiopods

Brachiopod valves and their fragments are dispersed throughout most of the Khuff B and C Reservoirs, but are especially concentrated within the foraminiferal and bryozoan-rich layers within the Khuff C. Brachiopods from the Khuff Formation in Oman have been documented by Angiolini *et al.* (1996, 1998, 2003).

Echinoids

Echinoid, probably crinoid, fragments are confined to selected levels within the subsurface Khuff C Reservoir carbonate, where they are typically rare, but locally common.

Ostracods

Simple, non-punctate, unornamented ostracod valves are present throughout most of the subsurface Khuff B and C Reservoir carbonate. They are especially concentrated within mudstones that are otherwise barren of microfossils and where they may represent adverse, possibly hypersaline, conditions. References of possible future value include Crasquin-Soleau *et al.* (1999) and Crasquin-Soleau (2003).

Gastropods

Low and high-coiled, non-costate gastropods are locally present within the subsurface Khuff C Reservoir carbonate, but rare within the subsurface Khuff B Reservoir carbonate. They are often present in close association with filamentous undifferentiated algae.

Polychaete worm tubes

These calcareous, coiled tubes are rare, and have been reported only from the upper part of the subsurface Khuff B Reservoir carbonate. Le Nindre *et al.* (1990) have recorded this form from the Khartam Member of the Khuff Formation, and use the species *Spirorbis phlyctaena* as evidence for an early Triassic age.

Sponge spicules

Simple monaxon sponge spicules are generally rare within the studied sections. They form local concentrations within the lower and middle parts of the subsurface Khuff C Reservoir carbonate at certain localities. Sponge spicules are usually either siliceous or calcareous. They, along with spongin, form the 'skeleton' of a sponge. Members of the group Calcarea are the only sponges that possess spicules composed of calcium carbonate. These spicules do not have hollow axial canals. Today, their diversity is greatest in the tropics, as is the case with most marine groups. They are predominantly found in shallow waters, though at least one species is known from a depth of 4000 metres.

Palaeoenvironmental aspects

The various biocomponents display vertical variations in both their distribution and abundance throughout both the Khuff B and C carbonates. Within the Khuff C carbonates and the basal part of the Khuff B carbonates, these variations mostly involve foraminifera. In the upper part of the B carbonate, significantly of Triassic age, only non-foraminiferal biocomponents, such as smooth ostracods and brachiopods, are present, and display local variations in their distribution.

These variations in the biofacies composition are related directly to variations in the depositional environment, except where dolomitization has completely destroyed any microfossils. The variations are cyclic, and a succession of ecologically determined cycles is evident, most of which are best represented, either as increase in abundance or diversity, within the maximum flooding zones. Biocomponent associations, or biofacies, develop as a result of the response of various organisms to particular depositional environmental conditions. The benthonic foraminifera can be conveniently subdivided into the eight simple morphotypes for potential palaeoenvironmental interpretation, an approach also proposed by Haynes (1981) and Gallagher (1998):

1. Coiled (*Agathammina*)
2. Discoid (*Hemigordius*)
3. Spherical (*Hemigordiopsis, Palaeoglomospira*)
4. Biserial (*Globivalvulina, Paraglobivalvulina, Dagmarita, Paradagmarita, Geinitzina*)
5. Uniserial (*Frondina, Ichthyolaria, Pachyphloia, Protonodosaria*)
6. Bulbous uniserial with internal partitions (*Colaniella*)
7. Larger fusulinids (subspherical) (*Nankinella caucasica, Staffella*)
8. Larger fusulinids (discoidal) (*Nankinella orbicularis*)
9. Smaller fusulinids (*Reichelina*)

These trends became evident during the author's studies of the measured surface sections in central Saudi Arabia (Hughes & Kamal 2000) and in the subsurface (Hughes & Kamal 1995), and have been used to establish the scheme illustrated in Table 3.

Conclusions

The Khuff Formation in outcrop consists of a basal mixed clastic and carbonate unit that is overlain by the Huqayl and Duhaysan carbonate members, the Midhnab clastic member and the Khartam carbonate member. In the subsurface of Ghawar, the Khuff Formation consists of four carbonate–evaporite couplets. A lithostratigraphical correlation between the outcrop and subcrop has been suggested, with the strongest link being based on a chronostratigraphical event that relates the top of the Midhnab Member to within the lower part of the Khuff B carbonate.

Micropalaeontological and palynological evidence provides a late Permian age for the Huqayl to Midhnab members in the outcrop

Table 3. *Microbiofacies and interpreted associated environmental parameters*

Microbiofacies	Energy level	Sediment type	Salinity	Depth range	Biotope
Spirorbis-ostracod-gastropod	Mod. low	Mudstone-wackestone	Elevated	Very shallow	Intertidal–shallow subtidal; ?hardgrounds for *Spirorbis*
Gastropod-brachiopod-ostracod	Mod. low	Mudstone-wackestone	Elevated		Intertidal–shallow subtidal
Ostracod-*Mizzia*-*Gymnocodium*	Mod. low	Mudstone-wackestone	Normal-elevated	Shallow	Subtidal lagoon
Agathammina-*Hemigordius*	Low	Mudstone	Normal-elevated	Mod. shallow	Subtidal lagoon
Agathammina-*Hemigordius*-*Globivalvulina*	Low	Mudstone	Normal	Mod. shallow	Inter-shoal deeps
Nankinella-*Stafella*	Mod. high	Packstone-Grainstone	Normal	Mod. shallow	Grainstone shoal flanks
Bryozoa	Mod. high	Grainstone	Normal	Shallow	Grainstone shoals
Pachyphloia-*Protonodosaria*	Low	Mudstone	Normal	Mod. deep	Platform, open marine

and also to the Khuff D, C and lower B carbonates in the subsurface. The Permo-Triassic boundary is placed at the Midhnab–Khartam contact in the outcrop, and within the lower Khuff B carbonate in the subsurface. The basal Triassic, Scythian, age is also based on micropalaeontological and palynological evidence and permits assignment of this age to the Khartam member in the outcrop and to the upper Khuff B carbonate and Khuff A carbonate in the subsurface.

The vertical stacking pattern, or tiering, of biofacies has permitted the following conclusions to be deduced for the various benthonic foraminifera and associated microfossils. Depositional cycles tend to commence with sparse foraminifera typically limited to *Agathammina* spp., or *Palaeoglomospira* spp. and these are considered to represent intertidal to shallow subtidal conditions within slightly hypersaline environments. Above this biofacies is one that contains *Agathammina* spp. with *Hemigordius* spp. and other planispiral forms, considered to represent slightly deeper conditions within the shallow subtidal, and with the associated near-normal salinity conditions. The dasyclad algae *Gymnocodium belerophontis* and *Mizzia* sp. are present within this and other biofacies. An increase in diversity is present in the overlying biofacies in which *Paraglobivalvulina* spp., *Pachyphloia* spp. and a few uniserial nodosarid forms are present, with only rare *Agathammina* spp. This assemblage is considered to represent progressively deeper subtidal conditions. Rare small fusulinids and bryozoan fragments testify to possible contemporaneous reworking of shoal-derived bioclasts. Oolitic sediments with small fusulinids such as *Nankinella* and *Stafella* spp. are interpreted as shoal biofacies, and are typically accompanied by bryozoan fragments. Muddy sediments without *Agathammina* or *Hemigordius* species, but with a moderately rich and moderately high diversity assemblage including *Ichthyolaria* spp., *Reichelina* spp., *Protonodosaria* spp., *Pachyphloia* spp., *Dagmarita* spp. and biserial forms are interpreted to represent the deepest assemblage with the most open marine facies. It is this last biofacies that may represent maximum flooding events within the depositional cycles. The distribution of these biofacies enables depositional cycles to be determined. It should be noted, however, that many cycles do not display all of these variations, depending upon the location of the studied well. The variations attributed to water depth may also be caused by salinity levels, with only slight palaeobathymetric control.

Micropalaeontological analysis has significantly contributed to am improved understanding of the depositional environment of the Khuff carbonates in Saudi Arabia. In addition, major regional flooding events have been identified by the microfauna that enhance inter-reservoir correlation in the subsurface carbonates.

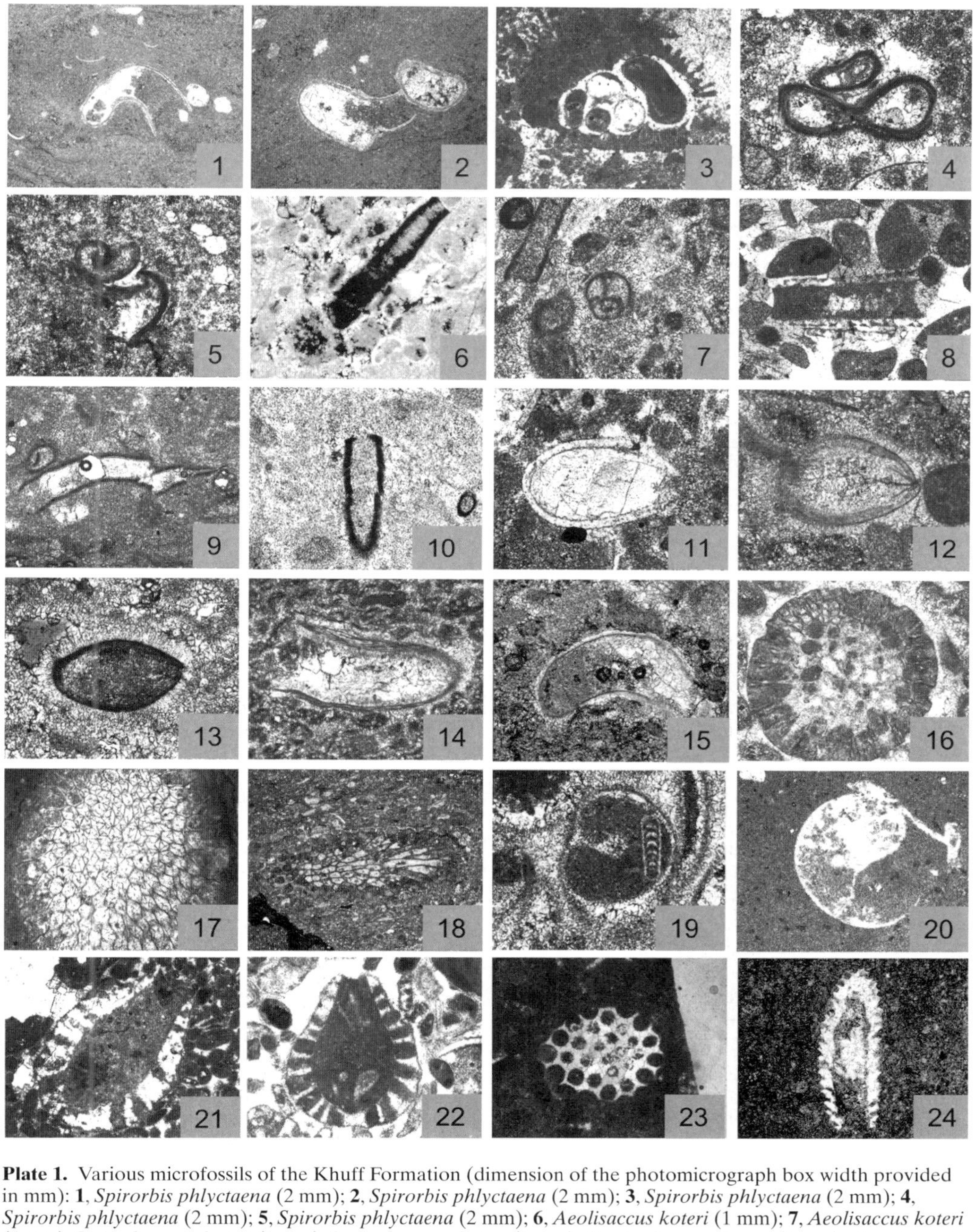

Plate 1. Various microfossils of the Khuff Formation (dimension of the photomicrograph box width provided in mm): **1**, *Spirorbis phlyctaena* (2 mm); **2**, *Spirorbis phlyctaena* (2 mm); **3**, *Spirorbis phlyctaena* (2 mm); **4**, *Spirorbis phlyctaena* (2 mm); **5**, *Spirorbis phlyctaena* (2 mm); **6**, *Aeolisaccus koteri* (1 mm); **7**, *Aeolisaccus koteri* (4 mm); **8**, *Aeolisaccus koteri* (1 mm); **9**, *Aeolisaccus koteri* (1 mm); **10**, *Aeolisaccus koteri* (1 mm); **11**, ostracod (1 mm); **12**, ostracod (1 mm); **13**, ostracod (1 mm); **14**, brachiopod (2 mm); **15**, brachiopod (2 mm); **16**, bryozoan *Dyscritella* cf. *tenuimuralis* (2 mm); **17**, bryozoan *Dyscritella* cf. *tenuimuralis* (2 mm); **18**, bryozoan *Dyscritella* cf. *tenuimuralis* (4 mm); **19**, gastropod (2 mm); **20**, gastropod (2 mm); **21**, *Mizzia velebitana* (2 mm); **22**, *Mizzia velebitana* (2 mm); **23**, *Mizzia velebitana* (4 mm); **24**, *Gymnocodium belerophontis* (4 mm).

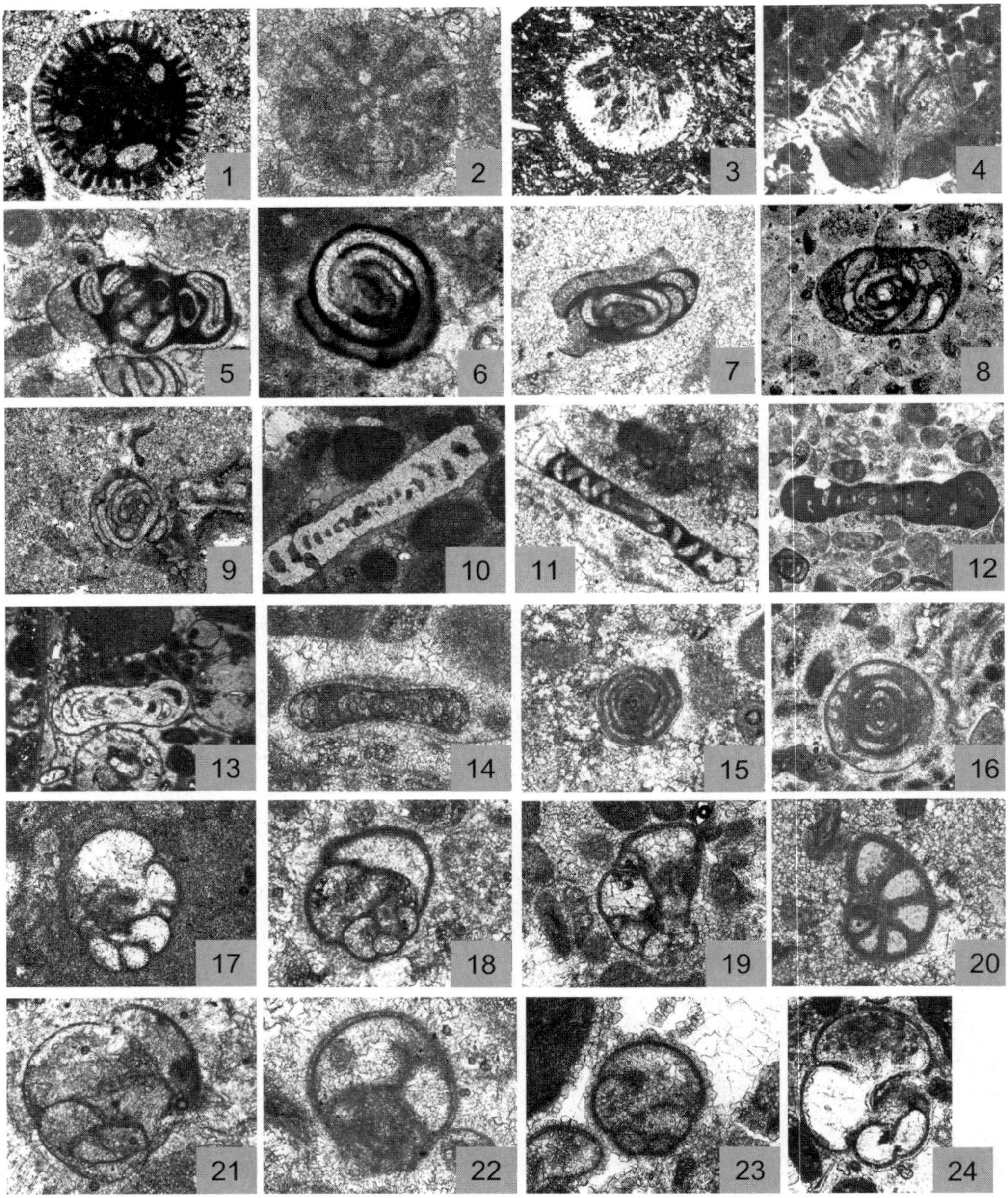

Plate 2. Various calcareous algae and foraminifera of the Khuff Formation (dimension of the photomicrograph box width provided in mm): **1**, *Mizzia velebitana* (2 mm); **2**, *Anatolipora* sp. (2 mm); **3**, *Rivularia* sp. (2 mm); **4**, *Rivularia* sp. (2 mm); **5**, *Agathammina pusilla* (1 mm); **6**, *Palaeoglomospira* sp. (2 mm); **7**, *Palaeoglomospira* sp. (2 mm); **8**, *Palaeoglomospira* sp. (2 mm); **9**, *Baisalina pulchra* (2 mm); **10**, *Hemigordius* sp. (1 mm); **11**, *Hemigordius* sp. (1 mm), **12**, *Brunsiella concava* (2 mm); **13**, *Hemigordius schlumbergeri* (2 mm); **14**, *Hemigordius* sp. (2 mm); **15**, *Hemigordius* sp. (axial section) (1 mm); **16**, *Hemigordiopsis irregularis* (2 mm); **17**, *Globivalvulina vonderschmitti* (1 mm); **18**, *Globivalvulina vonderschmitti* (1 mm); **19**, *Globivalvulina vonderschmitti* (1 mm); **20**, *Globivalvulina vonderschmitti* (1 mm); **21**, *Paraglobivalvulina globulosa* (1 mm); **22**, *Paraglobivalvulina globulosa* (1 mm); **23**, *Paraglobivalvulina globulosa* (1 mm); **24**, *Globivalvulina* sp. (1 mm).

Plate 3. Various foraminifera of the Khuff Formation (dimension of the photomicrograph box width provided in mm): **1**, *Pachyphloia* cf. *cukurkoyi* (1 mm); **2**, *Pachyphloia* cf. *cukurkoyi* (1 mm); **3**, *Cribrogenerina sumatrana* (1 mm); **4**, *Lingulina semilavata* (1 mm); **5**, *Nodosaria* cf. *sagitta* (1 mm); **6**, *Protonodosaria rauserae* (1 mm); **7**, *Protonodosaria rauserae* (1 mm); **8**, *Nodosaria ochotica* (1 mm); **9**, *Frondina permica* (1 mm); **10**, *Frondina permica* (1 mm), **11**, *Frondina permica* (1 mm); **12**, *Polarisella sagitta* (1 mm); **13**, *Nodosaria farcimeniformis* (1 mm); **14**, *Pachyphloia iranica* (1 mm); **15**, *Pseudolangella pulchra* (1 mm); **16**, *Frondina permica* (1 mm); **17**, *Frondina permica* (1 mm); **18**, *Frondina permica* (2 mm); **19**, *Nodosaria elabugae* (1 mm); **20**, *Nodosaria elabugae* (2 mm); **21**, *Pachyphloia iranica* (2 mm); **22**, *Pachyphloia iranica* (2 mm); **23**, *Nodosaria suchonensis* (1 mm); **24**, *Pachyphloia cukurkoyi* (1 mm); **25**, *Pachyphloia cukurkoyi* (1 mm).

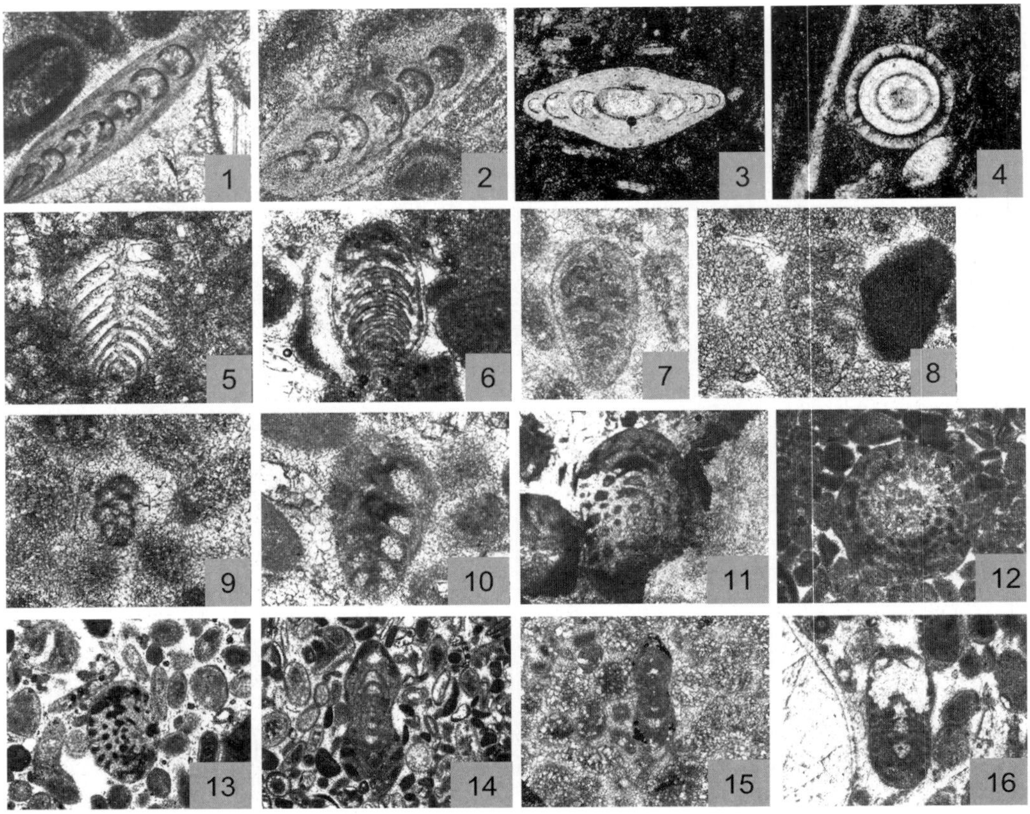

Plate 4. Various foraminifera of the Khuff Formation (dimension of the photomicrograph box width provided in mm): **1**, *Pachyphloia cukurkoyi* (axial section) (1 mm); **2**, *Pachyphloia iranica* (axial section) (1 mm); **3**, *Pachyphloia cukurkoyi* (transverse section) (1 mm); **4**, *Pachyphloia iranica* (transverse section) (1 mm); **5**, *Pachyphloia iranica* (1 mm); **6**, *Pachyphloia iranica colcheni* (1 mm); **7**, *Paradagmarita* sp. (1 mm); **8**, *Paradagmarita* sp. (1 mm); **9**, *Dagmarita chanakchiensis* (2 mm); **10**, *Paradagmarita* sp. (1 mm); **11**, *Nankinella* cf. *orbicularis* (transverse section) (2 mm); **12**, *Nankinella* cf. *orbicularis* (axial section) (2 mm); **13**, *Nankinella* cf. *orbicularis* (transverse section) (2 mm); **14**, *Stafella hupehensis* (transverse section) (2 mm); **15**, *Stafella hupehensis* (transverse section) (2 mm); **16**, *Stafella* sp. (2 mm).

This contribution is published with the permission of the Saudi Arabian Oil Company (Saudi Aramco) and the Ministry of Petroleum. The comments provided by the anonymous editors of the manuscript, together with the careful final editing by M. Hart, are gratefully acknowledged.

References

AL-JALLAL, I.A. 1989. *Depositional environments, diagenesis and reservoir characteristics of the Permian Khuff Formation in Eastern Saudi Arabia.* PhD Thesis, University of London.

AL-JALLAL, I.A. 1995. The Khuff Formation: its regional reservoir potential in Saudi Arabia and other Gulf countries, depositional and stratigraphic approach. *In*: AL-HUSSEINI, M.I. (ed.) *GEO 94*. Middle East Petroleum Geosciences Bahrain, 103–119.

ALSHARHAN, A.S. & NAIRN, A.E.M. 1994. The late Permian carbonates (Khuff Formation) in the western Arabian Gulf: its hydrocarbon parameters and paleogeographic aspects. *Carbonates and Evaporites*, **9**, 132–142.

ALSHARHAN, A.S. & NAIRN, A.E.M. 1997. *Sedimentary Basins and Petroleum Geology of the Middle East.* Elsevier, Amsterdam.

ANGIOLINI, L., NICORA, A. *ET AL*. 1996. Late Permian fauna from the Khuff Formation, South Oman. *Permophiles*, **29**, 62–64.

ANGIOLINI, L., NICORA, A. *ET AL*. 1998. Evidence of a Guadaloupian age for the Khuff Formation of southeastern Oman: preliminary report. *Rivista Paleontologia e Stratigrafia*, **104**, 329–340.

ANGIOLINI, L., BALIN, M., GARZANTI, E., NICORA, A., TINTORI, A., CRASQUIN, S. & MUTTONI, G. 2003. Permian climatic and paleogeographic changes in the Khuff Formation of interior Oman.

Palaeogeography, Palaeoclimatology, Palaeoecology, **191**, 269–300.

BANNER, F.T. & SIMMONS, M.D. 1994. Calcareous algae and foraminifera as water-depth indicators: an example from the Early Cretaceous carbonates of northeast Arabia. *In*: SIMMONS, M.D. (ed.) *Micropalaeontology and Hydrocarbon Exploration in the Middle East.* British Micropalaeontological Society Publications Series, Chapman & Hall, London, 243–252.

BAUD, A. 2003. The Permian-Triassic boundary and the Griesbachian substage. *Abstract, International Congress on carboniferous and Permian Stratigraphy, Utrecht, 10–16 August, 2003*, 43.

BAUD, A.M., MAGARITZ, M. & HOLSER, W.T. 1989. Permian-Triassic of the Tethys: carbon isotope studies. *Geologische Rundschau*, **78**, 649–677.

BEYDOUN, Z.R. 1988. *The Middle East: Regional Geology and Petroleum Resources*. Scientific Press, Beaconsfield.

BEYDOUN, Z.R. 1991. *Arabian Plate Hydrocarbon, Geology and Potential – a Plate Tectonic Approach*. American Association of Petroleum Geologists, Tulsa, Studies in Geology, **33**.

BOZORGNIA, F. 1967. *Microfacies and Microorganisms of Paleozoic through Tertiary Sediments of Some Parts of Iran*. National Iranian Oil Company, Tehran.

BOZORGNIA, F. 1973. *Paleozoic Foraminiferal Biostratigraphy of Central and East Alborz Mountains, Iran*. National Iranian Oil Company, Tehran.

BROOKFIELD, M.E., TWITCHETT, R.J. & GOODINGS, C. 2003. Palaeoenvironments of the Permo-Triassic transitions in Kashmir, India. *Palaeogeography, Palaeoclimatology, Palaeoecology*, **198**, 353–371.

BROUTIN, J., ROGER, J. *ET AL*. 1995. The Permian Pangea. Phytogeographic implications of new paleontological discoveries in Oman (Arabian peninsula). *Compte Rendu de l'Académie des Sciences, Paris*, **321**, 1069–1096.

CRASQUIN-SOLEAU, S. 2003. A new ostracod (Crustacea) species (*Geffenina wangi*) from the Middle Permian of Jebel Gharif (Central Oman). *Journal of Micropalaeontology*, **22**, 101–105.

CRASQUIN-SOLEAU, S., BROUTIN, J. *ET AL*. 1999. First ostracode fauna from the Upper Permian of Oman (Khuff Formation), *Micropaleontology*, **45**, 163–182.

DAWSON, O. & RACY, A. 1993. Fusuline-calcareous biofacies of the Permian Ratburi Limestone, Saraburi, Central Thailand. *Journal of Southeast Asian Earth Sciences*, **8**, 49–65.

DELOFRE, R. 1988. New taxonomy of dasyclad algae. *Bulletin des Centres de Recherches Exploration-Production Elf Aquitaine*, **12**, 165–217.

DELOFRE, R. 1992. Revision of the Gymnocodiaceae (Red Algae, Permian–Miocene) taxonomy, biostratigraphy, paleobiogeography, 3rd part. *Revue de Micropaléontologie*, **35**(1), 23–37.

EL-KHAYEL, A.A. & WAGNER, R.H. 1985. Upper Permian stratigraphy and megafloras of Saudi Arabia: palaeogeographic and climatic implications. *C.R.X. Congress International Proceedings of the 2nd International Congress on Carboniferous Stratigraphy and Geology, Strat. Geol. Carbonifère*, Madrid, 1983, 17–26.

EL-KHAYAL, A.A., CHALONER, W.G. & HILL, C.R. 1980. Palaeozoic plants from Saudi Arabia. *Nature*, London, **285**, 33–34.

ELLIOTT, G.F. 1955. The Permian calcareous alga *Gymnocodium*. *Micropaleontology*, **1**, 83–90.

ERNST, A. 2000. Permian Bryozoans of the NW-Tethys. *Facies*, **42**, 79–102.

ERNST, A. 2001. Bryozoa of Upper Permian Zechstein Formation of Germany. *Senckenbergiana Lethaea*, **81**, 135–181.

FLÜGEL, E. 1977. *Fossil Algae*. Springer, Berlin.

FONTAINE, H. & NGUYEN, D.T. 1989. Morphological study of *Rectostipula quadrata* Jenny-Deshusses (Upper Permian foraminifera) after its discovery in Thailand. *Revue de Micropaléontologie*, **32**, 118–125.

GALLAGHER, S.J. 1998. Controls on the distribution of calcareous foraminifera in the Lower Carboniferous of Ireland. *Marine Micropaleontology*, **34**, 197–211.

GALLOWAY, W.E. 1989. Genetic Stratigraphic Sequences in Basin Analysis 1: Architecture and genesis of flooding surface bounded depositional units. *AAPG Bulletin*, **73**, 125–142.

GLINTZBOECKEL, C. & RABATE, J. 1964. Microfaunes et microfacies du Permo-Carbonifère du Sud Tunisien. *In*: CUVILLIER, J. & SCHURMAN, H.M.E. (eds) *International Sedimentary Petrographical Series*, **7**. Brill, Leiden.

GOLUBIC, S., RADOJCIC, R. & LEE, S.J. 2003. Cyanobacteria on Modern and Ancient carbonate platforms. *In*: *Scytonema – Aeolisaccus. Abstracts of the 8th International Symposium on Fossil Algae, Granada, Spain, 18–20 September, 2003, Publicaciones de la Universidad de Granada, Spain*, 21.

GRANIER, B. & DELOFFRE, R. 1994. Inventaire critique des algues Dasyclades fossils. III partie – Les Algues Dasycladales du Permien et du Trias. *Revue de Paléobiologie*, **14**, 49–84.

GROVES, J.R. & BOARDMAN, D.R. 1999. Calcareous smaller foraminifers from the Lower permian Council Grove Group near Hooser, Kansas. *Journal of Foraminiferal Research*, **29**, 243–262.

GROVES, J.R. & WAHLMAN, G.P. 1997. Biostratigraphy and evolution of Late Carboniferous and Early Permian foraminifers from the Barents Sea (offshore Arctic Norway). *Journal of Paleontology*, **71**, 758–779.

GROVES, J.R., ALTINER, D. & RETTORI, R. 2003. Origin and early evolutionary radiation of the Order Lagenida (Foraminifera). *Journal of Paleontology*, **77**, 831–843.

HAYNES. J.R. 1981. *Foraminifera*. Wiley, Chichester.

HILL, C.R. & EL-KHAYEL, A.A. 1983. Late Permian plants including charophytes from the Khuff Formation of Saudi Arabia. *Bulletin British Museum, Natural History (Geology)*, **37**, 105–112.

HOLSER, W.T. & MAGARITZ, M. 1987. Events near the Permian–Triassic boundary, *Geology*, **11**, 155–180.

HUGHES, G.W. 2002. The stratigraphic significance of microbiofacies to selected Saudi Arabian reservoirs. GEO 2002 Abstracts. *GeoArabia*, **7**, 248–249.

HUGHES, G.W. 2003. Permian to Cretaceous foraminiferal–calcareous algal associations from Saudi Arabia. *In*: *Abstracts of the 8th International Symposium on Fossil Algae, Granad, Spain, 18–20 September, 2003, Publicaciones de la Universidad de Granada, Spain*, 30.

HUGHES, G.W. 2005. Calcareous algae of Saudi Arabian Permian to Cretaceous carbonates. *Revista Española de Micropaleontología*, **37**, 131–140.

HUGHES, G.W. & KAMAL, R.A. 1995. The Khuff Formation at BIRK-2: sedimentology, palaeoenvironments and depositional sequences. *Saudi Aramco Geological Research and Development Division*, Miscellaneous Report, **1027**, Dhahran.

HUGHES, G.W. & KAMAL, R.A. 2000. Recognition of cryptic, high-frequency depositional cycles within the Khuff-C carbonates in the subsurface of south-central Saudi Arabia. *GeoArabia,* **5**, 110.

HUGHES, G.W., MEYER, F.O. & SWART, P. 2004. The Permo-Triassic boundary in Saudi Arabia. Abstract and Poster, *32nd International Geological Congress. August 20–28, 2004, Florence, Abstracts Part 2, Session 206–18.*

JIN, Y., WARDLAW, B.R., GLENISTER B.F. & KOTLYAR, G.V. 1997. Permian chronostratigraphic subdivisions. *Episodes*, **20**, 10.

JIN, Y., SHANG, Q. & CAO, C. 2000. Late Permian magnetostratigraphy and its global correlation. *Chinese Science Bulletin* **45**, 698–704.

JOHNSON, B. 1981. Microfaunal biostratigraphy of the Dalan Formation (Permian) Zagros Basin, southwest Iran. *In*: NEALE, J.W. & BRASIER, M.D. (eds) *Microfossils from Recent and fossil shelf seas.* British Micropalaeontological Society Series, Ellis Horwood, Chichester, 52–60.

KALANTARI, A. 1982. Microbiostratigraphy of Paleozoic through Jurassic sediments of Ahmadi Anticline, southwest Iran. *Revista Española de Micropaleontología*, **14**, 263–290.

KASHFI, M.S. 2000. Greater Persian Gulf Permian-Triassic stratigraphic nomenclature requires study. *Oil and Gas Journal*, Nov.6. Also: *http://membres.lycos.fr/middle_east/Kashfi_2000.htm*.

KOBAYASHI, F. 1997. Upper Permian foraminifers from the Iwai-Kanyo area, west Tokyo, Japan. *Journal of Foraminiferal Research*, **27**, 186–195.

KOTLYAR, G.V., ZAKHAROV, Y.D. *ET AL.* 1983*a*. [*Dzhulfian and Dorashamian regional stages in the USSR.*] *Project 106 (Permo-Triassic stage of geological evolution) of the International Geological Correlation Programme*. Nauka, Leningrad, 129–191, plates 1, 6 and 7 [in Russian].

KOTLYAR, G.V., ZAKHAROV, Y.D., KOCZYRKEVICZ, B.V., ROSTOVTSEV, K.O., CHEDIJA, I.O., VUKS, G.P. & GUSEVA, E.A. 1983*b*. [Evolution of the Latest Permian Biota. *Dzhulfian and Dorashamian regional stages in the USSR.*] *Project 106 (Permo-Triassic stage of geological evolution) of the International Geological Correlation Programme*. Nauka, Leningrad [in Russian].

KOTLYAR, G.V., NESTELL, G.P., ZAKHAROV, Y.D. & NESTELL, M.K. 1999. Changhsinian of the Northwestern Caucasus, Southern Primorye and Southeastern Pamirs, *Permophiles, Newsletter of the Subcommision on Permian Stratigraphy, International Commission on Stratigraphy and International Union of Geological Sciences*, **35**, 18–22.

KRASSILOV, V.A. & AFONIN, S.A. 2000. Floristic changes over the Permian–Triassic boundary. *Abstract, American Association of Petroleum Geologists Conference*, Rio de Janeiro, 2000.

KRYSTYN, L., RICHOZ, S., BAUD, A. & TWITCHETT, R. 2003. A unique Permian-Triassic boundary section from the Neotethyan Hawasina Basin, Central Oman Mountains. *Palaeogeography, Palaeoclimatology, Palaeoecology*, **191**, 329–344.

LEMOIGNE, Y. 1981*a*. Presence d'une flore comprenant des elements cathaysiens dans le centre de l'Arabie Saoudite au Permien superioeur. *Compte Rendu Hebdomadaire des Séances de l'Académie des Sciences, Paris*, 3, **292**, 975–977.

LEMOIGNE, Y. 1981*b*. Flore mixte au Permien superieur en Arabie Saoudite. *Geobios*, Lyon, **14**(5), 611–635.

LE NINDRE, Y.-M., VASLET, D. & MANIVIT, J. 1990. Le Permo-Trias d'Arabie Centrale. *Histoire géologique de la bordure occidentale de la plateforme Arabe*, **3**, Editions du Bureau de Recherches Géologiques et Minières, Document 193.

LOOY, C.V., BRINKHUIS, H., VAN-KONIJNENBURG – VAN CITTERT, J.J.A., SEPHTON, M.A., TWITCHETT, R.J. & VISSCHER, H. 2003. *Abstract of the International Congress on Carboniferous and Permian Stratigraphy, Utrecht, Netherlands, 10–16 August, 2003*, University of Utrecht, 336.

LYS, M. 1984. Benthic foraminifera (non fusulinida) as a means of identifying biozones in the biostratigraphy of the Carboniferous and Permian. *In*: OERTLI, H.J. (ed.) *Benthos '83: 2nd International Symposium on Benthic Foraminifera (Pau, April 1983),* Elf Aquitaine, Esso REP and Total CFP, Pau and Bordeaux, 393–400.

LYS, M. 1988. *Biostratigraphie du Carbonifère et du Permien en Mesogee (Espagne, Afrique du Nord, regions Egenees, Proche-Orient), Etudes micropaleontologiques (Foraminifères), paleobiogeographie.* Thèses de doctorat d'état de l'université de Paris-Sud, Orsay, documents du Bureau de Recherches Geologiques et Minières (BRGM), Orleans, France, **147**.

LYS, M., COLCHEN, M., BASSOULLET, J.P., MARCOUX, J.& MASCLE, G. 1980. La Biozone a *Colaniella parva* du Permien Superieur et sa microfaune dans le bloc calcaire exotique de Lamayuru, Himalaya du Ladakh. *Revue de Micropaléontologie*. **20**, 78–108.

MADI, A., BOURQUE, P.A. & MARNET, B.L. 1996. Depth-related ecological zonation of a Carboniferous carbonate ramp – Upper Visean of Bechar Basin, Western Algeria. *Facies*, **35**, 59–79.

MAGARITZ, M. 1989. ^{13}C minima follow extinction events: a clue to faunal radiation. *Geology*, **17**, 337–340.

MAGARITZ, M., KRISHNAMURTHY, R.V. & HOLSER, W.T. 1992. Parallel trends in organic and inorganic carbon isotopes across the Permian–Triassic boundary. *American Journal of Science*, **292**, 727–739.

MEYER, F.O. & HUGHES, G.W. 2000. A multidisciplinary integration of selected Saudi Arabian carbonate unconformities. Abstract. *GeoArabia*, **5**, 144.

MEYER, F.O. & HUGHES, G.W. 2002. Application of high-resolution sequence stratigraphy in defining Khuff-reservoir facies partitioning, Uthmaniyah area, Ghawar field, Saudi Arabia. GEO 2002 Abstracts. *GeoArabia*, **7**, 279.

MURRIS, R.J. 1980. The Middle East: stratigraphic evolution and oil habitat. *AAPG Bulletin*, **64**, 597–618.

OKIMURA, Y. 1988. Primitive colaniellid foraminiferal assemblage from the Upper Permian Wargal Formation of the Salt Range, Pakistan. *Journal of Paleontology*, **52**, 715–723.

POWERS, R.W. 1968. *Saudi Arabia. Lexique Stratigraphique International, 3*, Centre National de la Recherche Scientifique, Paris.

POWERS, R.W., RAMIREZ, L.F., REDMOND, C.D. & ELBERG, E.L. 1966. *Geology of the Arabian Peninsula, Sedimentary Geology of Saudi Arabia*. Geological Survey Professional Paper 560-D, United States Government Printing Office, Washington.

PRONINA, G.P. 1988. [Late Permian miliolata of Transcaucasia.] *Proceedings of the Zoological Institute, Leningrad*, **184**, 49–63 [in Russian].

PRONINA, G.P. 1989. [Foraminifers of the *Paratritolites kittli* Zone of the Dorashamian of Late Permian of Transcaucasia.] *Eahegodnik Vsesoyuznogo Paleontologischeskogo, Leningrad*, **32**, 30–37 [in Russian].

PRONINA, G.P. 1995. Genus *Sphaerionia* and its stratigraphic significance. Reports of Shallow Tethys 4, International Symposium Albrechtsberg (Austria) 8–11 September, 1994. *Supplemento agli Annali dei Musei Civici di Rovereto Sezione Archeologia, Storia e Scienze Naturali*, **1**, 105–117.

PRONINA, G.P. 1996. Genus *Sphairionina* and its stratigraphic significance. *Annali dei Musei Civici di Rovereto Sezione Archeologia, Storia e Scienze Naturali, St. Petersberg, Russia*, **11**, 105–117.

PRONINA, G.P. 1999. [New species of small foraminifers from the Upper Permian of Transcaucasia.] *In*: STEPANOV, D.L. & KISILEV, G.N. (eds) *Vaprocy Paleontologii, Saint Petesburg State University*, 4–14 [in Russian].

PRONINA, G.P. & NESTELL, M.N. 1997. Middle and Late Pemian foraminifera from exotic limestone blocks of the Alma River basin, Crimea. *In*: ROSS, C.A., ROSS, J.R.P. & BRENCKLE, P.L. (eds) Late Paleozoic Foraminifera; their biostratigraphy, evolution and paleoecology and the Mid-Carboniferous boundary. *Cushman Foundation for Foraminiferal Research, Special Publications*, **36**, 111–114.

PRONINA-NESTELL, G.P. & NESTELL, M.K. 2001. Late Changsingian foraminifera of the northwestern Caucasus. *Micropaleontology*, **47**, 205–234.

RAZGHALLEH, S. & VACHARD, D. 1991. Systematics and biosedimentology of the Permian building calcareous algae Tubiphytes and Archaeolithoporella (Upper Murghabian, Jebel Tebaga, Tunisia). *Palaeontolographica*, **221**, 171–205.

RETALLACK, G.J. 1999. Postapocalyptic greenhouse paleoclimate revealed by earliest Triassic paleosols in the Sydney Basin, Australia. *Geological Society of America, Bulletin*, **III**, 52–70.

RICHOZ, S., BAUD, A. KOZUR, H.W. & MARCOUX, J. 2003. Deep-water records from Middle Permian to Lower Triassic of Oman: the demise of Permian biogenic chert and abnormal C isotope curve. *Abstract of the International Congress on Carboniferous and Permian Stratigraphy, Utrecht, Netherlands, 10–16 August, 2003, University of Utrecht*, 441.

ROSS, C.A. & HAMAN, D. 1989. Suprageneric ranges of Foraminiferida. *Journal of Foraminiferal Research*, **19**, 72–83.

ROUX, A. 1991. Revision of the Gymnocodiaceae (Red Algae, Permian–Miocene) taxonomy, biostratigraphy, paleobiogeography, 2nd part. *Revue de Micropaléontologie*, **34**, 136–173.

SALAJ, J., MAAMOURI, A.-L. & ZARBOUT, M. 1988. Permo-Triassic and Jurassic microbiofacies of Tunisia, Part 1. *In*: SALAJ, J. & MAMMOURI, A.-L. (eds) *Atlas of Tunisian microbiofacies and its relation to some circum-Tethyian types, Hodonin, Czech Republic. Zemny plyn a nafta,* **43**, 3–110.

SARTORIO, D. & ROZZA, R. 1991. The Permian of Amanda 1-bis well (Northern Adriatic Sea). *Giornale di geologia*, **53**, 187–196.

SENALP, M. & DUWAIJI, A. 1994. Stratigraphy and sedimentation of the Unayzah Reservoir, Central Saudi Arabia. *In*: AL-HUSSEINI, M.I. (ed.) *GEO 94*. Middle East Petroleum Geosciences, Bahrain, 837–847.

SHARLAND, P., ARCHER, R. *ET AL*. 2000. A proposed sequence stratigraphy for the Phanerozoic succession of the Arabian Plate: Mesozoic and Cenozoic sequences. *GeoArabia*, **5**, 181.

SHARLAND, P., ARCHER, R. *ET AL*. 2001. *Arabian Plate Sequence Stratigraphy*. GeoArabia Special Publication **2**, Gulf Petrolink, Bahrain.

STAMPFLI, G.M. & BOREL, G.D. 2002. A plate tectonic model for the Paleozoic and Mesozoic constrained by dynamic plate boundaries and restored synthetic oceanic isochrones. *Earth and Planetary Science Letters*, **196**, 17–33.

STEINER, M.B., ESHET, Y., RAMPINO, M.R. & SCHWINDT, D.M. 2003. Fungal abundance spike and the Permian-Triassic boundary in the Karroo Supergroup (South Africa). *Palaeogeography, Palaeoclimatology, Palaeoecology,* **194**, 405–414.

STEINEKE, M., BRAMKAMP, R.A. & SANDER, N.J. 1958. Stratigraphic relations of Arabian Jurassic oil. *American Association of Petroleum Geologists Symposium, Tulsa*, 1294–1329.

STEPHENSON, M.H., OSTERLOFF, P.L. & FILATOFF, J. 2003. Palynological biozonation of the Permian of Oman and Saudi Arabia: progress and challenges. *GeoArabia*, **8**, 467–496.

SWART, P.K. & MACKENZIE, G. 2001. *Geochemical and*

Petrographic Observations on the Diagenesis of the Khuff Formation, Ghawar Field, Saudi Arabia. Unpublished report for Saudi Aramco.

VACHARD, D. & FERRIERE, J. 1991. An assemblage with Yabeina (fusulinid foraminifer) from the Midian (Upper Permian) of Whangaroa area (Orua Bay, New Zealand). *Revue de Micropaléontologie*, **34**, 201–230.

VACHARD, D. & FOURCADE, E. 1997. Foraminifères et algues du Permien du Guatemala. *Geobios*, **30**, 745–784.

VACHARD, D., MARTINI R. & ZANIONETTI, L. 1994. Le Murghabien a fusulinoides des Iles d'Hydra, Crete et Mytilene (Permien Superieur de Grece). *Geobios*, **28**, 395–406.

VACHARD, D., HAUSER, M., MARTINI, R., ZANINETTI, L., MATTER A. & PETERS, T. 2002. Middle Permian (Midian) foraminiferal assemblages from the Batain Plain (Eastern Oman): their significance to Neotethyan paleogeography. *Journal of Foraminiferal Research*, **32**, 155–172.

VASLET, D., LE NINDRE, Y.-M., VACHARD, D., CRASQUIN-SOLEAU, S., BROUTIN, J., HALAWANI M. & AL-HUSSEINI, M. 2004. Stratigraphy of the central Saudi Arabian Khuff Formation. Poster in GEO 2004, 6th Middle East Geosciences Conference and Exhibition, 7–10 March, Kingdom of Bahrain, Conference Abstracts, *GeoArabia*, **9**, 139.

VIRGONE, A., MARRO, E., LAPOINTE, P., FRAISSE, C., CAMOIN, G., GHAEMI A. & KHALILI, A. 2000. The Permian/Triassic boundary, example from the South Pars Field, Iran. GEO 2000 Abstract, *GeoArabia*, **5**, 192–193.

WRAY, J.L. 1977. *Calcareous Algae. Developments in Sedimentology*. Elsevier, New York, **4**.

WU YA SHENG & FAN JIASONG 2003. Amplitude of end Permian (end Chiagxingian) eustatic sea-level drop: from reefs in China. *Acta Geologica Sinica*, **1**, 95–102.

WEI, W., MATSUMOTO, R., HOSEIN, G., OHDE S. & AKIHIRO, K. 2003. Chemostratigraphy of the Permo-Triassic boundary in the Middle Zagros Area, Iran. *Abstract of the International Congress on Carboniferous and Permian Stratigraphy, Utrecht, Netherlands, 10–16 August, 2003, University of Utrecht*, 574.

Foraminifera and dinoflagellate cyst zonation and stratigraphy of the Callovian to Volgian reference section in the Tyumenskaya superdeep well (West Siberia, Russia)

VERA I. ILYINA, BORIS L. NIKITENKO & LARISA A. GLINSKIKH
Institute of Petroleum Geology, Siberian Branch of RAS, Koptyug av. 3, Novosibirsk, 630090, Russia

Abstract: The Callovian to Volgian section in the Tyumenskaya superdeep well SDW-6 is very important for the development and improvement of Jurassic biostratigraphy in West Siberia. It was drilled with almost continuous recovery of core (about 200 m), which is characterized by rich assemblages of microfossils. In this paper the results of micropalaeontological and palynological studies of the most complete Jurassic succession in the north of West Siberia are discussed, taking into consideration the problems of bio- and lithostratigraphy of the region.

The comprehensive analysis of the stratigraphical distribution of the different microfossil groups (foraminifers, microforaminifers, ostracods, dinoflagellate cysts, acritarchs, prasinophytes, spores and pollen) allows the creation of a detailed zonation. For the first time, an almost continuous sequence of dinoflagellate cyst zones for the Callovian–Middle Volgian has been established in the north of West Siberia, providing the basis for development of the Jurassic dinoflagellate cyst zonation in West Siberia.

Distinctive features of microbenthos communities and microphytoplankton associations reflecting biofacies have been studied. The pattern of distribution of microbenthic communities and microphytoplankton associations, depending on changes of transgressive–regressive events and palaeoenvironments in the Callovian to Volgian, has been established.

To devise reliable geological models for the occurence of hydrocarbons in northern West Siberia, it is very important to refine the stratigraphy of the Middle (Callovian) and Upper Jurassic sequences. The new and revised palaeontological, biostratigraphical and lithostratigraphical data accumulated for more than the last ten years has changed the boundaries of some zonal and lithostratigraphical units (Shurygin *et al.* 2000). Macrofauna (ammonites and bivalves) is rare in cores of the Jurassic sequences in West Siberia, and some of the zones recognized by their macrofauna are often discontinuous in particular sections. As a result, their correlatability is poor, but their identification serves to establish datums, allowing us to link up the zonal units based on the microfossils. The Callovian–Upper Jurassic sections covered by core are rare to absent in West Siberia. The development of zonal schemes was, therefore, based on composite sections and the comparison of partial records of Jurassic strata from different wells. As a result, the exact position of zonal boundaries according to different groups of fossils relative to one another and relative to the general stratigraphical scale was often rather vague.

One of the sections where the Middle (Callovian) and Upper Jurassic sequence is almost completely cored is the section of the Tyumenskaya superdeep well (SDW-6) in the north of central West Siberia (Fig. 1). The continuous coring in this interval of the Jurassic succession helps to solve many problems of the Jurassic stratigraphy of West Siberia. It has been possible to determine precisely the relationship of lithostratigraphical and biostratigraphical boundaries, to refine the position of biostratigraphical boundaries in the zonations based on different groups of microfossils and, finally, to improve relevant zonal schemes. To solve the above problems, studies were undertaken on the Callovian–Upper Jurassic part of the SDW-6 section with a detailed lithological and taphonomical description of the core and detailed sampling, followed by micropalaeontological and palynological analysis of the resulting preparations from the Vasyugan, Georgievka and Bazhenovo formations. Rich assemblages of microfossils (foraminifers, microforaminifers, ostracods, dinoflagellate cysts, acritarchs and prasinophytes) have been identified, permitting a detailed zonation of this succession.

From: POWELL, A. J. & RIDING, J. B. (eds) 2005. *Recent Developments in Applied Biostratigraphy.* The Micropalaeontological Society, Special Publications, 109–144. 1747-602X/$15.00

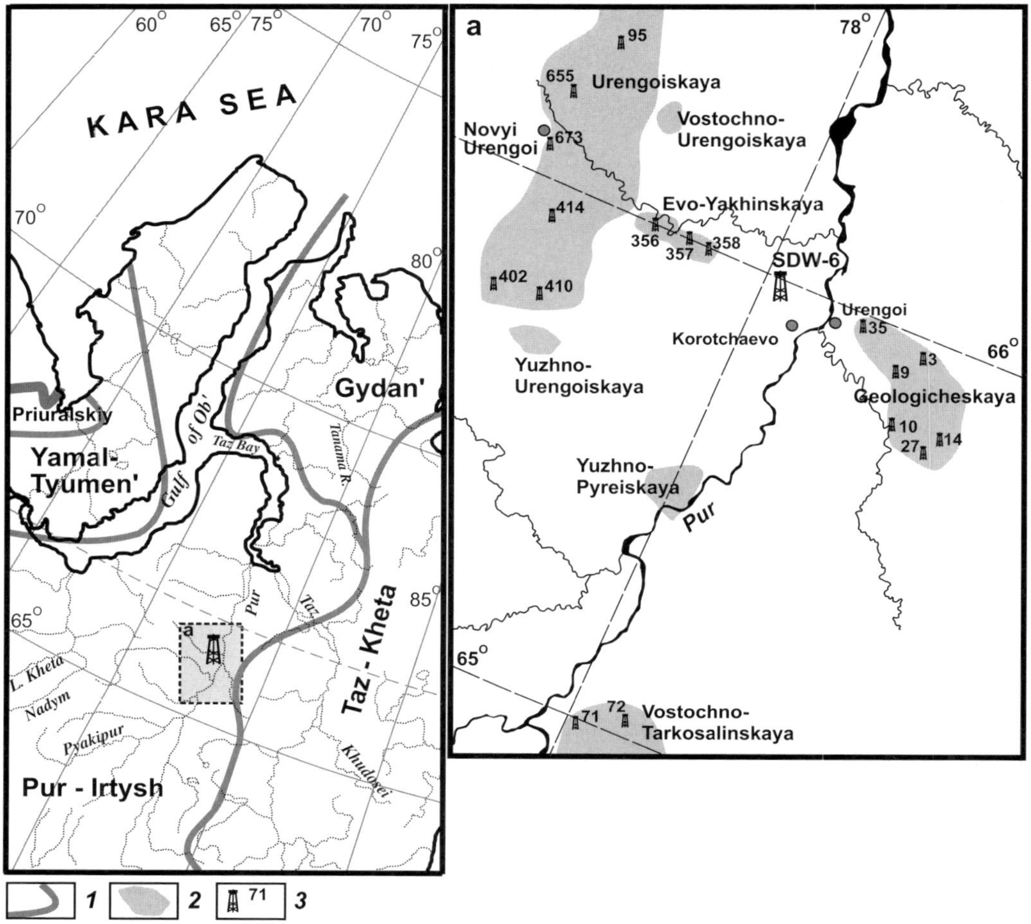

Fig. 1. Map showing facies zonation of Callovian and Upper Jurassic of northern West Siberia (Shurygin *et al.* 2000) and Tyumenskaya superdeep well (SDW-6) location map (a). 1, boundaries of facies districts; 2, drilling areas; 3, wells and their numbers.

Previous research on the Jurassic section of the SDW-6

Preliminary data on the lithological composition, geochemical characteristics of the section and stratigraphical differentiation of the Vasyugan, Georgievka and Bazhenovo formations in the SWD-6 have been published by Ekhlakov *et al.* (1991). Somewhat later, further information (Mazur 1996) was published on the detailed geological, geochemical, lithological, and other characteristics of this part of the SDW-6 section. However, biostratigraphical data on the Callovian to Volgian part of the section were absent from these publications.

In 1999, VNIGRI researchers published a paper (Kirichkova *et al.* 1999) on the biostratigraphy of the Mesozoic succession of the SWD-6 borehole, where they identified the Abalak and Bazhenovo formations within the Callovian–Upper Jurassic part of the core. From the results of micropalaeontological studies in the upper Tyumenskaya and lower Abalak formations, they identified the Callovian (intervals 3976.6–3989.7 and 3967–3976 m). A significantly greater range for the Callovian in this section (interval 4000–3888.8 m) was determined by spore and pollen analysis (Kirichkova *et al.* 1999). At the top of the Bazhenovo Formation and at the bottom of the Achimovka Member (3778.3–3798.6 m), a prasinophyte assemblage was found, which was interpreted as indicating the middle substage of the Volgian Stage and the Berriasian.

Thus the biostratigraphy of the Callovian–Upper Jurassic part of the SDW-6 section is inadequately understood. There is also no consensus on the lithostratigraphical differentiation of the section. Some researchers distinguish the Vasyugan and Georgievka formations (Ekhlakov *et al.* 1991; Mazur 1996; Dikovskii & Sedykh 1996; Ekhlakov & Ugryumov 1996), while others identify the Abalak Formation (Kirichkova *et al.* 1999).

Materials

During our investigation of this part of the Tyumenskaya superdeep well (SDW-6), the Vasyugan, Georgievka and Bazhenovo formations were studied in detail. When samples were chosen for micropalaeontological and palynological analyses, attention was given to the textural and structural features of the rocks, with nearly all lithological varieties explored. Simultaneously, taphonomic observations were undertaken. A collection of 70 core samples sampled throughout the Callovian and Upper Jurassic section of SDW-6 was examined in the laboratory. To carry out the microfaunal analysis, 200 g of core sample was disintegrated by washing in water through a 0.056 mm mesh sieve. The frequency of occurrence of each species was determined on a seven-point scale: very rare (1–2 specimens), rare (3–5 specimens), frequent (6–10 specimens), common (11–15 specimens), high quantity (few tens of specimens), abundant (many tens of specimens), and very abundant (hundreds of specimens) (Saks 1969). Palaeoecological and biofacies analyses were carried out to study the structure of the microfaunal communities, the dynamics of taxonomic diversity and the numbers of microfossils at the species and generic level and the quantitative relationships between different taxa.

The same samples were subjected to a combined palynological analysis, including the study of spores and pollen of terrestrial plants as well as the microphytoplankton (dinoflagellate cysts, acritarchs and prasinophytes). Emphasis was placed on the study of dinoflagellate cysts, which in the Vasyugan, Georgievka and the lower part of the Bazhenovo formations comprised the main part of palynological spectra. The dinoflagellate cysts as a key group of microphytofossils have been used for the zonation of the Callovian and Upper Jurassic section. Prasinophytes and acritarchs were used in the facies analysis of the deposits.

Geological setting

The Tyumenskaya superdeep well (SDW-6) is situated in the northern part of the Pur-Irtysh facies zone (Fig. 1). It is especially interesting that within this small area the Abalak and Bazhenovo formations laterally grade into the Vasyugan, Georgievka and Bazhenovo formations (in central areas), which in their turn pass laterally into the Tochino, Sigovoe and Yanov Stan formations further eastwards (Figs 2 and 3). To estimate the stratigraphical and lateral ranges of the formations and their thicknesses within this territory, we analysed the palaeontological, lithostratigraphical and well-log data on the sections in Urengoiskaya, Evo-Yakhinskaya, Geologicheskaya, Vostochno-Tarkosalinskaya and other areas. On the Central Urengoi megabar (Kontorovich *et al.* 2001), in the west part of this territory (Urengoi area), the high density of drilled wells permits observation of a gradual transition from the clay-dominated Abalak Formation to the clay–sand Vasyugan Formation and the clay-dominated Georgievka Formation (Fig. 4). In this region the Abalak Formation is very thin, varying from 55 to 75 m. In western areas, the thickness of the Vasyugan and Georgievka formations range from 70 to 87 m. In the central part of the Bol'shaya Kheta 'megasyncline' in the West Korotchaevo depression, within the Evo-Yakhinskaya area, and in SD-6, the Vasyugan and Georgievka formations vary in thickness from 100 to 150 m (see Fig. 4). In the east of the territory, in the northwest part of the Mid-Pur megatrough (east of the Geologicheskaya area), the Vasyugan Formation is replaced by the mudstones and siltstones of the Tochino Formation and the sandstones and siltstones of the lower member of the Sigovoe Formation. The mudstones of the Georgievka Formation grade into variably glauconitic sandstones and siltstones of the upper member of Sigovoe Formation and the clays and mudstones with sandstone beds of the lower part of Yanov Stan Formation (Fig. 3). Within this area, the Tochino, Sigovoe and the lower part of Yanov Stan formations are rather persistent in thickness, from 140 to 160 m. The bituminous mudstones of the Bazhenovo Formation gradually become non-bituminous and are replaced by clays and mudstones, with partings of siltstones and sandstones, in the upper part of the Yanov Stan Formation.

This variation in the thicknesses of the Abalak Formation in western areas, the Vasyugan and Georgievka formations in central areas, and the Tochino, Sigovoe and lower part of the Yanov Stan formations in eastern areas is in agreement with the data on the position of the base of the Abalak, Vasyugan and Tochino formations (Fig. 4) in the area. The lowest absolute depths are documented near the West Korotchaevo depression (Bol'shaya Kheta 'megasyncline'), on the Evo-Yakhinskaya area

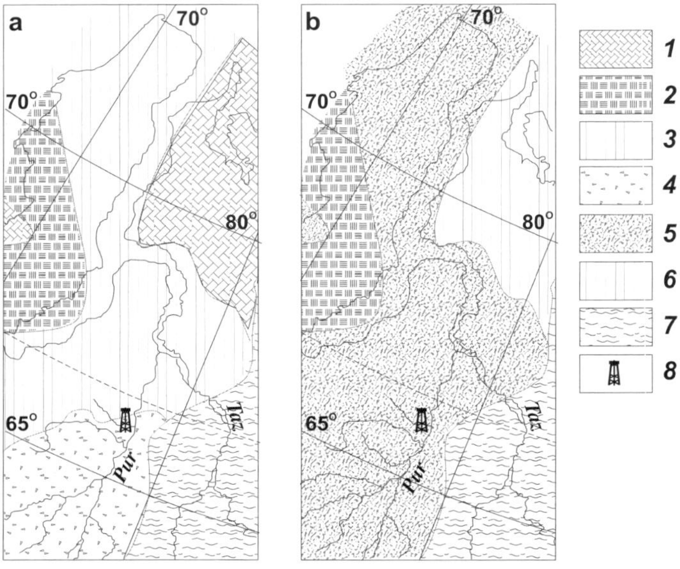

Fig. 2. Regional distributions of formations in (a) Callovian and Lower Volgian and (b) Lower–Upper Volgian in northern West Siberia. 1, Mauryn'ya Fm. and Lopsiya Fm. (a), Fedorovskiy Fm. (b); 2, Danilovskoe Fm (a, b); 3, Abalak Fm (a); 4, Vasyugan Fm and Georgievka Fm (a); 5, Bazhenovo Fm (b); 6, Gol'chikha Fm (a, b); 7, Tochino Fm and lower member of Sigovoe Fm (a), upper member of Sigovoe Fm and Yanov Stan Fm (b); 8, location of Tyumenskaya superdeep well (SDW-6).

and Tyumenskaya superdeep well SDW-6, gradually rising westward, to the Central Urengoi megabar and reaching its highest point in the east towards the Mid-Pur tilted 'megatrough' (Kontorovich *et al.* 2001) (Geologicheskaya wells area); see Figure 4.

The upper part of the Bathonian and Upper Jurassic sections in the Tyumenskaya superdeep well are represented by the Vasyugan, Georgievka and Bazhenovo formations.

The Vasyugan Formation, the lower member (interval 3994–3923.2 m) with a thickness of 70.8 m, is composed of dark-grey sometimes black argillites. In the basal bed of the Vasyugan Formation, interbedded siltstones, argillites with lenses and intercalations of sandstone, and beds of coaly argillites (the Pakhomovskaya Member) have been defined. In the lower member of Vasyugan Formation an Upper Callovian fauna is recorded: ammonites (? *Longaeviceras* sp. ind., *Quenstedtoceras* (*Soaniceras*) cf. *parvulum* (identified by S. V. Meledina)) as well as various and abundant foraminifera of the *Dorothia insperata*, *Trochammina rostovzevi* F15 Zone, *Kutsevella memorabilis*, *Guttulina tatarensis* F14 Zone, *Conorboides taimyrensis*, *Globulina paalzovi* F18 Zone and *Ammobaculites tobolskensis*, *Trochammina oxfordiana* F20 Zone. Marine microphytoplankton is represented by various dinoflagellate cysts of the *Fromea tornatilis* (Ft), *Impletosphaeridium* spp., *Lagenadinium callovianum* (I/Lc), *Gonyaulacysta jurassica* subsp. *adecta* var. *longicornis* (Gjl), *Wanaea thysanota* (Wth) and *Wanaea fimbriata* (Wf) dinoflagellate cyst Zones, as well as rare acritarchs (*Micrhystridium* spp.) and prasinophytes (*Leiosphaeridia*).

The upper member of the Vasyugan Formation (interval 3923.2–3855.1 m), with a thickness of 68.1 m is represented by intercalations of siltstones, sandy siltstones, fine-grained sandstones with abundant carbonized plant detritus, with rare beds of dark-grey claystones and sandstones. The lower part of the upper member in this section is characterized by foraminifers of *Ammobaculites tobolskensis*, *Trochammina oxfordiana* F20 Zone. In the upper member of the Vasyugan Formation, dinoflagellate cysts of the *Wanaea fimbriata* (Wf) (upper half) and *Rigaudella aemula* (Ram) dinoflagellate cyst Zones, as well as solitary *Micrhystridium* spp. and *Leiosphaeridia* have been found.

The Vasyugan Formation is transgressively overlapped by the dark-grey, greenish argillites of the Georgievka Formation (interval 3855.1–3844 m) with a thickness of 11.1 m. The

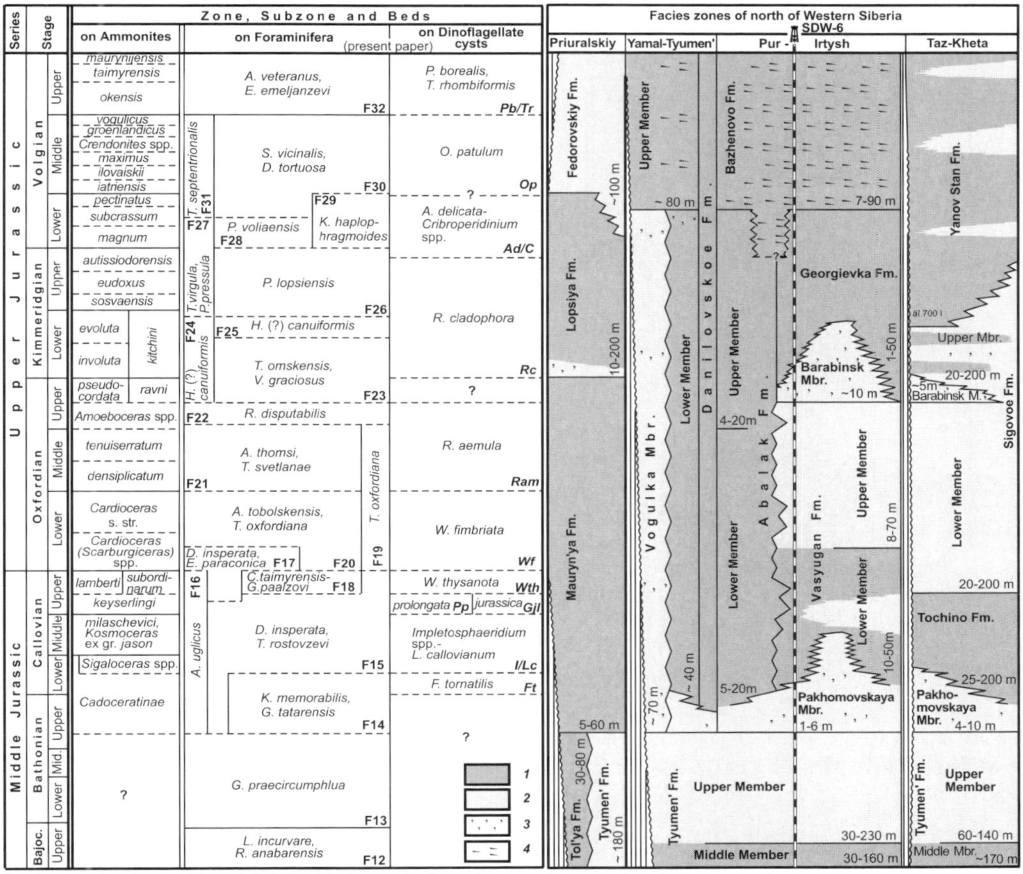

Fig. 3. Middle and Upper Jurassic stratigraphy. 1, mainly clayey deposits; 2, mainly sandy deposits; 3, glauconite-enriched beds; 4, bituminous deposits.

basal part of the Georgievka Formation (the Barabinsk Member) is represented by greenish-grey siltstones and brownish-grey sandstones. In the upper part of the Georgievka Formation rare foraminifers and rich assemblages of dinoflagellate cysts characteristic of *Rhynchodiniopsis cladophora* (Rc) and *Amphorula delicata, Cribroperidinium* spp. (Ad/C) dinoflagellate cyst Zones are recorded. Acritarchs are represented by rare *Veryhachium* sp. and *Micrhystridium* sp. In the lower part and at the top of the Georgievka Formation some prasinophytes (*Leiosphaeridia* and *Pterospermella* sp.) are found.

In the upper parts of the section the dark-grey argillites gradually become bituminous, allowing the identification of the Bazhenovo Formation which is composed of brownish bituminous argillites (interval 3844–3785 m) with a thickness of 59 m. In the lower part of the formation, the dinoflagellate cysts of *Amphorula delicata, Cribroperidinium* spp. (Ad/C) and *Oligosphaeridium patulum* (Op) dinoflagellate cyst Zones, as well as rare *Veryhachium, Solisphaeridium* and more frequent *Leiosphaeridia, Pterospermella* spp., and *Tasmanites* have been found. In the upper part of the Bazhenovo Formation some rare dinoflagellate cysts and abundant prasinophytes have been found.

Biostratigraphy

Callovian to Volgian sediments of the Tyumenskaya superdeep well (SDW-6) have been analyzed by micropalaeontological and palynological methods. The samples contained diverse and rather rich foraminiferal and microphytoplankton assemblages. Microphytoplankton is

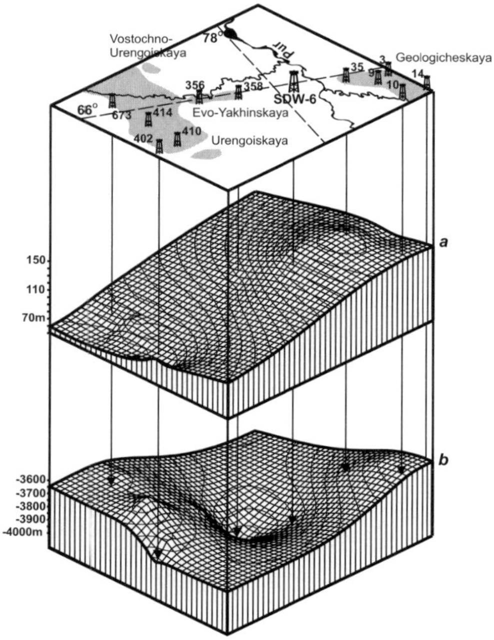

Fig. 4. Thickness of the Abalak, Vasyugan and Georgievka, Tochino, Sigovoe and Lower part of Yanov Stan Formations (a) and relief of basement of the Abalak, Vasyugan and Tochino Formations in the Urengoi district and adjacent territories (b).

represented by abundant dinoflagellate cysts and rare acritarchs, as well as by common prasinophytes, which sometimes reach very high percentages and dominate the microphytofossils in the upper part of the succession.

Foraminiferal zonation

The foraminiferal zonation is based on the stratigraphical range of important taxa, taxonomic changes in the succession of foraminiferal assemblages, and the characteristic features and variations within the foraminiferal assemblages during the Jurassic. As the basis for biostratigraphical subdivision, a composite set of foraminiferal zones has been used, including biozones, concurrent-range zones and ecozones or a series of parallel phylozones. Thus, to define the stratigraphical range of the zones, the combination of events of differing nature have been used (migrations, ecosystematics, changes of community structure, dominance shifts, acme of some taxa, phylogenetics and autochthonous appearance of new taxa). Widespread inter-regional correlations of these zonal subdivisions using the microbenthos can be considered as 'bioevents', with marker level horizons characterized by a unique succession of combinations of biological events of different nature (phylogenetic, migrations and ecological changes). At the same time, zones with narrow and wide ranges, characterized by different microfossil assemblages in various facies, have been defined. The lower boundaries of foraminiferal zones have been defined by the first appearance of new assemblages, new taxa (formed as a result of phylogenetic or migration processes) as well as by the acme of characteristic taxa.

Dorothia insperata, Trochammina rostovzevi Zone (F15)

Index-species: Dorothia insperata (Pl.1, **4**, **8**; Pl. 2, **11**, **12**), *Trochammina rostovzevi* (Pl.1, **5**; Pl. 2, **8**).

Characteristic species: Ammodiscus uglicus (Pl. 2, **1**), *Ammobaculites pungaensis* (Pl. 1, **1**), *Anmarginulina suprajurensis, Astacolus* ex gr. *lokossovensis* (Pl.1, **16**), *Dorothia insperata, Haplophragmoides magnus, Recurvoides scherkalyensis* (Pl. 1, **6**, **11**; Pl. 2, **6**), *Lenticulina elegia* (Pl. 2, **13**) and *Trochammina rostovzevi* (**5**).

In addition to these species, the foraminiferal assemblage contains *Ammodiscus* cf. *arangastachiensis, Saccammina compacta* (Pl. 1, **13**; Pl. 2, **2**), *Dentalina* sp. ind., *Globulina* ex gr. *oolithica, G. paalzowi, Kutsevella memorabilis, Lenticulina daschevskajae, L.* ex gr. *subpolonica, L.. subinvolvens* (Pl. 1, **9**; Pl. 2, **7**), *L. memorabilissima, Geinitzinita praenodulosa, Jaculella* sp., *Pseudonodosaria brandi, Recurvoides* ex gr. *anabarensis, Verneuilinoides* sp., *Planularia colligatiformis, P.* ex gr. *septentrionalis, Conorboides taimyrensis, Eomarsonella paraconica, Saracenella juganica, Ammobaculoides* ex gr. *primoris, Saracenaria* ex gr. *carzevae* and *Oolina borealis* (Fig. 5).

In different regions of West Siberia, the lowermost part of the F15 Zone contains ammonites *Cadoceratinae* and *Pseudocadoceras* sp., whereas the interval above contains abundant in *Rondiceras milaschevici, Longaeviceras* spp., *Eboraciceras* spp., *Quenstedtoceras* spp. (Bulynnikova *et al.* 1990). The oldest ammonites found at the bottom of the Vasyugan Formation (Pakhomovskaya Member) belong to beds with Cadoceratinae, which were previously regarded only as Lower Callovian strata (Bulynnikova *et al.* 1990). After a revision of the Cadoceratinae (Meledina 1994), the stratigraphical position of these beds, by analogy with sections of northern East Siberia, is defined as the uppermost Upper

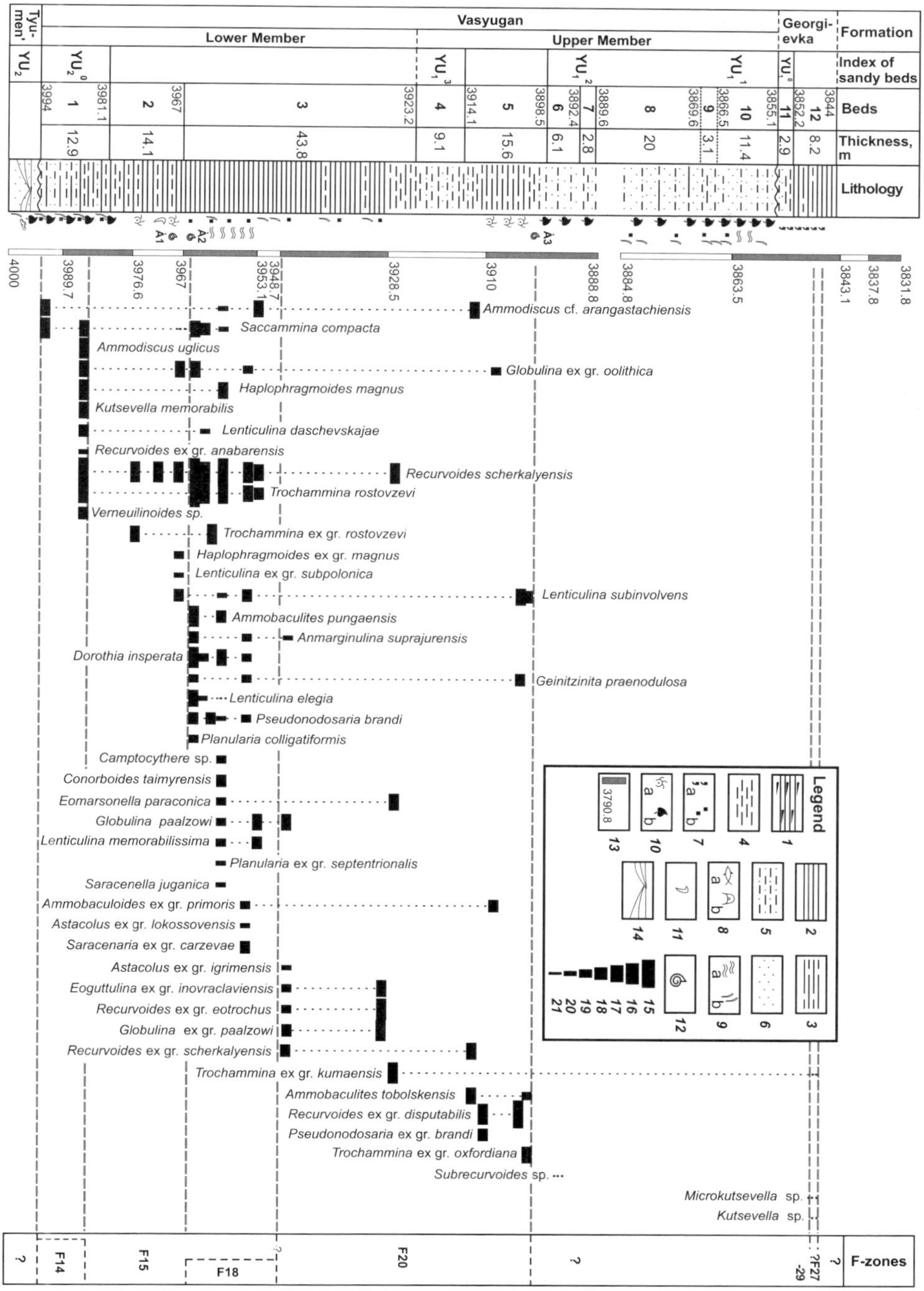

Fig. 5. Summary of the distribution of selected Callovian and Upper Jurassic foraminifers and foraminifer biozonation, Tyumenskaya superdeep well (SDW-6). 1, bituminous mudstone; 2, mudstone; 3, silty mudstone; 4, siltstone; 5, sandy siltstone; 6, sandstone; 7, glauconite (a), pyrite (b); 8, fish (a) and crustacean (b) remains; 9, *Planolites* isp. (a), *Skolithos* isp. (b); 10, small coquina detritus (a), charred plant debris (b); 11, onychite remains; 12, ammonites; 13, core sampling intervals, m; 14, cross bedding; 15, hundreds of specimens (very abundant); 16, many tens of specimens (abundant); 17, few tens of specimens (high quantity); 18, 11–15 specimens (common); 19, 6–10 specimens (frequent); 20, 3–5 specimens (rare); 21, 1–2 specimens (very rare).

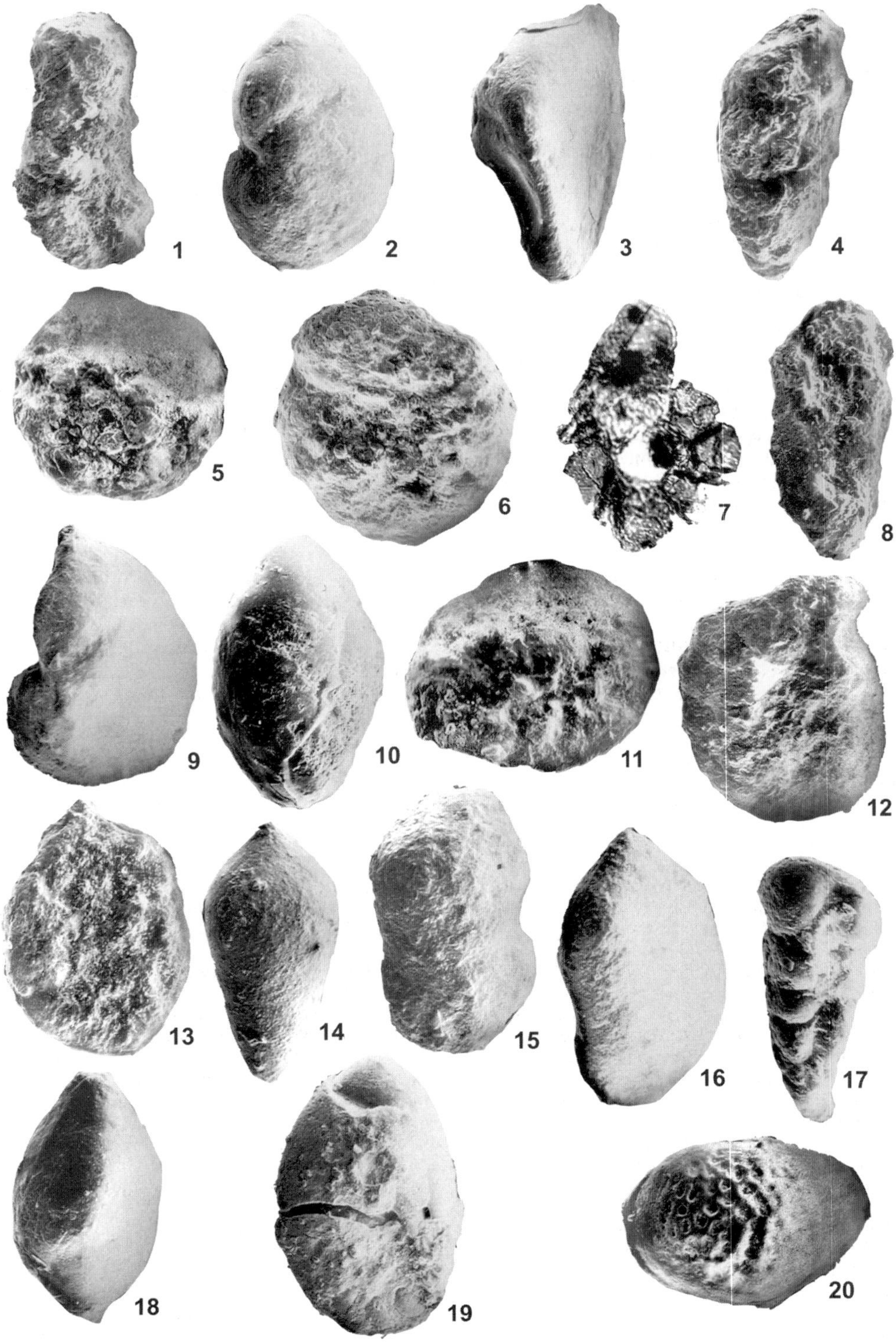
1
2
3
4
5
6
7
8
9
10
11
12
13
14
15
16
17
18
19
20

Bathonian–Lower Callovian. In the sections of the Ust'-Yenisei region, and in the northern East Siberia, the foraminiferal assemblage of the *Dorothia insperata, Trochammina rostovzevi* Zone coexist with ammonites of the *Cadoceras bamstoni, C. falsum, C. anabarense* and *C. emelianzevi* Zones (uppermost Upper Bathonian–Lower Callovian), *Rondiceras milaschevici* (Middle Callovian) in the Tochino Formation, and coexists with *Longaeviceras keyserlingi, Eboraciceras subordinarium* (Upper Callovian) in the Sigovoe Formation (Sharovskaya 1968; Lutova 1981; Meledina 1994).

Age: Latest Bathonian and Callovian.

Stratigraphical position: In the studied section, the foraminiferal assemblage of the *Dorothia insperata, Trochammina rostovzevi* F15 Zone, is found in the Pakhomovskaya Member and the lower part of the lower member of Vasyugan Formation, beds 1–2 and the lowermost 18.3m of bed 3 (interval 3993.5–3948.7 m) (Figs 5 and 6). The F15 Zone is also represented in the lowermost parts of the Danilovskoe, Abalak, Vasyugan and Naunak formations as well as the Tochino and lowermost Sigovoe formations (Figs 3 and 5).

Late Callovian ammonites identified by S. V. Meledina as ? *Longaeviceras* sp. ind. and *Quenstedtoceras (Soaniceras)* cf. *parvulum* were found at depths of 3969.2 and 3965.7 m, respectively in the studied section of SDW-6.

Geographical distribution: West and East Siberia, Arctic Canada, Svalbard, Franz Josef Land.

Kutsevella memorabilis, Guttulina tatarensis Zone (F14)

Index-species: Kutsevella memorabilis (Pl. 2, **4**), *Guttulina tatarensis*.

Characteristic species: Recurvoides ex gr. *anabarensis* (Pl. 2, **5**), *Kutsevella memorabilis, Verneuilinoides* sp., *Lenticulina daschevskajae, Trochammina rostovzevi, Dorothia insperata* and *Ammodiscus* cf. *arangastachiensis* (Fig. 5).

The foraminiferal assemblage of F14 Zone also contains *Ammodiscus uglicus, Dentalina* sp. ind., *Globulina* ex gr. *oolithica, Haplophragmoides magnus* (Pl. 1, **12**; Pl. 2, **9**), *Recurvoides scherkalyensis* and *Saccammina compacta* (Fig. 5).

The assemblage is usually dominated by *Recurvoides* (up to 60% at some levels). The subordinate position is occupied by *Ammodiscus* (1–40%), *Trochammina* (<8%), *Haplophragmoides* (2–8%), *Lenticulina* (<6%), *Globulina* (<5%) and *Saccammina* (2–50%).

Plate 1. In all the plates, the specimens are from the Tyumenskaya superdeep well (SDW-6), Pur-Irtysh facies zone, north of West Siberia. **1–6**, **8–20**, Vasyugan Fm: **1–6**, **8–14**, **16**, **18**, Lower member, **1–4**, **8**, **10–14**, **16**, **18**, **20**, Interval 3953.1–3967; lower part of bed 3, upper Callovian, *Dorothia insperata, Trochammina rostovzevi* F15 Zone; *Conorboides taimyrensis, Globulina paalzowi* F18 Zone. **1–4**, **10–12**, **20**, 7 m from top of core interval. **8**, **18**, Interval 3953.1–3967; 1.3 m from bottom of core interval. **5**, **6**, Interval 3976.6–3989.7; 8 m from top of core interval, Pakhomovskaya Mbr., bed 1, upper part of upper Bathonian–lower part of lower Callovian, *Dorothia insperata, Trochammina rostovzevi* F15 Zone, *Kutsevella memorabilis, Guttulina tatarensis* F14 Zone. **15**, **17**, **19**, upper member of Vasyugan Fm., lower Oxfordian, *Ammbaculites tobolskensis, Trochammina oxfordiana* F20 Zone.

1, *Ammobaculites pungaensis* Levina, 1990. lateral view, × 54,6; 1092/1; **2**, *Lenticulina memorabilissima* Gerke et Scharovskaja, 1961. lateral view, × 102; 1092/2; **3**, *Saracenella juganica* Kosyreva, 1975. × 70; 1092/3; **4**, *Dorothia insperata* (Bulynnikova 1962). × 83; 1092/4; **5**, *Trochammina rostovzevi* Levina, 1972. dorsal view, × 92; 1092/6; **6**, *Recurvoides scherkalyensis* Levina, 1962. dorsal view, × 65; 1092/7; **7**, *Microkutsevella* sp. × 672; 1092/9. Interval 3843.1–3863.5; 5 m from top of core interval, upper part of Georgievka Fm., bed 12, upper Kimmeridgian–lower Volgian, *Tolypammina virgula, Planularia pressula* F27 Zone and *Kutsevella haplophragmoides* F29 Zone; **8**, *Dorothia insperata* (Bulynnikova 1962). × 83; 1092/5; **9**, *Lenticulina subinvolvens* Scharovskaja, 1961. lateral view, × 74; 1092/10. Interval 3967.0–3976.6; 0.3 m from top of core interval, lower member of Vasyugan Fm., upper part of bed 2, upper Callovian, *Dorothia insperata, Trochammina rostovzevi* F15 Zone: **10**, *Globulina paalzowi* Myatlyuk, 1954. × 102; 1092/11: **11**, *Recurvoides scherkalyensis* Levina, 1962. dorsal view, × 77; 1092/8: **12**, *Haplophragmoides magnus* Bulynnikova, 1962. lateral view, × 37; 1092/12· **13**, *Saccammina compacta* Gerke, 1968. × 83; 1092/13. 1.5 m from bottom of core interval: **14**, *Pseudonodosaria brandi* (Tappan 1955). × 55; 1092/14. 10 m from top of core interval: **15**, *Ammobaculites tobolskensis* Levina, 1972. lateral view, × 62; 1092/15; Interval 3910–3928.5; 3 m from top of core interval, lower part of bed 5: **16**, *Astacolus* ex gr. *lokossovensis* Kosyreva, 1972. lateral view, × 109; 1092/16. 2.1 m from top of core interval: **17**, *Geinitzinita praenodulosa* Dain, 1972. × 92; 1092/17; Interval 3888.8–3910; 6 m from bottom of core interval, bed 5: **18**, *Anmarginulina suprajurensis* (Gerke et Scharovskaja 1961). × 129; 1092/18: **19**, *Astacolus* ex gr. *igrimensis* Levina, 1972. lateral view, × 67; 1092/19; Interval 3928.5–3948.7; 2.6 m from bottom of core interval, upper part of bed 3: **20**, *Camptocythere* sp. × 88; 1092/20; left valve.

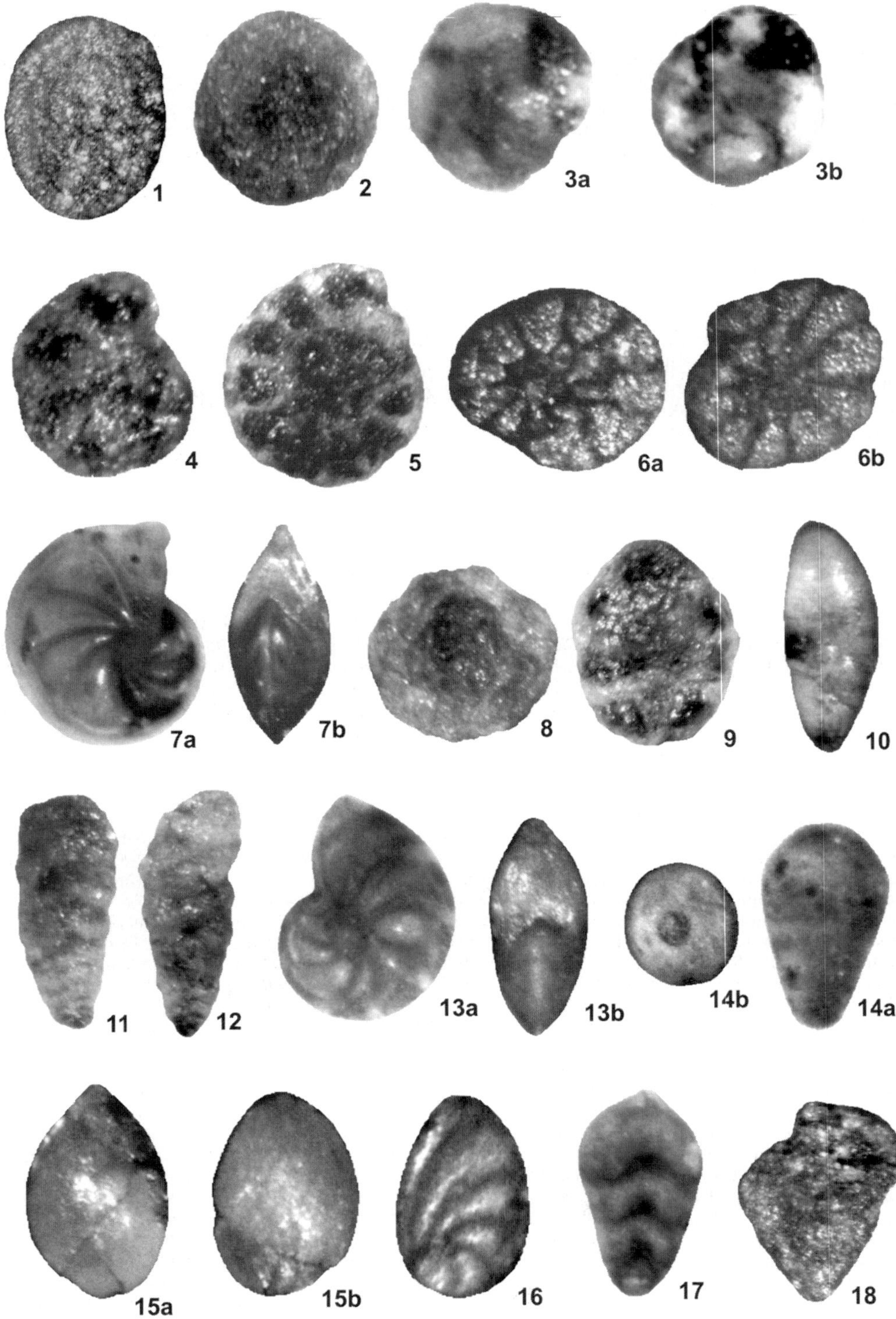
1
2
3a
3b
4
5
6a
6b
7a
7b
8
9
10
11
12
13a
13b
14b
14a
15a
15b
16
17
18

Agglutinated foraminifers dominate taxonomically (2–9 species) and quantitatively (88–100%). Calcareous foraminifers are much less diverse in the assemblage (<3 species) and rare (<12%). The foraminiferal assemblages of the F14 Zone in northern East Siberia are taxonomically richer (12 species of agglutinated and 12 species of calcareous foraminifers), but the agglutinated assemblages remain dominant over the calcareous assemblages (90–100%).

The foraminiferal assemblage of the overlying part of the *Dorothia insperata, Trochammina rostovzevi* Zone (between the *Kutsevella memorabilis, Guttulina tatarensis* Zone and the *Conorboides taimyrensis, Globulina paalzowi* Zone) is taxonomically poor. Eight species from seven genera (four species each of agglutinated and calcareous foraminifera) are identified in this assemblage. The representatives of *Recurvoides* continue to dominate (45–60%), while *Lenticulina* and *Trochammina* increase to 23% and 35% respectively.

Age: Latest Bathonian–Earliest Callovian.

Stratigraphical position: In the studied section, the foramiferal assemblage of the *Kutsevella memorabilis, Guttulina tatarensis* Zone occurs in the lowermost Pakhomov Member (lowermost Bed 1), interval 3993.5–3984.6 m (Figs 5 and 6). The foraminiferal assemblage of the *Kutsevella memorabilis, Guttulina tatarensis* Zone also occurs in the lowermost parts of the Abalak, Danilovskoe, Vasyugan and Tochino formations of Siberia (Fig. 3). The typical foraminifers in the West and central regions of West Siberia coexist with members of the *Cadoceratinae* and *Pseudocadoceras* sp. (Bulynnikova *et al.* 1990). In the Ust'-Yenisei region and in northern East Siberia, the stratigraphical range of the F14 Zone is determined by the occurrence of the ammonites *Cadoceras barstoni, C. falsum* and *C. anabarense* Zones (Sharovskaya 1968; Lutova 1981; Meledina 1994).

Geographical distribution: West and East Siberia.

Conorboides taimyrensis, Globulina paalzowi Zone (F18)

Index-species: Conorboides taimyrensis (Pl. 2, **3**), *Globulina paalzowi* (Pl. 2, **10**; Pl. 2, **15**).

Characteristic assemblage: Geinitzinita praenodulosa (Pl. 2, **17**), *Lenticulina elegia, Lenticulina* spp., *Planularia colligatiformis* (Pl. 2, **16**), *P.* ex gr. *septentrionalis, Eomarsonella paraconica, Saracenella juganica* (Pl. 1, **3**), *Conorboides taimyrensis, Ammobaculoides* ex gr. *primoris, Saracenaria* ex gr. *carzevae* and *Globulina paalzowi* (Fig. 5).

In addition to these species, the assemblage contains *Ammobaculites pungaensis, Anmarginulina suprajurensis* (Pl. 1, **18**; Pl. 2, **10**), *Dorothia insperata, Globulina* ex gr. *oolithica, Jaculella* sp. (Pl. 2, **18**), *Pseudonodosaria brandi* (Pl. 2, **14**), *Recurvoides scherkalyensis, Saccammina compacta, Trochammina rostovzevi,*

Plate 2. 1–18, Vasygan Fm, lower member, **1**, **2**, **4**, **5**, **8**, **9**, Interval 3976.6–3989.7; 8 m from top of core interval, Pakhomovskaya Mbr., bed 1, upper part of upper Bathonian–lower part of lower Callovian, *Dorothia insperata, Trochammina rostovzevi* F15 Zone, *Kutsevella memorabilis, Guttulina tatarensis* F14 Zone., **3**, **6**, **7**, **11**, **12**, Interval 3953.1–3967; 7 m from top of core interval, lower part of bed 3, upper Callovian, *Dorothia insperata, Trochammina rostovzevi* F15 Zone, *Conorboides taimyrensis, Globulina paalzowi* F18 Zone, **13**, **16**, Interval 3953.1–3967; 1.5 m from bottom of core interval, lower part of bed 3, upper Callovian, *Dorothia insperata, Trochammina rostovzevi* F15 Zone, *Conorboides taimyrensis, Globulina paalzowi* F18 Zone, **14**, **17**, **18**, **Interval** 3953.1–3967; 1.3 m from bottom of core interval, lower part of bed 3, upper Callovian, *Dorothia insperata, Trochammina rostovzevi* F15 Zone, *Conorboides taimyrensis, Globulina paalzowi* F18 Zone.

1, *Ammodiscus uglicus* Ehremeeva, 1972. × 76; 1092/21, **2**, *Saccammina compacta* Gerke, 1968. × 64; 1092/22, **3**, *Conorboides taimyrensis* Lutova, 1976. × 145; 1092/23; **3a**, dorsal view, **3b**, ventral view; **4**, *Kutsevella memorabilis* (Scharovskaja 1958). × 76; 1092/24; lateral view; **5**, *Recurvoides ex gr.anabarensis* Bassov et A.Sokolov, 1985. dorsal view, × 60; 1092/25; **6**, *Recurvoides scherkalyensis* Levina, 1962. × 73; 1092/26; **6a**, dorsal view, **6b**, ventral view; **7**, *Lenticulina subinvolvens* Scharovskaja, 1961. × 50; 1092/27; **7a**, lateral view; **7b**, apertural face view; **8**, *Trochammina rostovzevi* Levina, 1972. dorsal view, × 70; 1092/28; **9**, *Haplophragmoides magnus* Bulynnikova, 1962. × 55; 1092/29; **10**, *Anmarginulina suprajurensis* (Gerke et Scharovskaja 1961). × 86.5; 1092/30. Interval 3953.1–3967; 2.1 m from top of core interval, lower part of bed 3, upper Callovian, *Dorothia insperata, Trochammina rostovzevi* F15 Zone, *Conorboides taimyrensis, Globulina paalzowi* F18 Zone; **11**, *Dorothia insperata* (Bulynnikova 1962). × 71; 1092/31; **12**, *Dorothia insperata* (Bulynnikova 1962). × 67; 1092/32; **13**, *Lenticulina elegia* Kosyreva, 1972. × 69; 1092/33; **13a**, lateral view; **13b**, apertural face view; **14**, *Pseudonodosaria brandi* (Tappan 1955). × 54; 1092/34; **14a**, apertural view; **14b**, lateral view; **15**, *Globulina paalzowi* Myatlyuk, 1954. × 108; 1092/35; Interval 3928.5–3948.7; 2.6 m from bottom of core interval, middle part of bed 3, lower Oxfordian, *Ammbaculites tobolskensis, Trochammina oxfordiana* F20 Zone; **16**, *Planularia colligatiformis* Kosyreva, 1972. lateral view, × 108; 1092/36; **17**, *Geinitzinita praenodulosa* Dain, 1972. × 120; 1092/37; **18**, *Jaculella* sp. × 39; 1092/38.

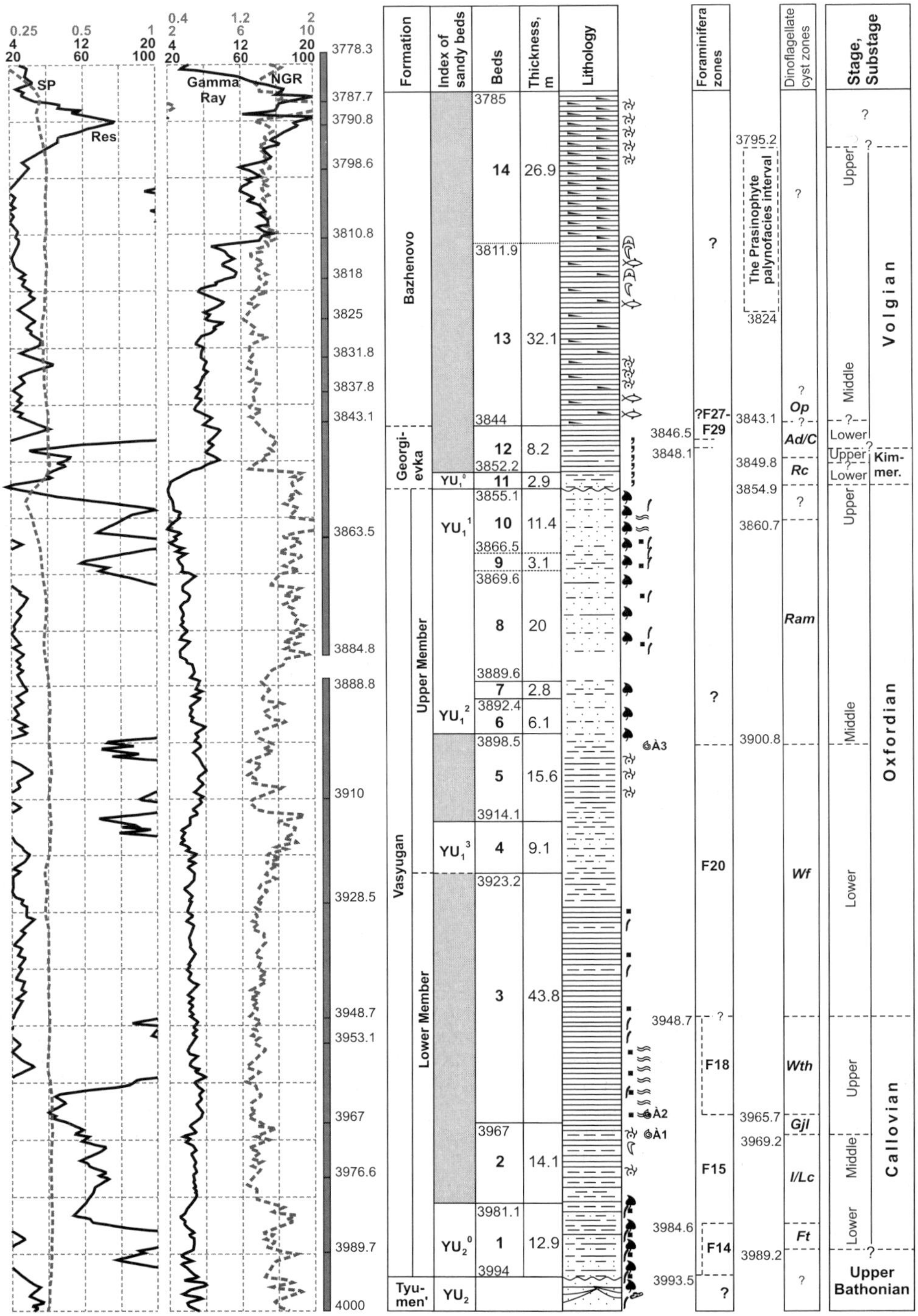

Fig. 6. Foraminiferal and dinoflagellate cyst biozonation and lithostratigraphical subdivision of the reference section of the Callovian and Upper Jurassic section in the Tyumenskaya superdeep well (SDW-6). Ammonites (identified by S. V. Meledina): A1, ? *Longaeviceras* sp.ind.; A2, *Quenstedtoceras* (*Soaniceras*) cf. *parvulum;* A3, *Cardioceratinae* gen. et sp. ind.

Dentalina sp. ind., *Lenticulina daschevskajae*, *L. memorabilissima* (Pl. 1, **2**), *L. subinvolvens*, *Oolina borealis*, *Ammodiscus* ex gr. *arangastachiensis* and *Haplophragmoides magnus* (Fig. 5).

The foraminiferal assemblage is composed of 26 species (10 and 16 species of agglutinated and calcareous foraminifers respectively) from a total of 21 genera. The genus *Recurvoides* (41–57%) dominates in this assemblage although there are considerable numbers of *Trochammina* (9–34%; locally <80%). *Ammobaculites* (<7%), *Lenticulina* (4–10%), *Globulina* (2–18%), *Pseudonodosaria* (1–10%) and *Dorothia* (2–7%) are also important in the foraminiferal assemblage. The upper half of the F18 Zone contains more specimens of *Globulina* (17–18%), which even becomes a dominant genus at this stratigraphical level in the central and southern parts of West Siberia. The agglutinated foraminifers are quantitatively predominant in the assemblage, representing from 50 to 88% of the fauna, but the calcareous foraminifera are taxonomically much more diverse.

In the north of East Siberia, the assemblage of the F18 Zone is characterized by 44 species of foraminifers (12 and 32 species of agglutinated and calcareous foraminifera respectively) that belong to 23 genera. The calcareous foraminifers make up a considerable part of the assemblage (5–40%). The assemblage of this Zone is characterized by abundant Polymorphinidae (*Globulina, Guttulina, Pyrulina*), that is less important in the underlying deposits (Lutova, 1981).

Age: Latest Callovian.

Stratigraphical position: In the studied section, the foramiferal assemblage of the *Conorboides taimyrensis, Globulina paalzowi* Zone has been established in the middle part of the lower member of Vasyugan Formation (the lower part of bed 3, interval 3965.7–3948.7 m) (Figs 5 and 6). The foraminiferal assemblages of the *Conorboides taimyrensis, Globulina paalzowi* Zone have been determined in the Tyumenskaya superdeep well, together with the Late Callovian ammonite (identified by S. V. Meledina) *Quenstedtoceras (Soaniceras)* cf. *parvulum* which was found at a depth of 3965.7 m (Fig. 6). Sediments of the F18 Zone represent a considerable part of the uppermost Upper Callovian stage and are widespread in the Abalak, Vasyugan, and Sigovoe formations of Siberia (Lebedev 1972) (Fig. 3). The typical foraminiferal assemblage of the F18 Zone in West and central regions of West Siberia coexists with the Late Callovian ammonites *Eboraciceras* spp. and *Quenstedtoceras* spp. (Bulynnikova *et al.* 1990).

Geographical distribution: West and East Siberia.

Ammobaculites tobolskensis, Trochammina oxfordiana Zone (F20)

Index-species: Ammobaculites tobolskensis (Pl. 1, **15**), *Trochammina oxfordiana*.

Characteristic species: Ammobaculites tobolskensis, Trochammina ex gr. *oxfordiana, Astacolus* ex gr. *igrimensis* (Pl. 1, **19**), *Eoguttulina* ex gr. *inovraclavensis* and *Recurvoides* ex gr. *eotrochus* (Fig. 5).

The foraminiferal assemblage of the F20 Zone also contains *Anmarginulina suprajurensis, Dentalina* sp. ind., *Geinitzinita praenodulosa, Globulina* ex gr. *paalzowi, Recurvoides scherkalyensis, R.* ex gr. *disputabilis, Eomarsonella paraconica, Trochammina* ex gr. *kumaensis, Ammodiscus* ex gr. *arangastachiensis, Pseudonodosaria* ex gr. *brandi, Ammobaculoides* ex gr. *primoris* and *Lenticulina subinvolvens* (Fig. 5).

Seventeen species of foraminifera are recorded from the F20 Zone (nine and eight species of agglutinated and calcareous foraminifers respectively) from a total of 14 genera. Different species of *Recurvoides* (33–79%) or *Trochammina* (<56%) dominate the assemblage. As in the assemblage of the F18 Zone, representatives of the genus *Globulina* have an important role (18–37%). Species of the genera *Ammobaculites* (<13%), *Lenticulina* (4–22%) and *Dentalina* (<11%) are also an important part of the fauna.

In northern East Siberia, the foraminiferal assemblages of the F20 Zone are represented by 13 species of agglutinated and 19 species of calcareous foraminifera from a total of 18 genera. Ratios of the content of agglutinated and calcareous foraminifera vary considerably depending on facies. Thus, in the Anabar region, the assemblage contains only agglutinated foraminifera, whereas the sections of East Taimyr show alternating assemblages dominated either by calcareous or by agglutinated foraminifera (Lutova 1981).

Age: Early Oxfordian.

Stratigraphical position: In the studied section, the foramiferal assemblage of the *Ammobaculites tobolskensis, Trochammina oxfordiana* Zone occurs in the upper part of the lower member and lower part of upper member of the

Vasyugan Formation (interval 3948.7–3908 m, upper 25 m of bed 3 of the lower member Vasyugan Formation and beds 4 and 5 of the upper member Vasyugan Formation) (Figs 5 and 6). In the Tyumenskaya Superdeep well (SDW-6), an Oxfordian–Early Kimmeridgian ammonite was found at a depth of 3900.8 m which was identified by S. V. Meledina as *Cardioceratinae* gen. et sp. ind. In some regions of West Siberia, the foraminiferal assemblage of the F20 Zone coexists with the Early Oxfordian ammonites of the *Cardioceras* (*Scarburgiceras*) spp. Zone and the *Cardioceras* s. str. Ammonite Zone (Nesterov 1991; Bulynnikova *et al.* 1990). The foraminiferal assemblages of the F20 Zone are widespread in the Abalak, Vasyugan and Sigovoe formations of Siberia.

Geographical distribution: West Siberia, Svalbard, Franz Josef Land and Arctic Alaska.

In the overlying part of the Tyumenskaya SDW-6 section, rare microforaminifers (*Subrecurvoides* sp.) were found in the middle part of the upper member of the Vasyugan Formation (bed 6, interval 3892.4–3898.5 m) (Figs 5 and 6). Depleted foraminiferal assemblages are also observed in the upper part of Georgievka Formation (interval 3846.5–3848.1 m). They are represented by rare *Trochammina* ex gr. *kumaensis, Lenticulina* sp. ind., *Kutsevella* sp. ind. and the microforaminifera *Microkutsevella* sp. (Pl. 1, **7**). The foraminifera and microforaminifera recovered from this interval typically represent the *Tolypammina virgula, Planularia pressula* F27 Zone and *Kutsevella haplophragmoides* F29 Zone (Upper Kimmeridgian–Lower Volgian) (Figs 3, 5 and 6).

Dinoflagellate cyst zonation

Analysis of the taxonomic composition of the dinoflagellate cysts (dinocyst) and their distribution in the cores of the Tyumenskaya superdeep well (SDW-6) provides a continuous sequence of dinocyst assemblages from the Lower Callovian to the Middle Volgian Substage for the first time in West Siberia (Figs 7 and 8). These assemblages have been used as a basis for a dinocyst zonation of the Middle (Callovian) and Upper Jurassic marine sediments in this section. The first appearance datum (FAD, the range base) and/or the last appearance datum (LAD, the range top) of dinocyst taxa and the acme of selected, stratigraphically significant species are recognized as bioevents in the dinocyst flora and serve as the criteria for formulating the zonation. The stratigraphical position of the dinocyst zones are calibrated with the regional ammonite succession and the foraminiferal assemblages found in this section. The Jurassic dinocyst zonations of the Russian Platform, the Pechora Basin and northern Siberia correlated with the Northwest European dinocyst standard zonation were also used as biostratigraphical reference points (Meledina *et al.* 1998; Riding *et al.* 1999; Shurygin *et al.* 2000). The following dinoflagellate cyst zonation is proposed for the Callovian to the Middle Volgian of the investigated section (Figs 9 and 10).

Fromea tornatilis Interval Zone (Ft)

Definition: The interval from the FADs of *Crussolia dalei, Chytroeisphaeridia hyalina, Gonyaulacysta jurassica* subsp. *adecta,* common *Fromea tornatilis,* FAD of *Impletosphaeridium varispinosum* in upper part to the range base of *Impletosphaeridium polytrichum* (Figs 7 and 9).

Characteristic assemblage: In addition to the species noted above, *Escharisphaeridia pocockii, Gonyaulacysta* sp., *Nannoceratopsis pellucida, Paragonyaulacysta* sp., *Pareodinia ceratophora* and *Pareodinia* sp. are recorded.

Age: Earliest Callovian.

Comments: This dinocyst zone is considered to be the equivalent to the *Fromea tornatilis* Interval Biozone erected for the lowermost Callovian (*herveyi* ammonite Zone of the European Standard) in the Russian Platform by Riding *et al.* (1999). The lower boundary of the *Fromea tornatilis* (Ft) Zone coincides with the base of the Callovian. In the SDW-6 section, this zone equates, approximately, with the upper half of the *Kutsevella memorabilis, Guttulina tatarensis* F14 Zone. The *Fromea tornatilis* Zone is analogous to the Boreal *Crussolia dalei–Paragonyaulacysta retiphragmata* dinocyst Zone found in the lowermost Callovian of East Siberia (Riding *et al.* 1999; Shurygin *et al.* 2000). Analysis of the characteristic assemblages of these dinocyst zones may help to correlate the lowermost Callovian of the Boreal and Subboreal provinces (Figs 6, 9 and 10).

Geographical distribution: The Russian Platform: the Moscow syncline and Pechora Basin, Lowermost Callovian; West Siberia: Shaim region, Abalak Formation, beds with ammonites *Cadoceratinae;* and Nyurolska Depression, the lower member of the Vasyugan Formation; the Tyumenskaya SDW-6 section: Pakhomovskaya Member of the Vasyugan Formation, middle part of bed 1, interval 3989.2–3984.6 m.

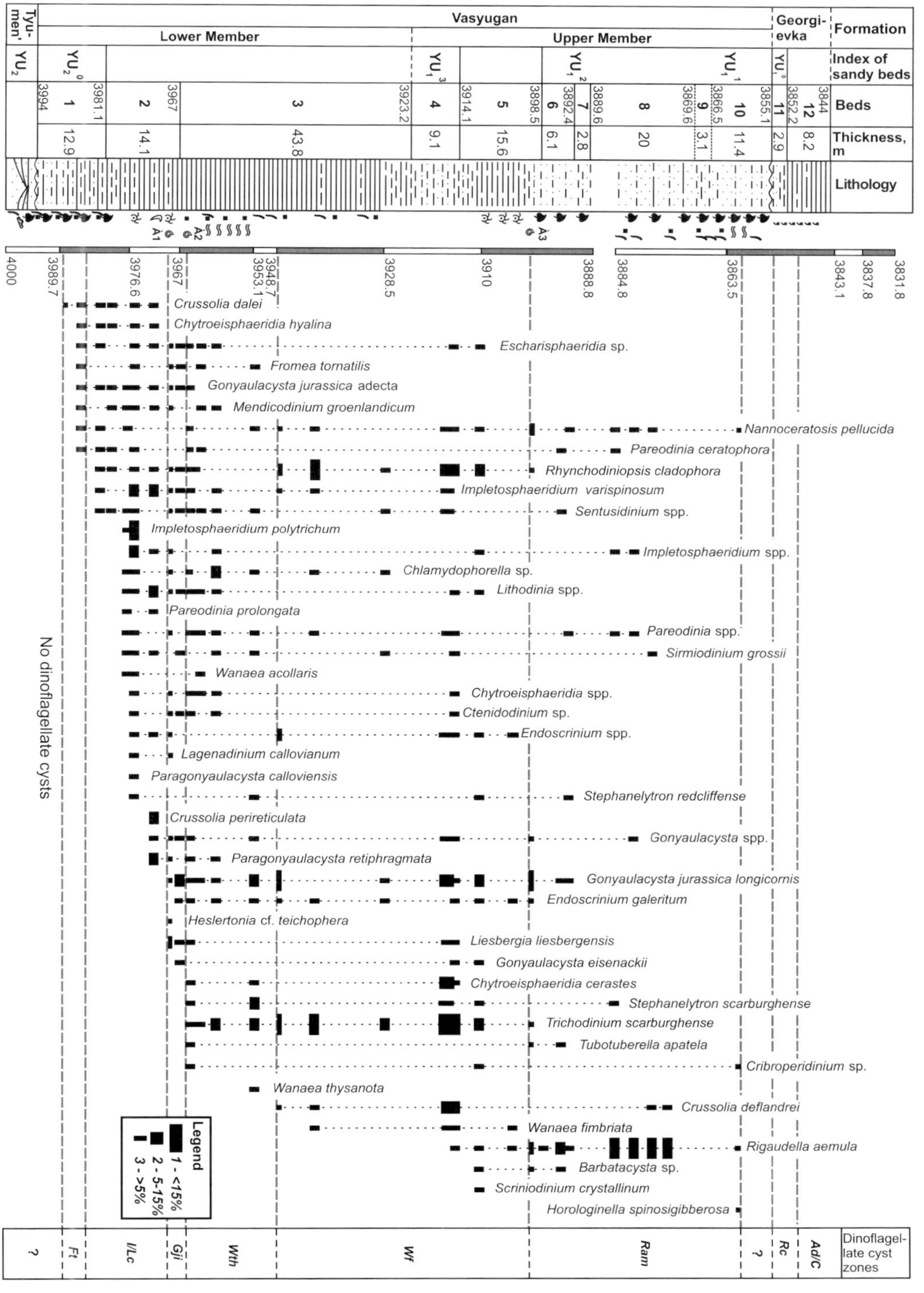

Fig. 7. Summary of the distribution of selected Callovian and Oxfordian dinoflagellate cysts and dinoflagellate cysts biozonation, Tyumenskaya superdeep well (SDW-6).

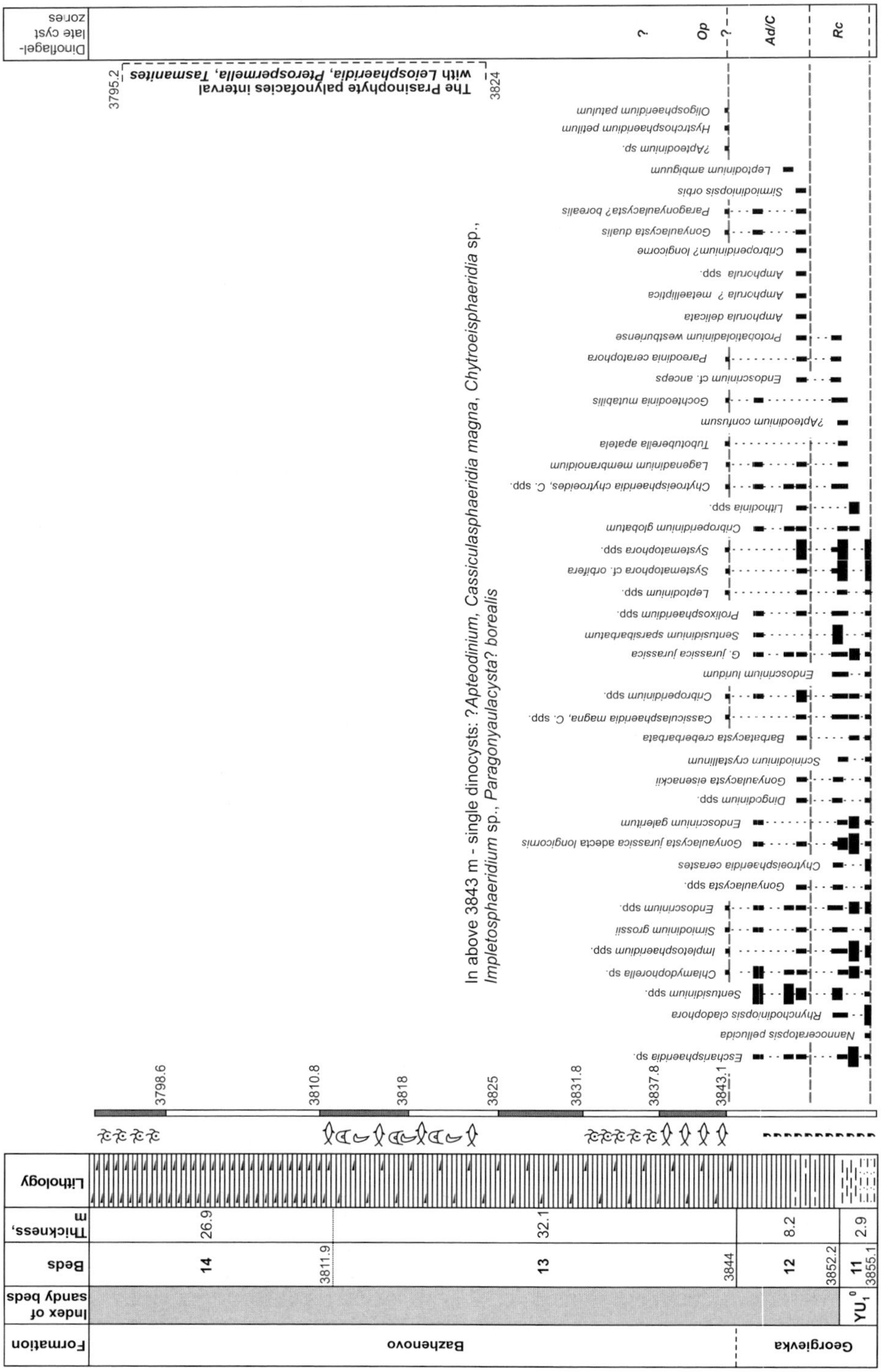

Fig. 8. Summary of the distribution of selected Kimmeridgian and Volgian dinoflagellate cysts and dinoflagellate cysts biozonation, Tyumenskaya superdeep well (SDW-6).

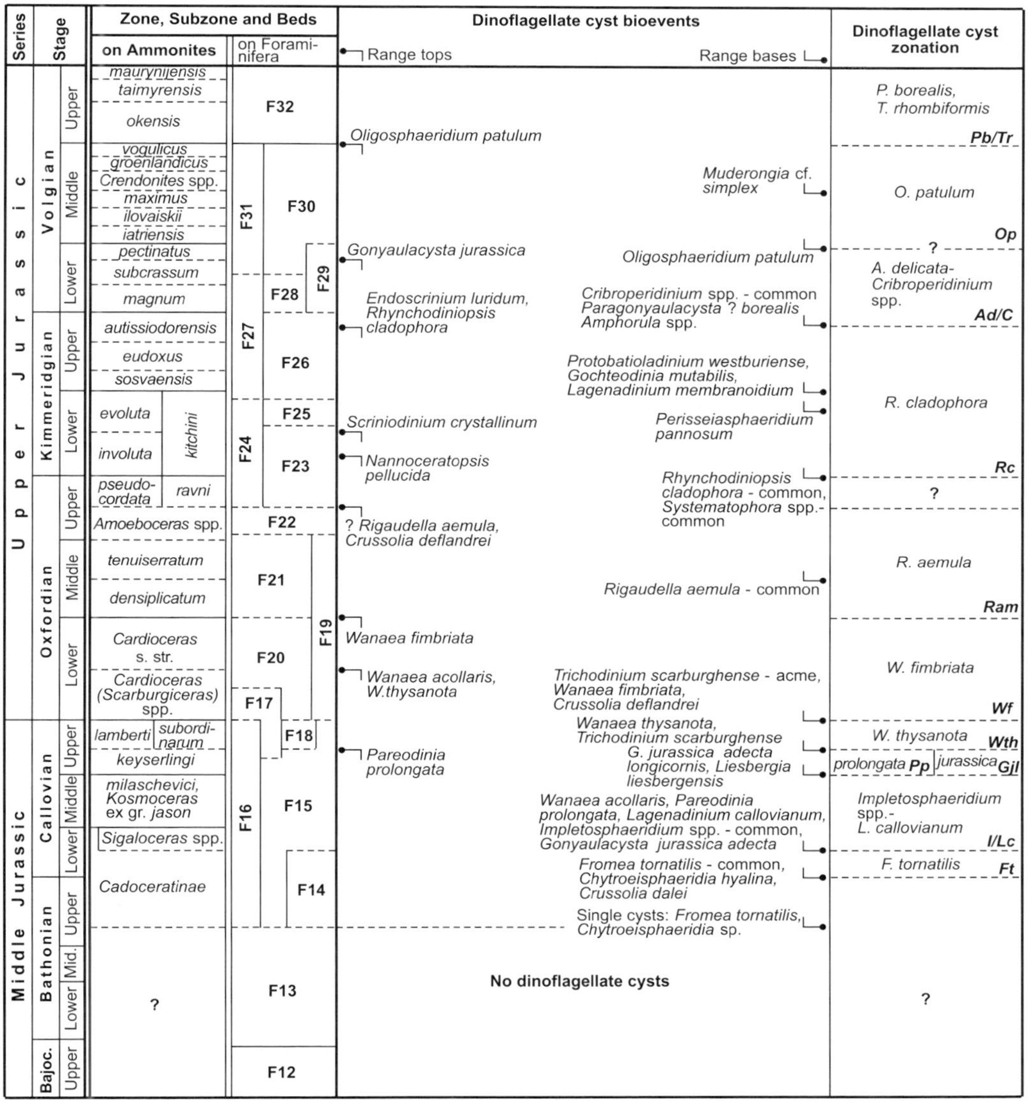

Fig. 9. Callovian to Volgian dinoflagellate cyst zonation and bioevents in West Siberia.

Impletosphaeridium spp., Lagenadinium callovianum Interval Zone (I/Lc)

Definition: The interval from the FAD of *Impletosphaeridium polytrichum* and *Lagenadinium callovianum* to the range base of *Gonyaulacysta jurassica* subsp. *adecta* var. *longicornis*. Within this dinocyst zone the abundance and diversity of *Impletosphaeridium* and *Sentusidinium* spp. increase while, in the upper half of the zone, *Mendicodinium groenlandicum* occurs regularly.

Characteristic assemblage: Chlamydophorella sp., *Chytroeisphaeridia hyalina, Crussolia dalei, Crussolia perireticulata*, *Endoscrinium* sp., *Escharisphaeridia* sp., *Gonyaulacysta jurassica* subsp. *adecta, Fromea tornatilis, Impletosphaeridium varispinosum* (common), *Impletosphaeridium polytrichum* (common), *Ctenidodinium* sp., *Lithodinia* sp., *Lagenadnium callovianum, Mendicodinium groenlandicum, Nannoceratopsis pellucida, Paragonyaulacysta retiphragmata, P. calloviensis, Pareodinia ceratophora, Pareodinia* cf. *prolongata, P.*

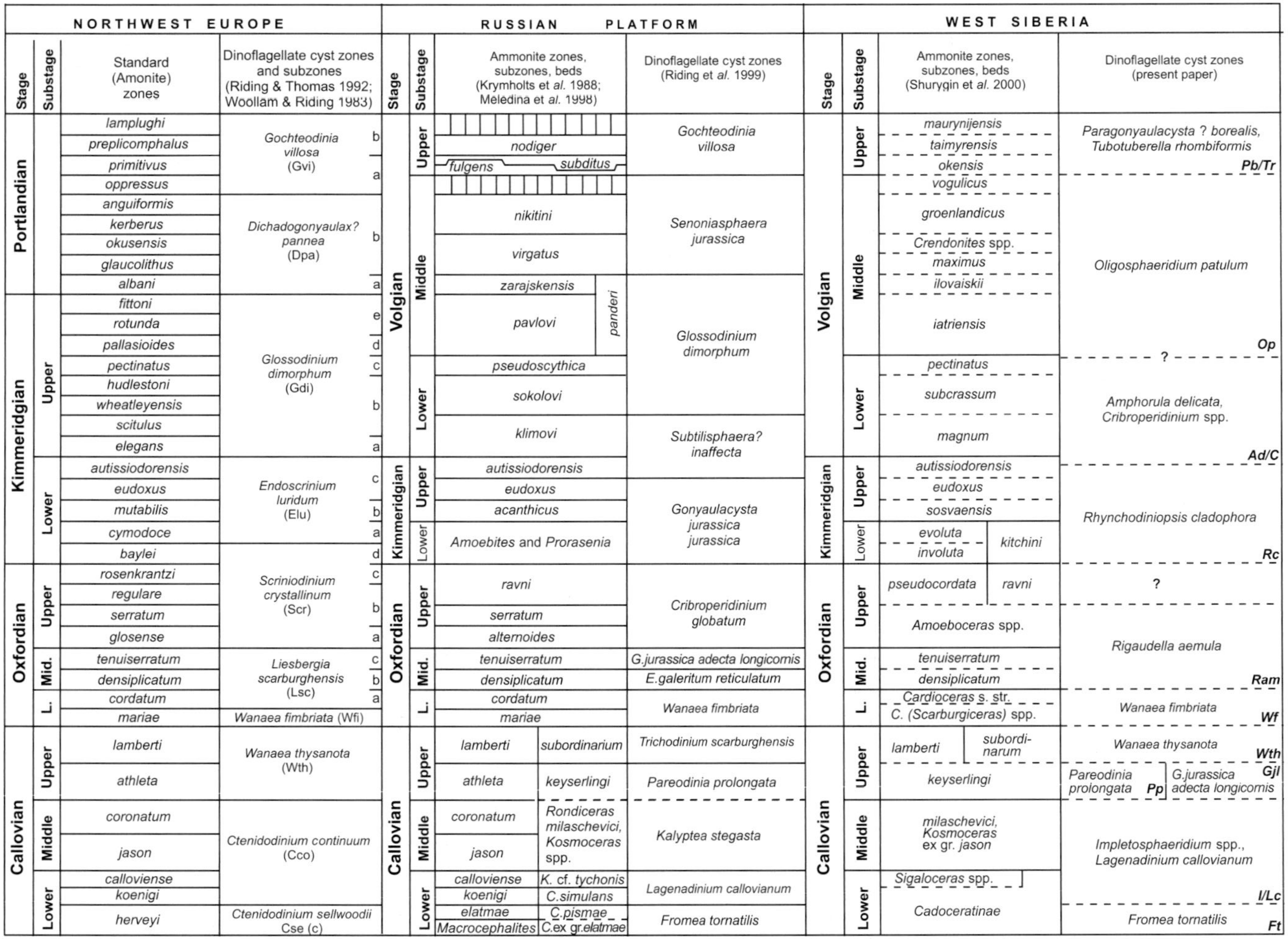

Fig. 10. Comparison and correlation of the Callovian to Volgian dinoflagellate cyst zonations of West Siberia, the Russian Platform and Northwest Europe.

halosa, *Rhynchodiniopsis cladophora*, *Sentusidinium* spp., *Sirmiodinium grossii*, *Stephanelytron redcliffense* and *Wanaea acollaris* (Figs 7, 9 and Pl. 3).

Age: Probably Early–Middle Callovian.

Comments: In the Russian Platform the first *Lagenadinium callovianum*, in combination with *Chlamydophorella*, *Impletosphaeridium*, and *Sentusidinium* spp., are determined from sediments with ammonites of the two upper zones (*koenigi* and *calloviense*) of the Lower Callovian. The first *Gonyaulacysta jurassica* subsp. *adecta* var. *longicornis* occurs (Fig. 10) in the base of the Upper Callovian (Ilyina 1991, 1998; Meledina *et al.* 1998; Riding *et al.* 1999). Based on this information, the *Impletosphaeridium* spp., *Lagenadinium callovianum* dinocyst Zone is thought to represent the upper part of the Lower Callovian (*koenigi* and *calloviense* ammonite Zones) and the Middle Callovian. The upper boundary of this dinocyst zone is placed by the discovery of the late Callovian ammonite *Longaeviceras* sp. ind. (identified by S. V. Meledina) and is tentatively correlated with the Middle–Upper Callovian boundary.

Geographical distribution: The *Impletosphaeridium* spp., *Lagenadinium callovianum* (I/Lc) Zone is established only in the Tyumenskaya SDW-6 section within the lower member of the Vasyugan Formation, top of the Pakhomovskaya Member, and most of the lower part of bed 2, interval 3984.6–3969.2 m (Figs 6 and 7).

Gonyaulacysta jurassica subsp. adecta var. longicornis Interval Zone (Gjl)

Definition: The interval from the range base and persistent occurrence of the index species to the FAD of *Trichodinium scarburghense* and/or *Wanaea thysanota*. Within this zone, *Ctenidodinium* cf. *continuum*, *Endoscrinium galeritum*, *Heslertonia* sp. and *Liesbergia liesbergensis* are recorded.

Characteristic assemblage: In addition to the above species, *Chlamydophorella* sp., *Chytroeisphaeridia*, spp., *Ctenidodinium* cf. *continuum*, *Escharisphaeridia* sp., *Fromea tornatilis*, *Gonyaulacysta eisenackii*, *Impletosphaeridium varispinosum*, *Lithodinia* sp., *Mendicodinium groenlandicum*, *Nannoceratopsis pellucida*, *Pareodinia* cf. *prolongata*, *Pareodinia* spp., *Rhynchodiniopsis cladophora*, *Sentusidinium* spp., *Sirmiodinium grossii*, *Tubotuberella apatela* are also recorded (Fig. 7).

Age: Late Callovian

Comments: The FAD of the index species has been reported from the lower Upper Callovian of some locations in northwest Europe and Arctic Canada (Davies 1983; Riding 1984; Riding & Thomas 1992; Smelror & Below 1993). The appearance of these dinocysts at the base of the Upper Callovian coincides with ammonites of the *athleta* Zone in the Moscow syncline and *keyserlingi* Zone in the Pechora Basin as reported by Ilyina (1991, 1998). This zone is the equivalent of the *Pareodinia prolongata* Interval Biozone erected for the Upper Callovian (*athleta* or *keyserlingi* ammonite Zones) on the Russian Platform by Riding *et al.* (1999). In the Tyumenskaya SDW-6 section the boundaries of the *Gonyaulacysta jurassica* subsp. *adecta* var. *longicornis* Zone in the lower part of the Upper Callovian are confirmed by the ammonite occurrences. The upper boundary of this zone is correlated with the lower boundary of the *Conorboides taimyrensis*, *Globulina paalzowi* F18 Zone (Figs 5, 6).

Geographical distribution: The Tyumenskaya SDW-6 section is the reference section of this dinocyst zone: the lower member of the Vasyugan Formation, top of bed 2 and the lowermost part of bed 3, interval 3969.2–3965.7 m (Fig. 6).

Wanaea thysanota Interval Zone (Wt)

Definition: The interval from the FAD and consistent occurrence of *Trichodinium scarburghense*, the appearance of *Wanaea thysanota* in base or within this zone to FAD of *Crussolia deflandrei* and/or *Wanaea fimbriata* at its top.

Characteristic assemblage: In addition to the species above, *Chlamydophorella* sp. (common), *Chytroeisphaeridia cerastes*, *Chytroeisphaeridia* spp., *Cribroperidinium* sp., *Crussolia* sp., *Ctenidodinium continuum*, *Endoscrinium galeritum*, *Escharisphaeridia* sp., *Fromea tornatilis*, *Gonyaulacysta jurassica* subsp. *adecta* var. *longicornis*, *Impletosphaeridium varispinosum*, *Impletosphaeridium polytrichum* (rarely), *Lithodinia* sp., *Liesbergia liesbergensis*, *Mendicodinium groenlandicum*, *Nannoceratopsis pellucida*, *Paragonyaulacysta retiphragmata*, *Pareodinia ceratophora*, *Rhynchodiniopsis cladophora*, *Sentusidinium* spp., *Stephanelytron redcliffense*, *S. scarburghense* and *Wanaea acollaris* are also recorded (Figs 7, 9 and Pl. 4).

Age: Late Callovian.

Comments: The *Wanaea thysanota* Zone is considered to be equivalent to the *Trichodinium scarburghense* Interval Biozone erected for the

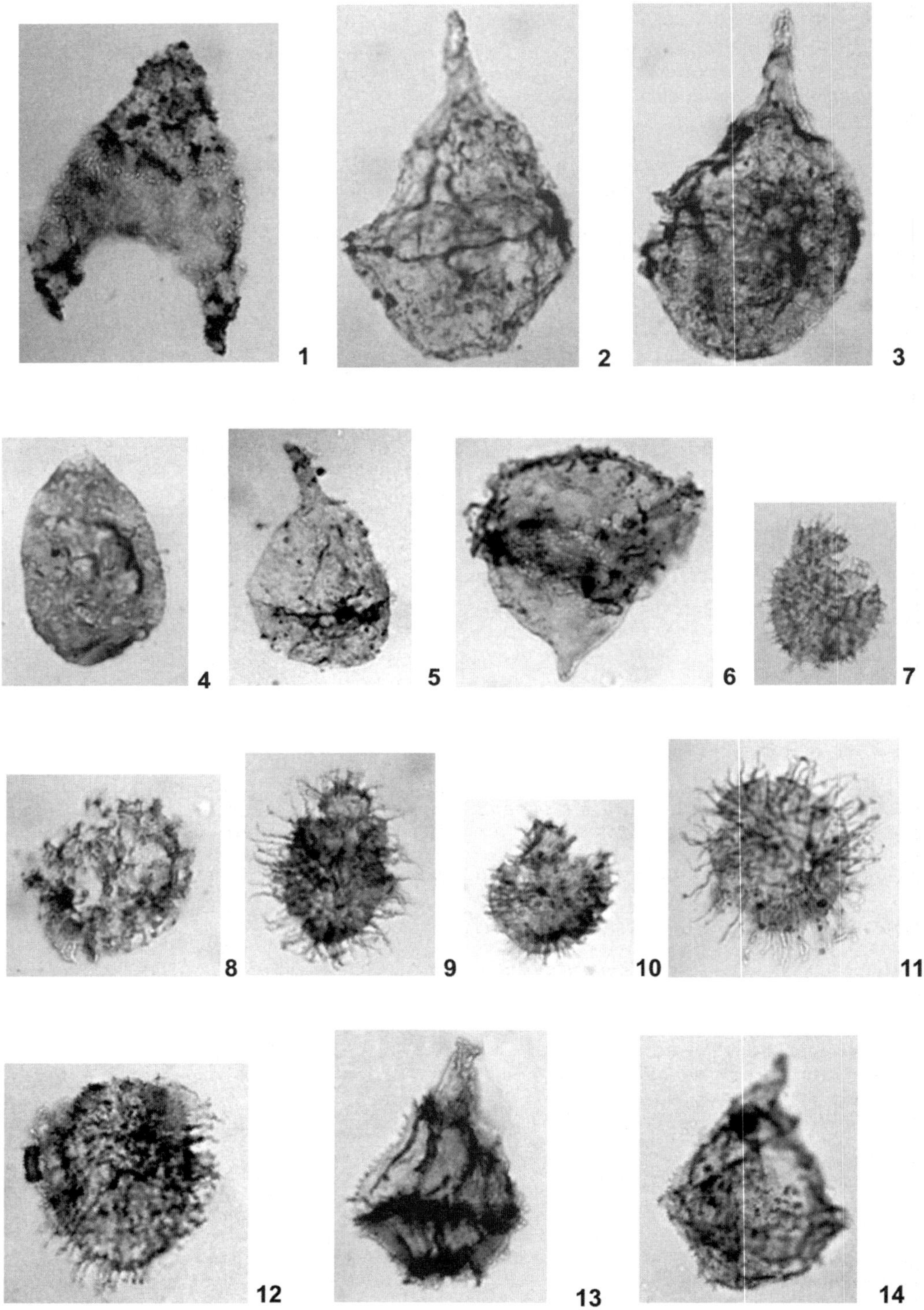
1
2
3
4
5
6
7
8
9
10
11
12
13
14

Upper Callovian (*lamberti* or *subordinarium* ammonite Zones) on the Russian Platform by Riding *et al.* (1999). *Trichodinium scarburghense* is used as the index species for an Upper Callovian dinocyst zone in Eastern Greenland, which is analogous to the *Trichodinium scarburghense*, *Wanaea thysanota* Zone in the Barents Sea shelf sections of Smelror & Below (1993). The *Wanaea thysanota* Zone has been tentatively recognized within the *subordinarium* ammonite Zone in the Upper Callovian sections of several wells in the north of West Siberia (Ilyina 1998; Shurygin *et al.* 2000). In the Tyumenskaya SDW-6 section this dinozone approximately corresponds with the *subordiinarium* ammonite Zone and *Conorboides taimyrensis*, *Globulina paalzowi* F18 Zone. The lower boundary of this dinocyst zone is dated by the discovery of the Late Callovian ammonite *Quenstedtoceras (Soaniceras)* cf. *parvulum* that has been identified by S. V. Meledina (Figs 6 and 10).

Geographical distribution: West Siberia: Zapadno-Novogodnaya-210 and Stakhanovskaya-910 wells as reported by Ilyina (1998). In the Tyumenskaya SDW-6 succession it is present in the lower member of the Vasyugan Formation, lower half of bed 3 (without the lowermost part), interval 3965.7–3948.7 m (Figs 6 and 7).

Wanaea fimbriata Range Zone (Wf)

Definition: The interval from the FAD of *Crussolia deflandrei*, acme of *Trichodinium scarburghense* and the range base to the range top of *Wanaea fimbriata*.

Characteristic assemblage: Recorded within this zone are *Adnatosphaeridium caulleryi*, *?Barbatacysta creberbarbata*, *Chlamydophorella* sp., *Chytroeisphaeridia cerastes*, *Cribroperidinium* sp., *Crussolia deflandrei*, *Ctenidodinium* sp., *Endoscrinium galeritum*, *Endoscrinium* sp., *Escharisphaeridia* sp., *Fromea tornatilis*, *Gonyaulacysta jurassica adecta* var. *longicomis*, *G. eisenackii*, *Gonyaulacysta* sp., *Impletosphaeridium varispinosum*, *Liesbergia liesbergensis*, *Lithodinia* spp., *Nannoceratopsis pellucida*, *Pareodinia* spp., *Prolixosphaeridium* sp., *Rhynchodiniopsis cladophora*, *Rigaudella aemula*, *Sentusidinium* spp., *Scriniodinium crystallinum*, *Sirmiodinium grossii*, *Stephanelytron scarburghense*, *S. redcliffense*, *Trichodinium scarburghense*, *Tubotuberella dangeardii* and *Wanaea fimbriata* (Figs 7, 9 and Pl. 4).

Age: Early Oxfordian

Geographical distribution: West Siberia: Zapadno-Novogodnaya, 210 wells; Stakhanovskaya-910 wells; Yarainerskaya, 23 wells; Zapadno-Purpeiskaya, 710 wells (Beisel *et al.* 2002). In the Tyumenskaya SDW-6 succession it is represented by the upper part of the lower member of the Vasyugan Formation and the bottom of its upper member, upper half of bed 3, beds 4 and 5 (without the uppermost part), interval 3948.7–3900.8 m (Figs 6 and 7).

Rigaudella aemula Interval Zone (Ram)

Definition: The interval from the LAD of *Wanaea fimbriata*, to the range top of *Rigaudella aemula*, including the last evolutionary peak of *Nannoceratopsis pellucida*.

Characteristic assemblage: The assemblage present within this zone includes *?Barbatacysta creberbarbata*, *Cassiculosphaeridia* sp., *Crussolia deflandrei*, *Cribroperidinium* sp., *Endoscrinium* sp., *Gonyaulacysta jurassica adecta* var. *longicornis*, *G. jurassica* subsp. *jurassica*,

Plate 3. Selected Callovian to Middle Volgian dinoflagellate cysts from the Tyumenskaya superdeep well (SDW-6), north of West Siberia. (All specimens ×550). **1–14**, Vasyugan Fm, Lower Member. **1–11**, Lower Callovian (upper part)–Middle Callovian, *Impletosphaeridium* spp., *Lagenadinium callovianum* (I/Lc) Zone. **1**, **5**, **9**, **11**, Interval 3976.6–3989 m; **1**, 4.8 m from top of core interval, sample 1832/8783; **5**, **9**, **11**, top of core interval, sample 1811/8785: **2–4**, **6–8**, **10**, Interval 3967–3976.6 m; **2**, **3**, **6**, 5 m from bottom of core interval, sample 1784/8787: **4**, **7**, **8**, **10**, 1 m from bottom of core interval, sample 1804/8786: **12–14**, Upper Callovian (*keyserlingi* ammonite Zone), *Gonyaulacysta jurassica* subsp. *adecta* var. *longicornis* (Gjl) Zone: **12–14**, Interval 3967–3976.6 m; **12**,**14**, top of core interval, sample 1752/8789; **13**, bottom of core interval, sample 1764/8788.

1, *Nannoceratopsis pellucida* Deflandre, 1939, specimen 169SDW-6; **2**, *Crussolia dalei* Smelror & Arhus 1989, specimen 144SDW-6; **3**, *Crussolia perireticulata* Arhus *et al.*, 1989, specimen 142SDW-6; **4**, *Fromea tornatilis* (Drugg 1978) Lentin & Williams, 1981, specimen 151SDW-6; **5**, *Pareodinia ceratophora* Deflandre, 1947, specimen 163SDW-6; **6**, *Wanaea acollaris* Dodekova, 1975, specimen 148SDW-6; **7**, *Sentusidinium* sp. or *Impletosphaeridium* sp., specimen 147SDW-6; **8**, *Lagenadinium callovianum* Piel, 1985, specimen 150SDW-6; **9**, *Impletosphaeridium polytrichum* (Valensi 1947) Islam, 1993, specimen 166SDW-6; **10**, *Sentusidinium* sp., specimen 160SDW-6; **11**, *Impletosphaeridium polytrichum* (Valensi 1947) Islam, 1993, specimen 167SDW-6; **12**, *Impletosphaeridium varispinosum* (Sarjeant 1959) Islam 1993, specimen 131SDW-6; **13**, *Gonyaulacysta jurassica* (Deflandre 1939) Norris & Sarjeant, 1965 subsp. *adecta* Sarjeant, 1982, specimen 138SDW-6; **14**, *Gonyaulacysta jurassica* subsp. *adecta* var. *longicornis* (Deflandre 1938) Downie & Sarjeant, 1965 emend. Sarjeant, 1982, specimen 132SDW-6.

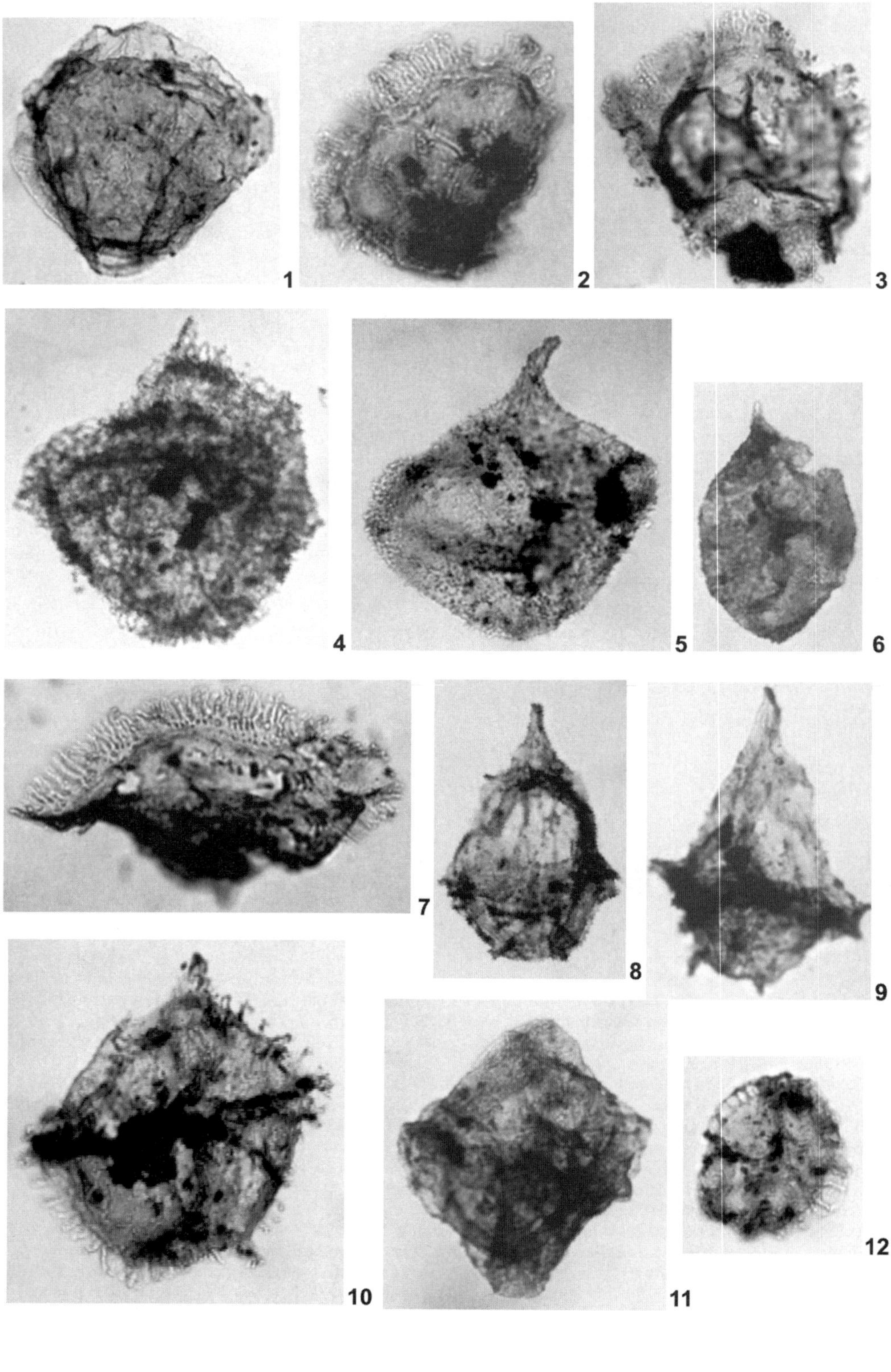
1
2
3
4
5
6
7
8
9
10
11
12

Gonyaulacysta sp., *Horologinella spinosigibberosa, Impletosphaeridium* sp., *Imbatodinium* sp., *Nannoceratopsis pellucida, Pareodinia ceratophora, Pareodinia* sp., *Rhynchodiniopsis cladophora, Rigaudella aemula* (Pl. 5, **1**; common), *Sentusidinium* sp., *Sirmiodinium grossii, Stephanelytron redcliffense* and *Tubotuberella apatela* (Figs 7 and 9).

Age: Middle and Late Oxfordian.

Comments: In northwest Europe the acme of *Rigaudella aemula* is characteristic of the Middle Oxfordian, the upper boundary of which limits the range top of this taxon (Riding 1987). The last peak in evolution of *Nannoceratopsis pellucida* also falls in the same interval. However, *Rigaudella aemula* has been reported from the transitional Upper Oxfordian–Kimmeridgian deposits in northern Yorkshire (Thomas & Cox 1988). In the north of West Siberia the stratigraphic interval represented by the *Rigaudella aemula* Zone is determined by the foraminiferal assemblages and was established within the Middle–lower Upper Oxfordian (Ilyina 1998). Recent studies of dinocyst and foraminifera distributions in the Oxfordian section of the Mykhpaiskaya well-723 has suggested an extension of the stratigraphical interval of this dinocyst zone to the upper boundary of the *Recurvoides disputabilis* F22 Zone in the Upper Oxfordian (Fig. 3). Based on this, the extent of the *Rigaudella aemula* Zone may be defined, with a degree of uncertainty, as the Middle and Upper Oxfordian (probably without the *ravni* ammonite Zone).

Geographical distribution: West Siberia: Zapadno-Novogodnyaya well-210, Stakhanovskaya well-910, Yarainerskaya well-23 and other wells in the north of the region; Novonadezhdinskaya-1 well in the southeast of this region. In the Tyumenskaya SDW-6 succession it is recorded within the upper member of the Vasyugan Formation, top of bed 5, beds 6–9, lower half of bed 10, interval 3900.8–3860.7 m (Figs 6 and 7).

Rhynchodiniopsis cladophora Interval Zone (Rc)

Definition: The interval from the last peak of the index species, and rare *Endoscrinium luridum*, to the range tops of these taxa. In the lower part of the (Rc) Zone, *Nannoceratopsis pellucida* and *Scriniodinium crystallinum* disappear. Within this zone *Gonyaulacysta jurassica* subsp. *jurassica* occurs regularly, whereas its upper part is marked by FADs of *Apteodinium confusum, Dingodinium jurassicum, Gochteodinia mutabilis, Lagenadinium ?membranoidium, Endoscrinium anceps* and *Protobatioladinium westburiense* and by markedly less frequent occurrences of the index species (Figs 8 and 9).

Characteristic assemblage: The dinocyst assemblage has a rich and diverse composition and includes *Apteodinium confusum, Barbatacysta creberbarbata, Cassiculosphaeridia* cf. *magna, Chlamydophorella* sp., *Chytroeisphaeridia chytroeides, Cribroperidinium* cf. *globatum, Cribroperidinium* spp., *Dingodinium jurassicum, Dingodinium* sp., *Endoscrinium luridum, Endoscrinium* cf. *anceps, Endoscrinium* spp., *Escharisphaeridia* sp., *Gochteodinia mutabilis, Gonyaulacysta jurassica subsp. jurassica, G. jurassica* subsp. *adecta* var. *longicornis, G. eisenackii, G.* cf. *fastigiata*, G. cf.

Plate 4. 1–12, Vasyugan Fm: **1–4**, **6**, **8**, **9**, **12**, Lower Member; **5**,**7**, **10**, **11**, Upper Member; **1**, **2**, **4**, **12**, Upper Callovian. **1**, *Gonyaulacysta jurassica* subsp. *adecta* var. *longicornis* (Gjl) Zone. Interval 3967–3976.6 m; 7.4 m from the bottom of core interval, sample 1764/8788: **2**, **4**, **12**, Upper Callovian (upper part), *Wanaea thysanota* (Wth) Zone: **4**, **12**, Interval 3953.1–3967 m; **12**, 1.5 m from the bottom of core interval, sample 1732/8791; **4**, 7 m from the top of core interval, sample 1708/8793; **2**, Interval 3948.7–3953.1 m; 0.5 m from the bottom of core interval, sample 1658/8794; **3**, **5–11**, Lower Oxfordian, *Wanaea fimbriata* (Wf) Zone; **3**, **6**, **9**, Interval 3928.5–3948.7 m; **9**, bottom of core interval, sample 1652/8795; **3**, **6**, **7**, 4 m from the bottom of core interval, sample 1615/8796; **5**, **7**, **8**,**10**, **11**, Interval 3910–3928.5 m; **5**, **7**, **10**, 5.3 m from the top of core interval, sample 1543/8799; **8**, 0.6 m from the top of core interval, sample 1582/8797; **11**, 0.2 m from the top of core interval, sample 1515/8800.

1, *Endoscrinium luridum* (Deflandre 1939) Gocht, 1970, specimen 134SDW-6; **2**, *Wanaea thysanota* Woollam, 1982, specimen 68SDW-6; **3**, *Wanaea fimbriata* Sarjeant, 1961, specimen100SDW-6; **4**, *Trichodinium scarburghense* (Sarjeant 1964) Williams *et al.*, 1993, specimen 119SDW-6; **5**, *Liesbergia liesbergensis* Berger, 1986, specimen 86SDW-6; **6**, *Crussolia deflandrei* Wolfard & Van Erve, 1981, specimen 101SDW-6; **7**, *Wanaea fimbriata* Sarjeant, 1961, specimen 9SDW-6; **8–9** *Gonyaulacysta jurassica* subsp. *adecta* var. *longicornis* (Deflandre 1938) Downie & Sarjeant, 1965 emend. Sarjeant, 1982; **8**, specimen 99SDW-6; **9**, specimen 102SDW-6; **10**, *Rhynchodiniopsis cladophora* (Deflandre 1939) Below, 1981, specimen 83SDW-6; **11**, *Endoscrinium galeritum* (Deflandre 1939) Vozzhennikova, 1967, specimen 96SDW-6; **12**, *Stephanelytron scarburghense* Sarjeant, 1961, specimen 127SDW-6.

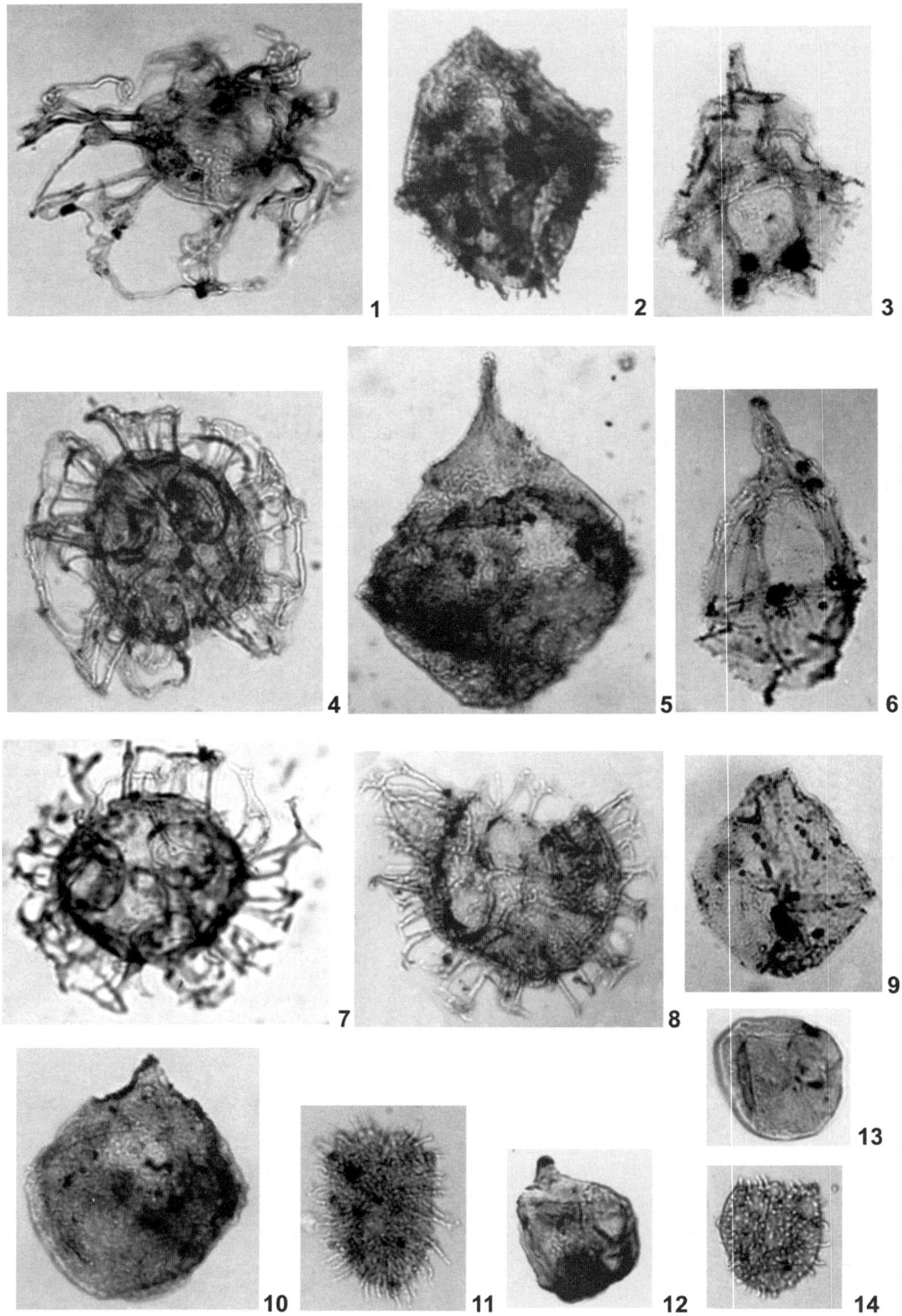
1
2
3
4
5
6
7
8
9
10
11
12
13
14

pectinigera, *Heslertonia* sp., *Imbatodinium kondratjevii*, *Impletosphaeridium* spp., *Lagenadinium ?membranoidium*, *Leptodinium* sp., *Lithodinia* spp., *Nannoceratopsis pellucida*, *Pareodinia* sp., *Prolixosphaeridium anasillum*, *Prolixosphaeridium* spp., *Protobatioladinium westburiense*, *Rhynchodiniopsis cladophora*, *Scriniodinium crystallinum*, *Sentusidinium sparsibarbatum*, *Sentusidinium* spp., *Sirmiodinium grossii*, *Stephanelytron* sp., *?Subtilisphaera* sp., *Systematophora* cf. *areolata*, *S.* cf. *davey*, *Systematophora* cf. *orbifera* and *Systematophora* spp. (common) and *Tubotuberella apatela*. (Figs 8, 9 and Pl. 5).

Age: Early and Late Kimmeridgian (probably without the latest part).

Comments: A similar dinocyst zone has been recognized in the Kimmeridgian (*baylei-eudoxus* ammonite Zones) for the Boreal regions of NW Europe (Fisher & Riley 1980). In the Boreal Kimmeridgian of the Russian Platform this dinocyst zone corresponds to the *Gonyaulacysta jurassica* subsp. *jurassica* Interval Biozone of Riding *et al.* (1999). In West Siberia, the *Rhynchodiniopsis cladophora* Zone was first defined in the Georgievka Formation of the Novonadezhdinskay-1 well for the Kimmeridgian Stage with a proposed upper boundary at the top of the *eudoxus* ammonite Zone (Shurygin *et al.* 2000). However, the range tops of *Endoscrinium luridum* and *Rhynchodiniopsis cladophora* also are considered as characteristic bioevents of the dinoflagellate flora in the *autissiodorensis* ammonite Zone in the Lower Kimmeridgian (*sensu anglico*) of England, Denmark and other regions of NW Europe (Riding & Thomas 1992; Poulsen 1993; Riding *et al.* 1999); Figures 9 and 10. Based of this, the disappearance of *Rhynchodiniopsis cladophora* and *Endoscrinium luridum* at the top of this dinocyst zone in the SDW-6 succession allows the suggestion that the upper boundary of the zone may lie within the *autissiodorensis* ammonite Zone of the Upper Kimmeridgian (according to the Jurassic Boreal zonal scheme of Zakharov *et al.* 1997). The latter corresponds to the equivalent ammonite zone of the Kimmeridgian in the European standard. As a result, the *Rhynchodiniopsi cladophora* dinocyst Zone may be defined with a degree of uncertainty in West Siberia as the Lower and Upper Kimmeridgian (without the uppermost beds). In the Kimmeridgian of the Russian Platform, it corresponds to the *Gonyaulacysta jurassica* subsp. *jurassica* biozone and to a part of the *Subtilisphaera inaffecta* biozone (until LAD of *Rhynchodiniopsis cladophora*) of Riding *et al.* (1999); see Figures 9 and 10.

Geographical distribution: West Siberia: SE of region, Novonadezhdinskaya-1 well. In the Tyumenskaya SDW-6 succession it is represented by the Georgievka Formation, upper part of the Barabinsk Member, bed 11 (the largest upper part), and the bottom of bed 12, interval, 3854.9–3849.8 m (Figs 6 and 8).

Amphorula delicata, Cribroperidinium spp. Interval Zone (Ad/C)

Definition: The zone is represented by the interval from the LADs of *Rhynchodiniopsis cladophora* and *Endoscrinium luridum*, and the FAD of *Amphorula delicata* to the range base of *Oligosphaeridium patulum*. Within this dinocyst zone *Cribroperidinium* spp. are regularly recorded.

Characteristic assemblage: The assemblage contains *Amphorula delicata*, *Amphorula*

Plate 5. **1**, Vasyugan Fm, Upper Member. Middle Oxfordian, *Rigaudella aemula* (Ram) Zone. Interval 3863.5–3884.8 m; 7 m from the bottom of core interval, sample 1350/8808; **2–14**, Georgievka Fm. Kimmeridgian, *Rhynchodiniopsis cladophora* (Rc) Zone. Interval 3843.1–3863.5 m; **2**,**12**, 12.5 m from bottom of core interval, sample 1174/8812; **3**, **8**, **9**, **13**, 9.5 m from the top of core interval, sample 1777/8816; **4**, **5**, **7**, **14**, 11.5 m from the top of core interval, sample 1199/8814; **6**, 10.2 m from the top of core interval, sample 1189/8815; **10**, 9 m from the top of core interval, sample 1169/8817; **11**, 11.8 m from the top of core interval, sample 1182/8813.

1, *Rigaudella aemula* (Deflandre 1939) Below, 1982, specimen 27SDW-6; **2**, *Rhynchodiniopsis cladophora* (Deflandre 1939) Below, 1981, specimen 6SDW-6; **3**, *Gonyaulacysta jurassica* subsp. *jurassica* Deflandre, 1939 emend. Poulsen, 1991, specimen 64SDW6; **4**, *Systematophora* cf. *orbifera* Klement, 1960, specimen 21SDW-6; **5**, *Cribroperidinium* sp., specimen 19SDW-6; **6**, *Gonyaulacysta jurassica* subsp. *adecta* var. *longicornis* (Deflandre 1938) Downie & Sarjeant, 1965 emend. Sarjeant, 1982, specimen 25SDW-6; **7**, *Systematophora* sp., specimen 20SDW-6; **8**, *Systematophora* sp., specimen 59SDW-6; **9**, *Cribroperidinium* sp., specimen 61SDW-6; **10**, *Cribroperidinium globatum* (Gitmes & Sarjeant) Helenes, 1984, specimen 82SDW-6; **11**, *Prolixosphaeridium anasillum* Erkmen & Sarjeant, 1980, specimen 16SDW-6; **12**, *Protobatioladinium westburiense* Nohr-Hansen, 1986, specimen 3SDW-6; **13**, *Chytroeisphaeridia chytroeides* (Sarjeant 1962) Downie & Sarjeant, 1965, specimen 60SDW-6; **14**, *Sentusidinium* sp., specimen 24SDW-6.

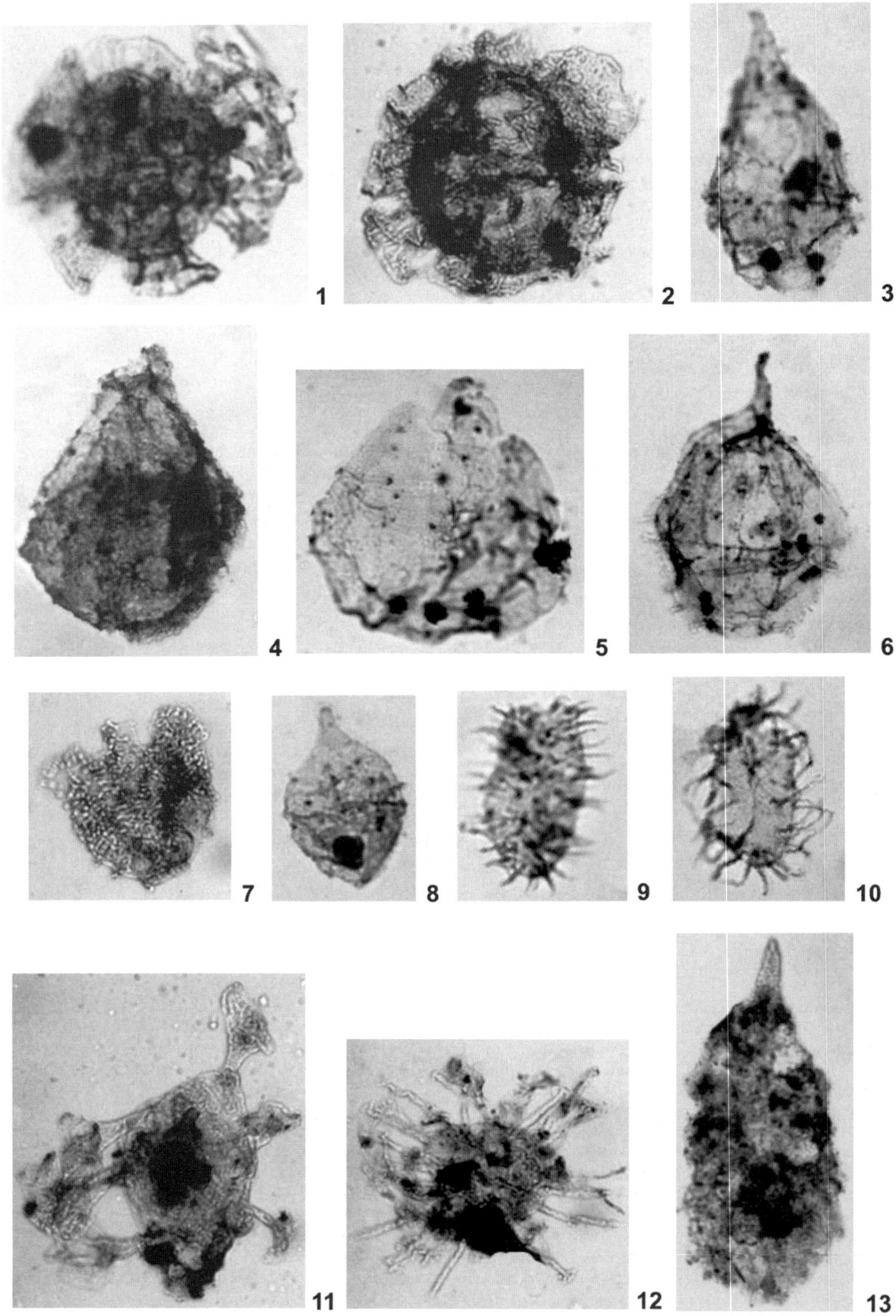

metaelliptica, *Amphorula* sp., *Barbatacysta creberbarbata*, *Cassiculosphaeridia* sp., *Chlamydophorella* sp. (common), *Chytroeisphaeridia chytroeides*, *Chytroeisphaeridia* sp., *Cribroperidinium globatum*, *Cribroperidinium longicorne*, *Cribroperidinium* spp. (common), *Dingodinium* sp., *Endoscrinium anceps*, *Endoscrinium* sp., *Escharisphaeridia*, *Gochteodinia mutabilis*, *Gonyaulacysta jurassica* subsp. *adecta* var. *longicomis*, *G. jurassica* subsp. cf. *jurassica*, *Gonyaulacysta* cf. *dualis*, *Gonyaulacysta pectinigera*, *Gonyaulacysta* spp., *Impletosphaeridium polytrichum*, *Impletosphaeridium tribuliferum*, *Lagenadinium ?membranoidium*, *Leptodinium ambiguum*, *Lithodinia* sp., *Paragonyaulacysta ?borealis*, *Pareodinia ceratophora*, *Prolixosphaeridinium* spp., *Protobatioladinium westburiense*, *Sentusidinium* spp. (common), *Sirmiodinium grossii*, *Sirmiodiniopsis orbis*, *Subtilisphaera? inaffecta*, *Systematophora* cf. *areolata*, *Systematophora* cf. *orbifera*, *Systematophora* spp. (common) and *Tubotuberella apatela* (Figs 8, 9 and Pl. 6).

Age: Latest Kimmeridgian–Early Volgian.

Comments: This dinocyst zone is only recognized in the Tyumenskaya SDW-6 succession and may be regarded as a provisional zonal unit. The diverse taxonomic composition of the assemblage, the base of which is represented by dinocysts typical of the Kimmeridgian and the Lower Volgian of the Russian Platform and NW Europe (Riding *et al.* 1999), tentatively allows this dinocyst zone to be placed in the uppermost Kimmeridgian to Lower Volgian interval. The boundaries of this zone have been defined informally.

Geographical distribution: West Siberia, the SDW-6 section: the upper part of the Georgievka Formation and the lowermost part of the Bazhenovo Formation, top of bed 12 and the lower part of bed 13 interval, 3849.8–3843.1 m (Figs 6 and 8).

Oligosphaeridium patulum Range Zone (Op)

Definition: The interval from the range base to the range top of *Oligosphaeridium patulum*.

Characteristic assemblage: In the Tyumenskaya SDW-6 section, the typical assemblage of this dinocyst zone, in addition to the presence of *Oligosphaeridium patulum* (Pl. 6, **11**), includes *Apteodinium* sp., *Cassiculosphaeridia magna*, *Chlamydophorella* sp., *Circulodinium* sp., *Chytroeisphaeridia chytroeides*, *Cribroperidinium* sp., *Endoscrinium* sp., *Gochteodinia mutabilis*, *Gonyaulacysta* cf. *dualis*, *Hystrichosphaeridium petilum*, *Impletosphaeridium tribuliferum*, *Impletosphaeridium* spp., *Lagenadinium ?membranoidium*, *Paragonyaulacysta ?borealis*, *Pareodinia ceratophora*, *Sirmiodinium grossii* and *Tubotuberella apatela*. This assemblage is typical for the base of the *Oligosphaeridium patulum* Zone. Higher in the SDW-6 section, single specimens of *?Apteodinium* sp., *Cassiculosphaeridia magna*, *Chytroeisphaeridia* sp, *Impletosphaeridium* spp., *Paragonyaulacysta ?borealis* occur together with some prasinophytes, including *Leiosphaeridia*, *Tasmanites* and *Pterospermella* spp. (Fig. 8).

Age: Middle Volgian (approximately).

Comments: The *Oligosphaeridium patulum* Zone has been recognized in some sections of the Bazhenovo Formation in the SE of West Siberia for the first time and is approximately defined within the limits of the Middle Volgian Substage (Shurygin *et al.* 2000). In the Tyumenskaya SDW-6 section this dinocyst zone is established by the FAD of the index species at a depth of 3843.1 m and by the general composition of the assemblage. In the stratotype sections of the Volgian stage on the River Volga (Gorodishce and Kashpir), *Oligosphaeridium patulum* appears in the *klimovi* ammonite Zone (= *elegans*-*scitulus* ammonite Zones in NW

Plate 6. 1–10, Georgievka Fm. Uppermost Kimmeridgian–Lower Volgian, *Amphorula delicata*, *Cribroperidinium* spp. (Ad/C) Zone. Interval 3843.1–3863.5 m: **1**, **2**,**5**,**7**,**9**, **10**, 6 m from the top of core interval, sample 1148/8819: **3**, **8**, 2.5 m from the top of core interval, sample 1120/8822: **4**, **6**, 6.7 m from the top of core interval, sample 1153/8818: **11–13**, Bazhenovo Fm (base), Middle Volgian (lowermost beds), *Oligosphaeridium patulum* (Op) Zone, Interval 3843.1–3863.5 m; the top of core interval, sample 1105/8827.

1, *Amphorula delicata* Van Helden, 1986, specimen 14SDW-6; **2**, *Amphorula* sp., specimen 11SDW-6; **3**, *Gonyaulacysta jurassica* subsp. *adecta* var. *longicornis* (Deflandre 1938) Downie & Sarjeant, 1965 emend. Sarjeant, 1982 specimen 76SDW-6; **4**, *Cribroperidinium* sp., specimen 81SDW-6; **5**, *Sirmiodinium grossii* Alberti, 1961 emend. Warren, 1973, specimen 49SDW-6; **6**, *Gonyaulacysta jurassica* subsp. *jurassica* Deflandre, 1939 emend. Poulsen, 1991, specimen 80SDW-6; **7**, *Lagenadinium ?membranoidium* (Vozzhennikova 1967) Lentin & Vozzhennikova, 1990, specimen 44SDW-6; **8**, *Paragonyaulacysta ?borealis* (Brideaux & Fisher 1976) Stover & Evitt, 1978, specimen 77SDW-6; **9**, *Prolixosphaeridium* sp., specimen 50SDW-6; **10**, *Prolixosphaeridium* sp., specimen 51SDW-6; **11**, *Oligosphaeridium patulum* Riding & Thomas, 1988, specimen 38SDW-6; **12**, *Hystrichosphaeridium petilum* Gitmez, 1970, specimen 35SDW-6; **13**, *Gochteodinia mutabilis* (Riley *in* Fisher & Riley 1980) Fisher & Riley, 1982 emend. Below, 1990, specimen 34SDW-6.

Europe), reaches its acme in the *panderi* ammonite Zone and does not cross the upper boundary of the Middle Volgian (Riding *et al.* 1999). In the boreal sections of Poland, *Oligosphaeridium patulum* is found in the same stratigraphic interval of the Volgian, but it occurs sporadically in the Lower Volgian and is most widespread in the Middle Volgian. The stratigraphic position of the assemblage with *Oligosphaeridium patulum* in the Tyumenskaya SDW-6 succession is within the Volgian Stage although subdivisions are not presently defined. The overlying deposits (in the interval of 3831–3825 m) are characterized by the Middle Volgian ammonites and bivalves (Kirichkova *et al.* 1999) were not studied because of a lack of samples. As a result, the lower boundary of the *Oligosphaeridium patulum* Zone may be approximately correlated with the boundary between the Lower and Middle Volgian. The upper boundary of this dinocyst zone in the investigated section is not established.

Geographical distribution: West Siberia, in the SE: Vesdeckodnaya-4 well, Myldzhinskaya-62 well, Dedovskaya-3 well, Bazhenovo Formation; in the NW, Shaim region, Danilovskaya-10554 well, Severo-Danilovskaya-10161 well, Bazhenovo Formation, the Middle Volgian. In Tyumenskaya SDW-6 section: Bazhenovo Formation, bed 13 at a depth of 3843.1 m.

In the upper part of the studied succession in SDW-6 (3824.0–3795 m interval) dinocysts are very rarely found (Fig. 8). Prasinophytes are common to abundant throughout this interval of the ssuccession and include *Leiosphaeridia* spp., *Tasmanites* and *Pterospermella* spp., associated with large amounts of amorphogen. This represents a 'prasinophyte palynofacies' typical of bituminous beds. In West Siberia, bituminous beds are developed in the upper part of the Bazhenovo Formation, especially in sediments of the Upper Volgian. The *Paragonyaulacysta borealis, Tubotuberella rhombiformis* (Pb/Tr) Zone was established in the same sediments in the north of East Siberia and traced throughout some regions of West Siberia and the Pripolar Ural district (Ilyina 1988; Lebedeva & Nikitenko 1999; Shurygin *et al.* 2000). Furthermore, repeated alternations of *Paragonyaulacysta borealis, Tubotuberella rhombiformis* and prasinophycean associations were found to be linked to changes in the geochemical environment during the latest Volgian and earliest Berriasian in the central part of the Khatanga marine basin (Kaplan *et al.* 1973; Ilyina 1985).

Microbenthos and microphytoplankton associations in the Callovian to Volgian Sea of the Urengoi District (West Siberia)

The succession of the Vasyugan, Georgievka and Bazhenovo formations penetrated by the Tyumenskaya superdeep well is quite informative for analysis of the biofacies characteristics of microbenthos communities and microphytoplankton associations. The section features an uninterrupted sequence of facies deposits, ranging from sub-continental through coastal to relatively deep water. At the beginning of the Late Bathonian the central part of the Urengoi district was occupied by a very shallow coastal plain, regularly flooded by the sea, which penetrated southwards during transgressions. At such times, a shallow marine regime, probably with unstable salinity, was temporarily established in the region, inhabited by rare bivalves and impoverished foraminiferal communities.

A typical marine regime was established in the Urengoi district, as in the whole of West Siberia, only from the end of the latest Bathonian with the beginning of the global Boreal *'Cadoceras'* transgression (Saks 1976; Glinskikh *et al.* 1999; Pinous *et al.* 1999; Shurygin *et al.* 2000). From the beginning of the transgression (when deposition of the Pakhomovskaya Member was intitiated), the SDW-6 region was covered by a shallow sea, probably of unstable salinity, inhabited by abundant eurybiontic *Saccammina* and *Ammodiscus* (Figs 11 and 12). The taxa constituting the foraminiferal community appear to have been adapted to unstable environmental conditions. Together with the foraminifera, rare remains of bivalves are found. Also typical of this time are the vertical trace fossils of *Skolithos* isp., remains of charred wood, and abundant coalified plant detritus. Microphytoplankton in the water column was represented only by *Leiosphaeridia.* Within this general transgressive regime, secondary regressive events can be observed, with the formation of fine-grained, cross-laminated and lenticular-bedded sandstones, containing charred wood but neither macro- or microfaunal remains. This regressive, and probably sub-continental stage, was probably of regional importance; evidence of it is found in the lower part of the basal layers of the Abalak Formation in the Shaim district of West Siberia.

With the development of the Callovian transgression the composition of the foraminiferal communities gradually became more complex. The community at that time already included ten genera, but was overwhelmingly dominated

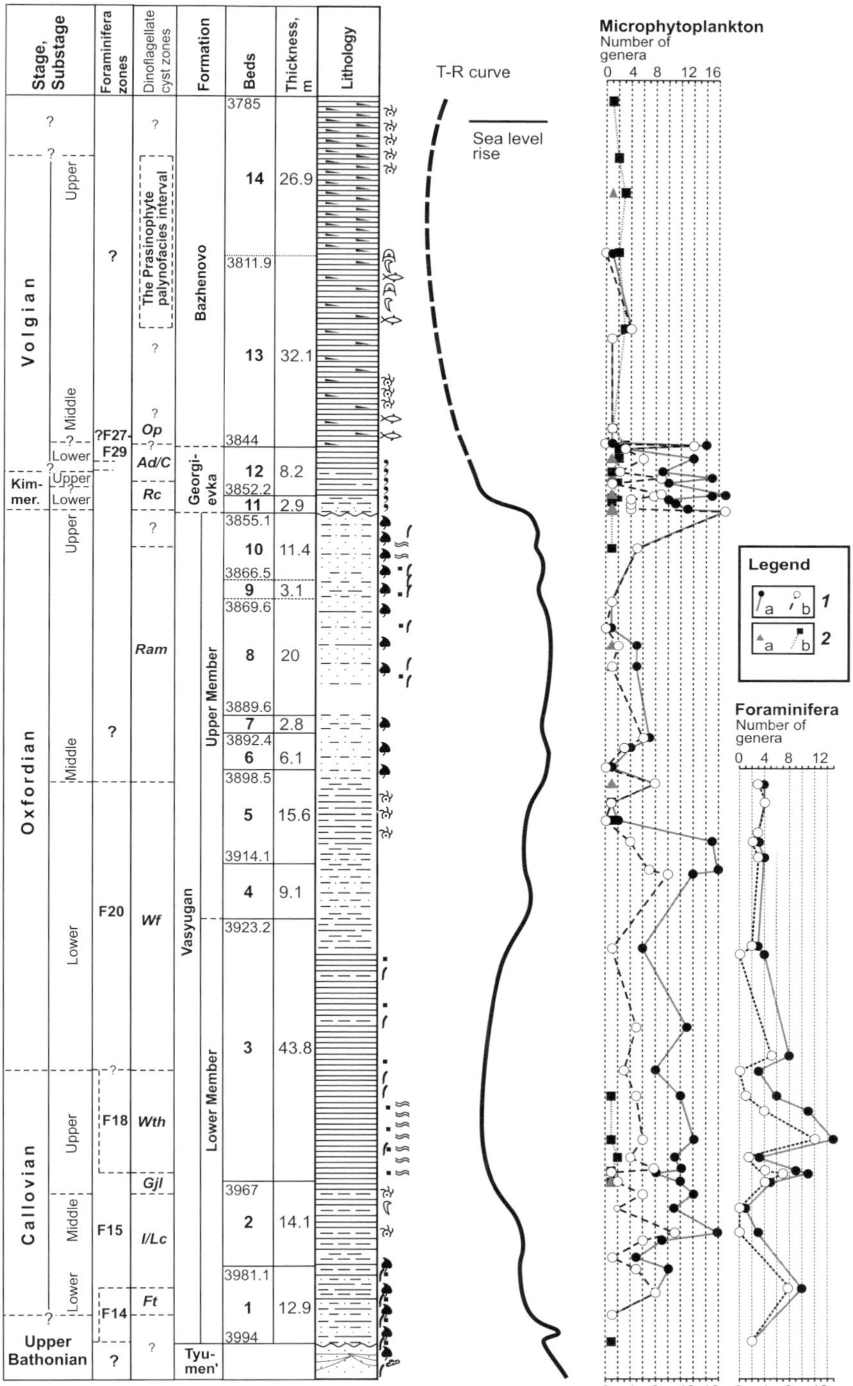

Fig. 11. Taxonomic diversity of microfossils and transgressive–regressive events.
1, number of foraminiferal and dinoflagellate cyst genera (a), number of foraminifera and dinoflagellate cyst genera (b); 2, number of acritarch (a) and prasinophyte (b) genera.

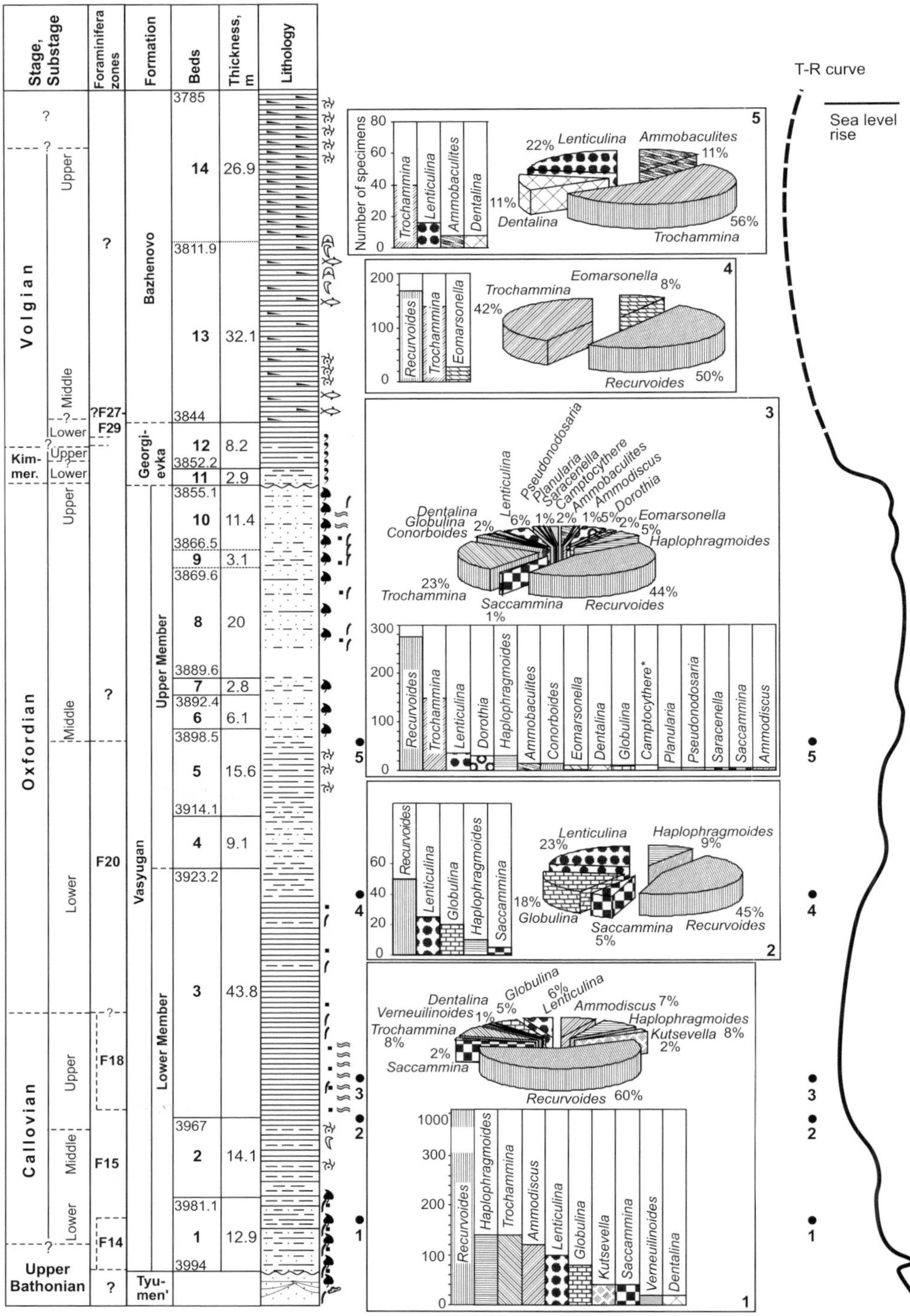

Fig. 12. Taxonomic composition of foraminifera associations and the fluctuation of taxonomic groups in the relation with transgressive–regressive events.

by *Recurvoides*. The pioneer community representatives *Ammodiscus* and *Saccammina* lost their importance. Stenobiotic *Lenticulina* and *Globulina* became relatively abundant for the first time (Figs 11 and 12). A similar situation is observed in the microphytoplankton associations; the prasinophyte *Leiosphaeridia* disappeared, while dinocysts of the genera *Chytroeisphaeridia, Escharisphaeridia, Fromea, Gonyaulacysta, Impletosphaeridium, Nannoceratopsis, Pareodinia* and *Rhynchodiniopsis* became abundant.

At the end of the Middle Callovian and in the earliest Late Callovian, the taxonomic diversity of the foraminiferal communities somewhat decreased, probably because of partial shallowing of the basin, but were still similar to the Early Callovian communities in their composition. These communities are dominated by *Recurvoides*, while *Lenticulina* and *Globulina* form subsidiary components. At the same time, the taxonomic diversity of microphytoplankton associations increased, probably as a result of increased warming of the water column. The number of dinoflagellate cyst genera varies from 11 to 16, with a high degree of taxonomic turnover in the associations (Figs 11 and 12). The most abundant are *Impletosphaeridium, Sentusidinium* and *Crussolia*. Prasinophytes (*Leiosphaeridia*) and acritarchs (*Leiofusa, Micrhystridium*) are rare.

At the end of the late Callovian, the northern hemisphere was affected by a global transgression (Mickey *et al.* 1998; Pinous *et al.* 1999; Shurygin *et al.* 2000). Fine clay deposits were formed at that time in the Tyumenskaya superdeep well area. Remains of nektonic fossils (ammonites, belemnites) first appear and the typical macrobenthos are small and thin-shelled gastropods and bivalves. Horizontal trace fossils of *Planolites* isp. are often found, whereas the vertical traces of *Skolithos* isp. are rare. Foraminiferal communities at this level are the richest and most diverse, represented by 10–16 genera (Figs 11 and 12). *Recurvoides* still dominates, but the formerly subordinate taxa are now replaced by *Trochammina*. Agglutinated foraminifera (2–3 species) predominate, representing up to 86% of the fauna, but calcareous foraminifera are the most taxonomically diverse, numbering up to nine genera (including *Lenticulina, Globulina, Saracenella, Conorboides, Pseudonodosaria* and *Geinitzinita*). The ostracod genus *Camptocythere* first appears at this level. Unlike the microbenthos, the taxonomic diversity of microphytoplankton decreased at the beginning of the transgression (Fig. 11). The dinocyst associations are represented only by *Impletosphaeridium, Escharisphaeridia, Gonyaulacysta, Lithodinia, Pareodinia* and *Sentusidinium*. Later, these dinocyst taxa are joined by *Chlamydophorella, Chytroeisphaeridia, Ctenidodinium, Nannoceratopsis, Rhynchodiniopsis, Stephanelytron, Trichodinium* and *Wanaea*. Prasinophytes are represented by rare *Pterosphaeridia* and *Leiosphaeridia*.

With the development of the transgression in the early Oxfordian, the basin probably deepened, with a change in foraminiferal communities. The taxonomic diversity decreased, and eurybiotic *Ammodiscus* became subordinate. In the second half of the early Oxfordian, at the beginning of the regressive stage, the environment became more favourable for microbenthos. Clay deposits still continued to accumulate, but the contribution of silt is evident. Abundant *Recurvoides, Dentalina, Globulina, Eoguttulina* and *Marginulina* are found in foraminiferal communities. Calcareous foraminifera began to dominate (<77%) the community structures (Figs 11 and 12).

The development of the regression is marked with alternating sandy and silty sediments, and the taxonomic diversity of foraminifera reduced to 3–4 species. Alternating communities with dominant calcareous and agglutinated foraminifera are observed. The dynamics of evolution of microphytoplankton associations was similar: dinocysts were represented by 12 genera at the beginning of the regression, whereas only six genera are found later, with *Trichodinium* and *Gonyaulacysta* species most common (Fig. 11).

At the end of the early Oxfordian, a brief secondary transgression probably occurred within the background of a general regression. Fine clay began to be deposited again. Macrofaunal remains are represented by nektonic forms (ammonites) and thin-shelled benthos (bivalves). Frequent environmental alterations were evidently unfavorable for the microbenthos. Foraminiferal communities were taxonomically impoverished (3–4 genera only) with *Recurvoides, Globulina* and *Trochammina* repeatedly dominant, and the eurybiotic *Ammodiscus* remaining important. *Ammobaculites, Lenticulina,* and *Dentalina* are also found. Microphytoplankton associations were also variable, the taxonomic diversity of dinocysts varying from 16 to 2 genera. The most abundant dinocyst species in the associations belong to *Gonyaulacysta* while *Escharisphaeridia, Endoscrinium, Liesbergia, Pareodinia, Rhynchodiniopsis, Rigaudella, Scriniodinium, Trichodinium* and *Wanaea* are rarely found. Prasinophytes and acritarchs are represented by *Leiosphaeridia* and *Micrhystridium,* respectively (Fig. 11).

The Middle and Late Oxfordian are characterized by the development of a regression in the basin, with several transgressive episodes (Figs 11 and 12). Sediments accumulated at this time are silty and fine-grained sandy, and contain abundant coalified detritus and remains of charred wood. The environment was unfavourable for benthos and microphytoplankton, probably the result of the extreme shallowness of the basin. No macrofaunal remains are found in these deposits; microbenthos is represented only by *Subrecurvoides.* Vertical trace fossils of *Skolithos* isp. are often found. In the microphytoplankton associations, dinocysts are represented by only 5–7 genera, and sometimes their diversity drops to a single genus. The most typical are the chorate cysts of *Rigaudella. Nannoceratopsis* is also recorded. In addition to dinocysts, rare *Micrhystridium* acritarchs are found. During brief secondary transgressive intervals, the dinoflagellate cyst associations become more diverse, and their taxonomic composition is almost completely renewed compared to the regressive stages. The prasinophyte *Leiosphaeridia* is also recorded (Fig. 11). Silty-sandy sediments were inhabited by *Subrecurvoides* microforaminifera. Horizontal traces of *Planolites* isp. are relatively common, whereas the vertical traces of *Skolithos* isp. are rare.

At the end of the Late Oxfordian and in the earliest Kimmeridgian, West Siberia was affected by a global Boreal transgression, observed in a number of sections in Boreal basins (Mickey *et al.* 1998; Shurygin *et al.* 2000). A stable marine regime was established in the central part of northwest Siberia. Deposits enriched in glauconite were formed at the beginning of the transgression (the Barabinsk Member of the Georgievka Formation), and with its development, fine clays with glauconite grains accumulated (seen in the Georgievka Formation). Macrobenthos is represented by remains of bivalves and gastropods. Silty environments were inhabited by rare microbenthic forms (foraminifera and microforaminifera) comprising *Trochammina, Lenticulina, Kutsevella* and *Microkutsevella.* Kimmeridgian microphytoplankton associations are renewed almost completely compared with those recorded in the Middle and Late Oxfordian (Fig. 11). The assemblages are taxonomically very diverse, containing 10–19 dinocyst genera. The most common genera in the Urengoi district were *Rhynchodiniopsis, Systematophora, Chlamydophorella* and *Sentusidinium. Dingodinium, Cribroperidinium, Endoscrinium, Gonyaulacysta, Cassiculosphaeridia, Amphorula and Tubotuberella* are often found. Other microphytoplankton forms, such as acritarchs (*Veryhachium* and *Solisphaeridium*) and prasinophytes (*Leiosphaeridia, Tasmanites* and *Pterospermella*) are also present. The Kimmeridgian microphytoplankton associations are stable, and their variability is rather low, ranging between 1 and 9 genera.

In the Kimmeridgian, the regional tectonical submergence of the West Siberian Plate began (Kontorovich *et al.* 1975; Pinous *et al.* 1999) which, in the Volgian, resulted in the formation of the black bituminous mudstone of the Bazhenovo Formation. At the beginning of the stagnant conditions, dinocysts in microphytoplankton associations are still quite diverse and represented by 14 genera, significantly more diverse than in the Kimmeridgian. *Apteodinium, Chlamydophorella, Chytroeisphaeridia, Impletosphaeridium, Cribroperidinium, Hystrichosphaeridium, Lagenadinium, Oligosphaeridium, Paragonyaulacysta* and *Pareodinia* are the most typical genera. However, in the progressively more stagnant environment, the abundance and taxonomic diversity of dinocysts in the associations sharply decreased to between 1 and 4 genera (Fig. 11). Only rare *Chlamydophorella, Paragonyaulacysta, Impletosphaeridium, Apteodinium* and *Cassiculosphaeridia* are found at this level. At the same time, prasinophytes became abundant and rather diverse, represented by *Leiosphaeridia, Tasmanites, ?Tytthodiscus,* and *Pterospermella,* which formed the so-called prasinophyte palynofacies by the end of the middle and beginning of the Late Volgian (Fig. 11).

Microbenthos is either very rare or absent from this environment. Macrobenthos is represented by bivalves on occasional horizons, together with rare crustaceans. The water column was inhabited by nektonic forms (ammonites and belemnites) and fish are common.

Conclusions

This investigation has facilitated a detailed description of members of the reference section of the Vasyugan, Georgievka, and Bazhenovo formations in the Tyumenskaya superdeep well (SDW-6), and their complete micropalaeontological and palynological characterization. Stratigraphical analysis of the distribution of these microfossils has allowed a first detailed zonation of this reference (and the most complete) section in the north of West Siberia

(Figs 5–8). An almost continuous sequence of dinoflagellate cyst zones for the Callovian to Middle Volgian in a single section has been established for the first time. Dinocyst zones have been correlated with the regional foraminiferal and ammonite zones of West Siberia as well as the Jurassic dinocyst zonations of the Russian Platform and northwest Europe (Figs 9 and 10). The Callovian to Middle Volgian dinocyst zonation of West Siberia is significantly refined and improved as a result of this study of the Tyumenskaya SDW-6.

Some specific features of facies distribution of the microbenthos communities and microphytoplankton associations have been established in this reference section for the Middle and Late Jurassic sea of West Siberia must be considered in the development of the regional biostratigraphical scales. The composition of microbenthos communities and microphytoplankton associations, and the peculiarities of their distribution, have been described, taking into consideration the palaeoenvironmental changes and transgressive–regressive events during the Callovian and late Jurassic (Figs 11 and 12). It is apparent that the microbenthos communities in the earliest stages of a transgression are dominated eurybiotic forms. The most diverse and stable communities are typical of transgressive and early regressive stages (Figs 11 and 12). Depleted microbenthos associations are observed during the peaks of the transgressive and regressive events. Unstable environments and frequent transgressive–regressive changes resulted in alternating communities with different dominant and subordinate forms. Similar trends are observed for microphytoplankton associations, though with some differences. In the Callovian, the most diverse associations have been found in the regressive stage, whereas at the beginning of the Late Callovian transgression their diversity somewhat decreases (Fig. 11). Unstable environments also caused the great dinocyst taxon turnover in the microphytoplankton associations. Dinocysts were extremely scarce in the anomalous stagnation conditions where the microphytoplankton associations were dominated by prasinophytes of quite monotonous composition, forming the so-called 'prasinophyte palynofacies' (Figs 8 and 11).

The authors are very grateful to S. V. Meledina for the identification of the Callovian and Oxfordian ammonites, found in core SDW-6, and valuable advice on biostratigraphy. N. K. Lebedeva is thanked for the many helpful discussions on this project; S. Feist-Burkhardt, S. Payne, M. Hart and J. B. Riding for the comments, review and suggestions on the improvement of our manuscript. The investigation has been carried out with the support of grant RFBR 03-05-64391.

Appendix 1. List of dinoflagellate cyst species with author citations

In this Appendix, only dinoflagellate cyst taxa identified from the investigated samples of the Tyumenskaya superdeep well (SDW-6) at species level and below are listed. Bibliographic references to the dinoflagellate cyst taxa may be found in Lentin & Williams (1993) and Williams *et al.* (1998). For convenience, certain taxa (e.g. the genus *Crussolia* proposed as taxonomic senior synonym of *Evansia* according to Below (1990) and *Fromea* the dinoflagellate affinity of which has been questioned by some authors (Fensome *et al.* 1993)), are regarded as unequivocal dinoflagellate cysts in this study.

Adnatosphaeridium caulleryi (Deflandre 1938) Williams & Downie 1969
Amphorula delicata Van Helden 1986
Amphorula metaelliptica Dodekova 1969 emend. Monteil 1990
Apteodinium confusum (Vozzhennikova 1967) Helenes 1984
Barbatacysta creberbarbata (Erkmen & Sarjeant 1980) Courtinat 1989
Cassiculosphaeridia magna Davey 1974 emend. Harding, 1990
Chytroeisphaeridia cerastes Davey 1979
Chytroeisphaeridia chytroeides (Sarjeant 1962) Downie & Sarjeant 1965
Chytroeisphaeridia hyalina (Raynaud 1978) Lentin & Williams 1981
Cribroperidinium globatum (Gitmez & Sarjeant 1972) Helenes 1984
Cribroperidinium longicorne (Downie 1957) Lentin & Williams 1985
Crussolia dalei Smelror & Arhus 1989
Crussolia deflandrei Wolfard & Van Erve 1981
Crussolia perireticulata Arhus *et al.* 1989
Ctenidodinium continuum Gocht 1970
Dingodinium jurassicum Cookson & Eisenack 1958
Endoscrnium anceps Raynaud 1978
Endoscrinium galeritum (Deflandre 1939) Vozzhennikova 1967
Endoscrinium luridum (Deflandre 1939) Gocht 1970
Escharisphaeridia pocockii (Sarjeant 1968) Erkman & Sarjeant 1980
Fromea tornatilis (Drugg 1978) Lentin & Williams 1981
Gochteodinia mutabilis (Riley *in* Fisher & Riley 1980) Fisher & Riley 1982 emend. Below 1990

Gonyaulacysta dualis (Brideaux & Fisher 1976) Stover & Evitt, 1978
Gonyaulacysta eisenackii (Deflandre 1938) Gorka 1965
Gonyaulacysta fastigiata Duxbury 1977
Gonyaulacysta jurassica (Deflandre 1939) Norris & Sarjeant 1965 subsp. *adecta* Sarjeant 1982
Gonyaulacysta jurassica subsp. *adecta* var. *longicornis* (Deflandre 1938) Downie & Sarjeant 1965 emend. Sarjeant 1982
Gonyaulacysta jurassica subsp. *jurassica* Deflandre, 1939 emend. Poulsen 1991
Gonyaulacysta pectinigera (Gocht 1970) Fensome 1979
Hologinella spinosigibberosa Brideaux & Fisher 1976
Hystrichosphaeridium petilum Gitmez 1970
Imbatodinium kondratjevii (Vozzhennikova 1967) Lentin & Vozzhennikova 1990
Impletosphaeridium polytrichum (Valensi 1947) Islam 1993
Impletosphaeridium tribuliferum (Sarjeant 1962) Islam 1993
Impletosphaeridium varispinosum (Sarjeant 1959) Islam 1993
Lagenadinium callovianum Piel 1985
Lagenadinium ?membranoidium (Vozzhennikova 1967) Lentin & Vozzhennikova 1990
Leptodinium ambiguum (Deflandre 1939) Helenes 1984
Liesbergia liesbergensis Berger 1986
Mendicodinium groenlandicum (Pocock & Sarjeant 1972) Davey 1979
Nannoceratopsis pellucida Deflandre 1939
Oligosphaeridium patulum Riding & Thomas 1988
Paragonyaulacysta ?borealis (Brideaux & Fisher 1976) Stover & Evitt 1978
Paragonyaulacysta calloviensis (Johnson & Hills 1973) Dörhofer & Davies 1980
Paragonyaulacysta retiphragmata Dörhofer & Davies 1980
Pareodinia ceratophora Deflandre 1947
Pareodinia prolongata Sarjeant 1959
Prolixosphaeridium anasillum Erkman & Sarjeant 1980
Protobatioladinium westburiense Nohr-Hansen 1986
Rhynchodiniopsis cladophora (Deflandre 1939) Below 1981
Rigaudella aemula (Deflandre 1939) Below 1982
Scriniodinium crystallinum (Deflandre 1939) Klement 1960
Sentusidinium sparsibarbatum Erkman & Sarjeant 1980
Sirmiodiniopsis orbis Drugg 1978
Sirmiodinium grossii Alberti 1961 emend. Warren 1973
Stephanelytron redcliffense Sarjeant 1961
Stephanelytron scarburghense Sarjeant 1961
Subtilisphaera ?inaffecta (Drugg 1978) Bujak & Davies 1983
Systematophora areolata Klement 1960
Systematophora daveyi Riding & Thomas 1988
Systematophora orbifera Klement 1960
Trichodinium scarburghense (Sarjeant 1964) Williams *et al.* 1993
Tubotuberella apatela (Cookson & Eisenack 1960) Ioannides *et al.* 1977
Tubotuberella dangeardii (Sarjeant 1968) Stover & Evitt 1978
Wanaea acollaris Dodekova 1975
Wanaea fimbriata Sarjeant 1961
Wanaea thysanota Woollam 1982

References

BEISEL, A.L., ZANIN, Y.N. *ET AL.* 2002. Reference section for the Upper Jurassic and Callovian in nortern West Siberia. *Russian Geology and Geophysics*, **43 (9)**, 763–777.

BELOW, R. 1990. Evolution und Systematik von Dinoflagellaten-Zysten aus der Ordnung Peridiniales. III. Familie Pareodiniaceae. *Palaeontographica,* Abteilung B, **220**, 1–96.

BULYNNIKOVA, S.P., KOMISSARENKO, V.K. & BELOUSOVA, N.A. (eds) 1990. *Atlas mollyuskov i foraminifer morskih otlozhenij verhnej yury i neokoma Zapadno-Sibirskoj neftegazonosnoj oblasti* (Atlas of molluscs and foraminifers of marine deposits of Upper Jurassic and Neocomian of West Siberian petroleum-bearing region). **1–2**, Nedra, Moscow, 3–359.

DAVIES, E.H. 1983. The dinoflagellate Oppel-Zonation of the Jurassic-Lower Cretaceous sequence in the Sverdrup Basin, Arctic Canada. *Geological Survey of Canada Bulletin*, **359**, 3–59.

DIKOVSKII, A.A. & SEDYKH, A.V. 1996. O sostoyanii vypolneniya programmy issledovanij Tyumenskoj sverhglubokoj skvazhiny (Preliminary reports on the Tyumenskaya superdeep well project). *In*: MAZUR, V.B. (ed.) *Tyumenskaya sverhglubokaya skvazhina (interval 0–7502m). Resul'taty bureniya i issledovaniya (Tyumenskaya superdeep well (interval 0–7502 m). Results of drilling and study).* Perm, Kamskij nauchno-issledovatel'skij institut kompleksnyh issledovanij glubokih i sverhglubokih skvazhin, 63–70.

EKHLAKOV, YU.A. & UGRYUMOV, A.N. 1996. Triasovye i yurskie otlozheniya v razreze Tyumenskoj sverhglubokoj skvazhiny (Triassic and Jurassic deposits in section of the Tyumenskaya superdeep well). *In*: MAZUR, V.B. (ed.) *Tyumenskaya sverhglubokaya skvazhina (interval 0–7502m). Resul'taty bureniya i issledovaniya (Tyumenskaya superdeep well (interval 0–7502m). Results of drilling and study).* Perm, Kamskij nauchno-issledovatel'skij institut kompleksnyh issledovanij glubokih i sverhglubokih skvazhin, 79–94.

EKHLAKOV, YU.A., DIKOVSKII, A.A., FRIK, M.G.,

SIROTENKO, L.V. & BELOKON', T.V. 1991. Osobennosti glubokozalegayushih mezozojskih otlozhenij severa Zapadno-Sibirskoj plity (Particular features of deep-seated Mesozoic deposits of northern West-Siberian Plate). *Sovetskaya Geologiya* (*Soviet Geology)*, **8**, 80–85.

FENSOME, R.A., TAYLOR, F.J.R., NORRIS, G., SARJEANT, W.A.S., WHARTON, D.I. & WILLIAMS, G.L. 1993. A classification of fossil and living dinoflagellates. *Micropaleontology, Special Publication*, **7**, 3–351.

FISHER, M.J. & RILEY, L.A. 1980. The stratigraphic distribution of dinoflagellate cysts at the boreal Jurassic-Cretaceous boundary. *Proceedings of the 4th International Palynology Conference,* Lucknow, **2**, 313–329.

GLINSKIKH, L.A., NIKITENKO, B.L. & SHURYGIN, B.N. 1999. The Jurassic Abalak Formation in West Siberia (palaeontology, lithostratigraphy and biostratigraphy). *Russian Geology and Geophysics*, **40 (7)**, 1043–1061.

ILYINA, V.I. 1985. *Palinologiya yury Sibiri.* (Jurassic palynology of Siberia.) Nauka, Moscow, 3–237.

ILYINA, V.I. 1988. Mikrofitoplankton pogranichnyh otlozheniy yury i mela na myse Urdyuk-Khaya (poluostrov Paksa) (Microphytoplankton of transitional Jurassic-Cretaceosus deposits in Urdyuk-Khaya Cape (Paksa Peninsular). *In*: KHLONOVA, A.F. (ed.) *Palinologiya v SSSR (Palynology in USSR)*. Novosibirsk, Nauka, 103–107.

ILYINA, V.I. 1991. Raschlenenie bat-oksfordskih otlozhenij Russkoj platformy po dinotsystam (Subdivision of the Bathonian to Oxfordian sediments of the Russian Platform on dinoflagellate cyst evidence). *In*: KIRICHKOVA, A.I. & CHIRVA, S.A. (eds) *Stratigrafiya i paleogeografiya osadochnyh tolsh neftegazonosnyh bassejnov SSSR (Stratigraphy and Paleogeography of the terrigenous successions of the oil-bearing basins of the USSR.* Leningrad, Vsesoyuznyj neftyanoj naucho-issledovatel'skij geologorazvedochnyj institut, 42–64.

ILYINA, V.I. 1998. Zonal'noe raschlenenie verhov kelloveya i oksforda Pur-Tazovskogo mezhdurech'ya v Zapadnoj Sibiri po cistam dinoflagellat (Zonal subdivision of the Upper Callovian to Oxfobdian of Pur-Taz interfluve of West Siberia on dinoflagellatc cysts evidance). *In*: VILZAN, I.A. (ed.) *Aktual'nye voprosy geologii i geografii Sibiri (Actual problems of geology and geography of Siberia)*. **1.** Tomsk, Tomskij gosudarstvennyj universitet, 215–218.

JOHNSON, C.D. & HILLS, L.V. 1973. Microplankton zones of the Savik Formation (Jurassic), Axel Heiberg and Ellesmere Islands, District of Franklin. *Bulletin of Canadian Petroleum Geology*, **21**, 179–218.

KAPLAN, M.E., YUDOVNY, E.G., ZAKHAROV, V.A., BASOV, V.A. & VOYTCEKHOVSKAYA, A.G. 1973. Usloviya nakopleniya morskikh otlozheniy poluostrova Paksa, perekhodnykh mezdu yuroy i melom (Anabarskiy zaliv) (The conditions of transitional Jurassic-Cretaceous marine sedimentation in Paksa Peninsular (Anabar Bay). *Doklady Akademii nauk SSSR (Reports of Reasearch Academia USSR)*, **209 (3)**, 691–694.

KIRICHKOVA, A.I., KULIKOVA, N.K., OVCHINNIKOVA, L.L., TIMOSHINA, N.A., TRAVINA, T.A. & FEDOROVA, V.A. 1999. Biostratigraficheskoe raschlenenie mezozojskih otlozhenij, vskrytyh Tyumenskoj sverhglubokoj skvazhinoj (Biostratigraphic subdivision of the Mesozoic deposits exposed by the Tyumenskaya superdeep well). *Stratigrafiya. Geologicheskaya Korrelyatsiya (Stratigraphy. Geological correlation),* **7 (1)**, 71–85.

KONTOROVICH, A.E., NESTEROV, I.I., SALMANOV, F.K., SURKOV, V.S., TROFIMUK, A.A. & ERV'E, YU.G. 1975. *Geologiya nefti i gaza Zapadnoj Sibiri* (Geology of oil and gas in West Siberia). Nedra, Moscow, 3–679.

KONTOROVICH, A.E., BELYAEV, S.YU., KONTOROVICH, V.A., KRASAVCHIKOV, V.O. & AGALAKOV, S.E. 2001. Tektonika osadochnogo chehla Zapadno-Sibirskoj neftegazonosnoj provintsii (Tectonics of sedimentary cover of West-Siberian petroleum-bearing province). *In*: LEONOV A.G. (ed.) *Materialy soveshaniya: Tektonika neogeya: obshie i regional'nye aspekty* (*Materials of Meeting: Tectonics of Neogean: general and regional aspects)*. **1**. Moscow, Geos, 307–310.

LEBEDEV, I.V. (ed.) 1972. *Stratigrafo-paleontologicheskaya osnova detal'noj korrelyatsii neftegazonosnyh otlozhenij Zapadno-Sibirskoj nizmennosti* (Stratigraphy-palaeontological basement for detailed correlation of petroleum-bearing deposits of West-Siberian Lowland). Tyumen, Zapadno-Sibirskij nauchno-issledovatel'skij geologo-razvedochnyj neftyanoj institut, 3–149.

LEBEDEVA, N.K. & NIKITENKO, B.L. 1999. Dinoflagellate cysts and microforaminifera of the Lower Cretaceous Yatria River section, Subarctic Ural, NW Siberia (Russia). Biostratigraphy, palaeoenvironmental and palaeogeographic discussion. *Grana,* **38**, 134–143.

LENTIN, J.K. & WILLIAMS, G.L. 1993. Fossil dinoflagellates: index to genera and species 1993 edition. *American Association of Stratigraphic Palynologists Contributions Series*, **28**, 3–856.

LUTOVA, Z.V. 1981. *Stratigrafiya i foraminifery kelloveya severa Srednej Sibiri* (Stratigraphy and foraminifers of Callovian of northern Middle Siberia). Nauka, Novosibirsk, 3–235.

MAZUR, V.B. (ed.) 1996. *Tyumenskaya sverhglubokaya skvazhina (interval 0–7502 m). Resul'taty bureniya i issledovaniya* (Tyumenskaya superdeep well (interval 0–7502 m). Results of drilling and study). Perm, Kamskij nauchno-issledovatel'skij institut kompleksnyh issledovanij glubokih i sverhglubokih skvazhin, 3–375.

MELEDINA, S.V. 1994. *Boreal'naya srednyaya yura Rossii (ammonity i zonal'naya stratigrafiya bajosa, bata i kelloveya* (Boreal Middle Jurassic of Russia (ammonites and zonal stratigraphy of the Bajocian, Bathonian, and Callovian). Nauka, Novosibirsk, 1–182.

MELEDINA, S.V., ILYINA, V.I. & NALNJAEVA, T.I. 1998. Parallel'nye biostratigraficheskie shkaly boreal'nogo bata I kelloveya Pechorskogo Severa kak instrument dlya mezhregional'nyh korrelyatsij

(Parallel biostratigraphic scales of boreal Bathonian and Callovian of northern Pechora as standard for interregional correlations). *Stratigrafiya. Geologicheskaya Korrelyatsiya (Stratigraphy. Geological correlation),* **6**, 29–42.

MICKEY, M., NIKITENKO, B.L. & SHURYGIN, B.N. 1998. Petroliferous Upper Jurassic correlated across W. Siberia, N. Alaska, Arctic Islands. *Oil & Gas Journal*, **96 (50)**, 84–87.

NESTEROV, I.I. (ed.) 1991. *Resheniya V Mezhvedomstvennogo regional'nogo stratigraficheskogo soveshaniya po mezozojskim otlozheniyam Zapadno-Sibirskoj ravniny* (Resolutions of the Fifth Interdepartmental regional stratigraphic meeting on Mesozoic deposits of the Siberian Plain). Tyumen, Zapadno-Sibirskij nauchno-issledovatel'skij geologo-razvedochnyj neftyanoj institut, 3–54.

PINOUS, O.V., SAHAGIAN, D.L., SHURYGIN, B.N. & NIKITENKO, B.L. 1999. High-resolution sequence stratigraphic analysis and sea-level interpretation of the Middle and Upper Jurassic strata of the Nyurolskaya depression and vicinity (Southeastern Siberia, Russia). *Marine and Petroleum Geology,* **16**, 245–257.

POULSEN, N.E. 1993. Dinoflagellate cyst biostratigraphy of the Oxfordian and Kimmeridgian of Poland. *Acta Geologica Polonica*, **43**, 251–272.

RIDING, J.B. 1984. Dinoflagellate range top biostratigraphy of the uppermost Triassic to lowermost Cretaceous of northwest Europe. *Palynology*, **8**, 195–210.

RIDING, J.B. 1987. Dinoflagellate cyst stratigraphy of the Nettleten Bottom Borehole (Jurassic, Hettangian to Kimmeridgian), Lincolshire, England. *Proceedings of the Yorkshire Geological Society*, **46**, 231–266.

RIDING, J.B. & THOMAS, J.E. 1992. Dinoflagellate cysts of the Jurassic System. *In*: POWELL A.J. (ed.) *A stratigraphic index of the dinoflagellate cysts*. British Micropalaeontological Society Publications Series, Chapman & Hall, London, 7–97.

RIDING, J.B., FEDOROVA, V.A. & ILYINA, V.I. 1999. Jurassic and Lowermost Cretaceous dinoflagellate cyst biostratigraphy of the Russian Platform and Northern Siberia, Russia. *American Association of Stratigraphic Palynologists Foundation, Contributions Series,* **36**, 1–184.

RILEY, L.A. & FENTON, J.P.G. 1982. A dinocyst zonation for the Callovian to Middle Oxfordian succession (Jurassic) of northwest Europe. *Palynology*, **6**, 193–202.

SAKS, V.N. (ed.) 1969. *Opornyj razrez verhneyurskih otlozhenij bassejna r. Heta* (Reference section of Upper-Jurassic deposits of Kheta River). Nauka, Leningrad, 3–124.

SAKS, V.N. (ed.) 1976. *Stratigrafiya yurskoj sistemy severa SSSR* (Stratigraphy of Jurassic of the northern USSR). Nauka, Moscow, 3–436.

SHAROVSKAYA, N.V. 1968. Kompleksy foraminifer iz yurskih i nizhnemelovyh otlozhenij Ust'-Enisejskogo i Turuhan-Ermakovskogo rajonov (Foraminiferal assemblages from Jurassic and Lower-Cretaceous deposits of Ust'- Yeniseisk and Turukhan-Ermakovo regions). *Uchenye zapiski. Paleontologiya i biostratigrafiya (Scientific reports. Paleontology and biostratigraphy).* Nauchno-issledovatel'skij institut geologii Arktiki, Leningrad, **23**, 106–117.

SHURYGIN, B.N., NIKITENKO, B.L. *et al.* 2000. *Stratigrafiya neftegazonosnyh bassejnov Sibiri. Yurskaja sistema* (Stratigraphy of petroleum-bearing basins of Siberia. Jurassic). Publishing House of Siberian Branch of the RAS, "GEO" Branch, Novosibirsk, 3–480.

SMELROR, M. 1988*a*. Bathonian and Early Oxfordian dinoflagellate cysts and acritarchs from Kong Karls Land, Svalbard. *Review of Paleobotany and Palynology*, **56**, 275–304.

SMELROR, M. 1988*b*. Late Bathonian and Early Oxfordian dinoflagellate cysts stratigraphy of Jameson Land and Milne Land, East Greenland. *Gronlands Geologiske Undersogelse Rapport,* **137**, 135–159.

SMELROR, M. & BELOW, R. 1993. Dinoflagellate biostratigraphy of the Toarcian to Lower Oxfordian (Jurassic) of the Barents Sea region. *In*: VORREN, T.O., BERGSAGER, E., DAHL-STAMNES, O. A., HOLTER, E., JOHANSEN, B., LIE, E. & LUND, T.B. (eds) *Arctic Geology and Petroleum Potential.* Norwegian Petroleum Society, Special Publication 2. Elsevier, Amsterdam, 495–513.

THOMAS, J.E. & COX, B.M. 1988. The Oxfordian–Kimmeridgian stage boundary (Upper Jurassic): dinoflagellate cysts assemblages from the Harome borehole, North Yorkshire, England. *Review of Paleobotany and Palynology*, **56**, 313–326.

WILLIAMS, G.L., LENTIN, J.K. & FENSOME, R.A. (eds) 1998. The Lentin and Williams index of fossil dinoflagellates 1998 edition. *American Association of Stratigraphic Palynologists Foundation, Contributions Series*, **34**, 3–817.

WOOLLAM, R. & RIDING, J.B. 1983. Dinoflagellate cyst zonation of the English Jurassic. *Institute of Geological Sciences Report,* London, **83/2**.

ZAKHAROV, V.A., BOGOMOLOV, Y.I. *ET AL.* 1997. Boreal zonal standard and biostratigraphy of the Siberian Mesozoic. *Russian Geology and Geophysics,* **38 (5)**, 965–993.

Quantitative biostratigraphy for the Cuervos Formation (Paleocene) of the Llanos foothills, Colombia: improving palynological resolution for oil exploration

C. A. JARAMILLO[1], F. MUÑOZ[2], M. COGOLLO[2] & F. DE LA PARRA[2]

[1]*Smithsonian Tropical Research Institute, Unit 0948, APO AA 34002-0948, USA*

[2]*Biostratigraphy team, Instituto Colombiano del Petróleo-Ecopetrol, AA 4185, Bucaramanga, Colombia (e-mail: carlos@flmnh.ufl.edu)*

Abstract: Oil exploration in Colombia has traditionally taken place in areas with relatively few structural complexities. However, in the last decade, exploration has moved to regions characterized by complex structural deformation, poor seismic resolution, and many stratigraphic problems, such as in the Llanos foothills. In this region, the major reservoirs occur in mostly continental Paleogene sequences, where palynomorphs are usually the only fossil group found. Thus, palynology has become an important tool in controlling the stratigraphic position of a well during drilling, in testing diverse seismic and structural interpretations, and correlating reservoirs. This study provides a biostratigraphic framework for the Cuervos Formation (Late Paleocene to earliest Eocene) of the Llanos and Llanos foothills, making use of graphic correlation. We used 14 sections with palynological information from outcrops, well cores, and well ditch cuttings. Five informal palynological zones are proposed. Based on the biostratigraphic model produced by this analysis, we reinterpreted a recently drilled well in the Llanos foothills upon which several previous interpretations had been made. A sidetrack of that well was subsequently drilled, validating the proposed model. This is a positive test that the biostratigraphic framework developed for the Cuervos Formation is reliable and can be successfully applied to exploration in the Llanos foothills.

Oil exploration in Colombia has taken place since the 1920s. In 2002, the country produced about 650 000 barrels per day. One of the main targets of Colombian oil exploration has been the Llanos and Llanos foothills basin, located on the eastern side of Colombia (Fig. 1). The stratigraphy of the basin is composed of about 2 km of Upper Cretaceous to Tertiary sediments (Fig. 2). The Carbonera Formation has been generally considered the seal, Mirador, Barco and Guadalupe Formations the reservoirs, and the black shales of the Upper Cretaceous, Gacheta-Luna Formation are the source (Cazier *et al.* 1995; Dzou *et al.* 1999). Most of the Tertiary sediments in the Llanos and Llanos foothills basin are continental with some discrete episodes of marine influence. Toward the Llanos foothills, the complexity of the stratigraphy increases, with large lateral changes in sediment thickness occurring in all units (Fajardo *et al.* 2000).

A cross section of the basin shows that the structural complexity of the basin significantly increases toward the foothills (Linares 1999, Fig. 3). Over the years, the trend of exploration has been moving from east to west (from areas with low structural complexity) to areas with more structural complexities (where larger reservoirs are expected, but the difficulty of finding them increases). Additionally, not only are the structures more complex, the exploration wells are now deeper. There has been a significant trend toward increasing the depth of exploratory wells in the Llanos through the years (Fig. 4). The rough topography of the Llanos foothills and strata dipping at high angles in the foothills have made good seismic resolution difficult to acquire. The level of uncertainty during the drilling phase therefore has increased through the years.

The present resolution of the biostratigraphy being used in Colombia is low compared to the needs of modern exploration in the Llanos foothills (Rubio 1997). For example, only one biostratigraphic zone exists for the Oligocene. Consequently it is not possible to divide the Carbonera C6 to C8 intervals in biostratigraphic terms (Figure 2, C1 to C8 units are subdivisions of the Carbonera Formation, C1 being the youngest). A similar problem occurs with the Carbonera C1 to C5 intervals, all in the lower Miocene, and the Paleocene Barco and Cuervos Formations (Fig. 2). Also, it is not possible to differentiate the lowermost Carbonera C8

From: Powell, A. J. & Riding, J. B. (eds) 2005. *Recent Developments in Applied Biostratigraphy*. The Micropalaeontological Society, Special Publications, 145–159. 1747-602X/$15.00

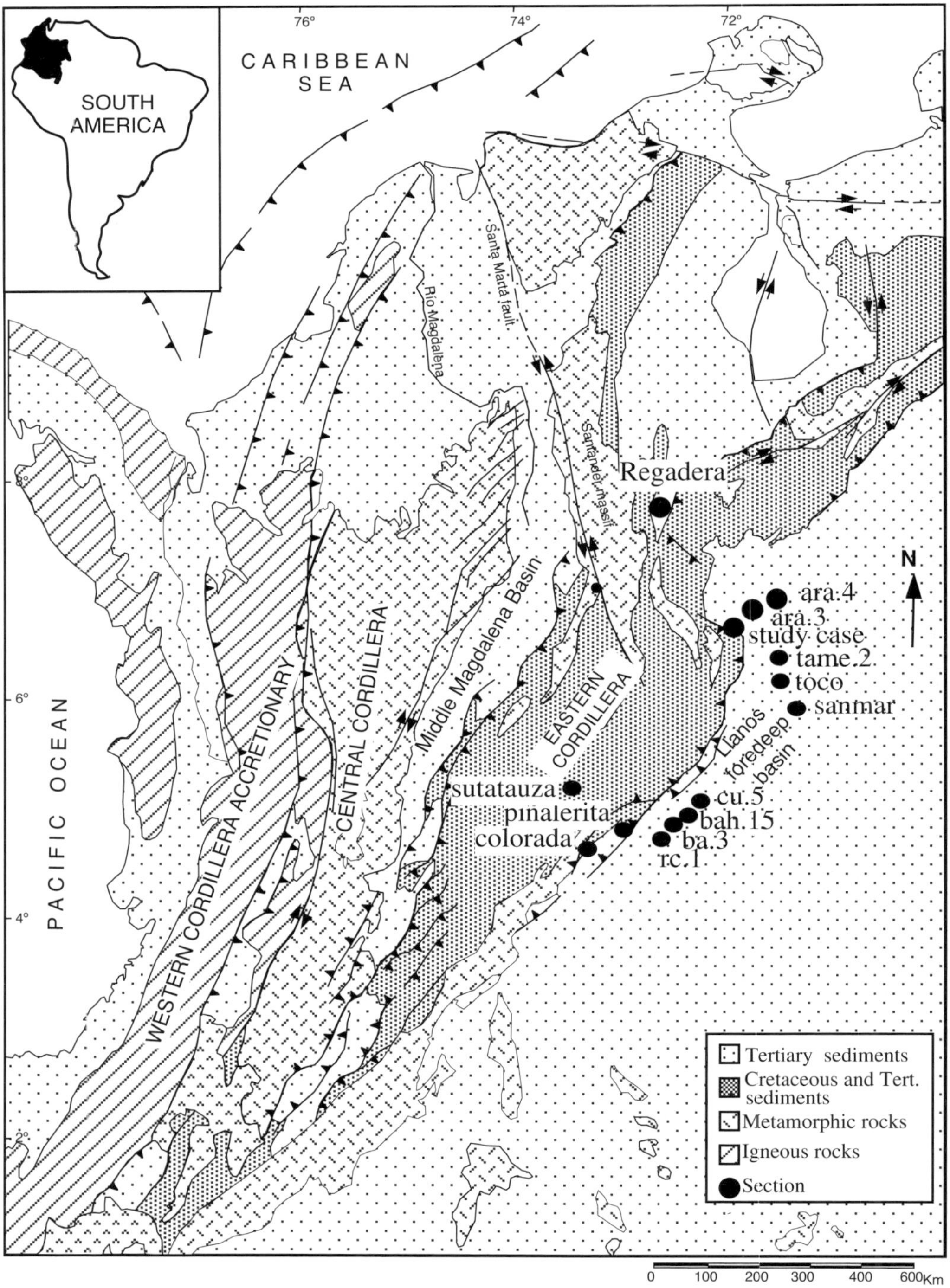

Fig. 1. Geological map of Colombia showing the 14 sections used for the analysis and the case study well (modified from Dengo & Covey 1993).

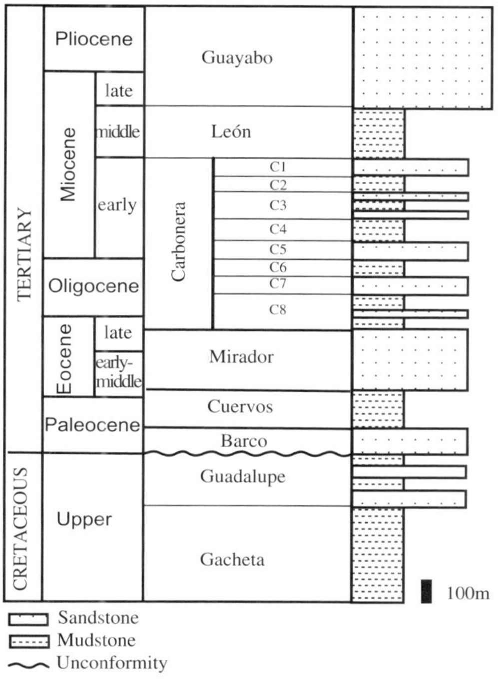

Fig. 2. Generalized stratigraphic column for the Llanos and Llanos foothills basin (after Ecopetrol 1996).

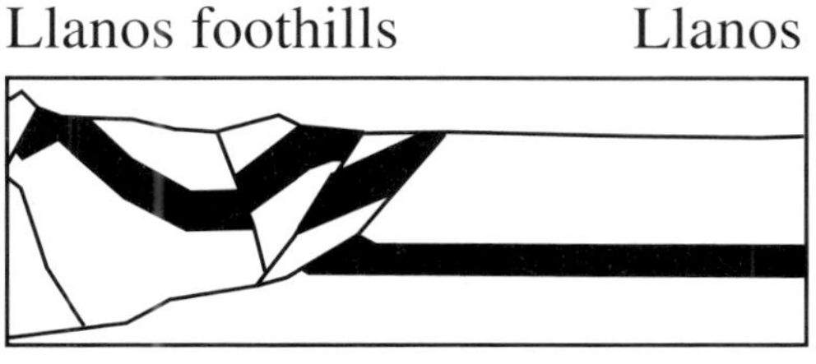

Fig. 3. Schematic structural cross section of the Llanos and Llanos foothills basin (after Linares 1999).

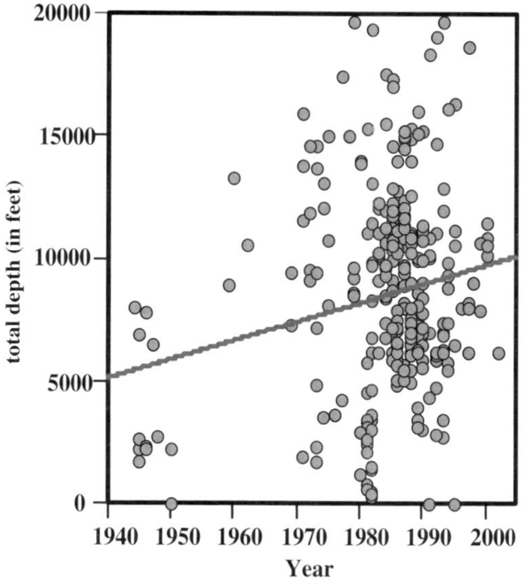

Fig. 4. Bivariate plot of depth of exploratory wells versus drilling year. There is a trend toward increasing the depth of exploratory wells through the years (trend calculated by linear regression, $r = 0.18$, $p < .005$).

versus the top of the Mirador Formation in biostratigraphic terms. This relatively low level of biostratigraphic resolution did not matter for wells drilled during early exploration of the Llanos, as biostratigraphy was mainly used as a rough control during drilling. This scenario has now changed. Exploration has moved into complex areas where more refined biostratigraphy is very important. Thus, biostratigraphy has become an important tool in assessing the stratigraphic position of a well during drilling, in testing diverse seismic, structural and stratigraphic interpretations, and in correlating to understand reservoir continuity.

In this study, a reliable biostratigraphic framework is presented that was developed for the Cuervos and equivalent formations (late Paleocene to earliest Eocene) of the Llanos and Llanos foothills basins based on palynological data and graphic correlation. This framework is then tested to solve a problem that originated during the drilling of an exploratory well in the northern Llanos foothills.

Geological setting

The Cuervos Formation, as it is known by the petroleum industry in the Llanos foothills of Colombia, is a fine-grained, coaly unit located in a stratigraphic position between the two main reservoirs in the region, the Mirador and Barco Formations (Ecopetrol 1996).

The Cuervos Formation was named by Notestein *et al.* (1944) after the Quebrada Los Cuervos in the Catatumbo area. The formation is composed mainly of claystones and shales, with some coal beds in its lower part; the upper part consists of grey and greenish-grey claystones with locally abundant red, yellow, and purple mottling (Notestein *et al.* 1944). Its thickness ranges from 282 to 490 m and increases northwards. The contact with the overlying Mirador Formation is sharp, and locally unconformable (Notestein *et al.* 1944). In the area of La Victoria creek, Germeraad *et al.* (1968) dated the Cuervos Formation as Paleocene (pollen zone of *Retidiporites magdalenensis*). In the Tibú anticline, Gonzalez (1967) dated the Upper Cuervos as Late Paleocene–Early Eocene based on pollen.

The Cuervos lithostratigraphic unit in the Llanos foothills has also been named the 'Arcillas de El Limbo' Formation, defined by Hubach in 1941 (in a Shell internal report, according to Van der Hammen 1958). The type locality is near El Limbo, Cravo Sur River, Colombia. It has been dated as Early Eocene (Cazier *et al.* 1995) or Late Paleocene (Cooper *et al.* 1995) although no biostratigraphic data was published to support the dating.

Methods

The most reliable biostratigraphic tools in the Paleocene of the Llanos and Llanos foothills basin are pollen and spores (Rubio 1997). Fourteen sections with palynological information from outcrops (4), well cores (4), and well ditch cuttings (6) were used for this study (Appendix 1). The outcrop data comes from published information: Piñalerita and Regadera (Jaramillo 2002; Jaramillo & Dilcher 2001), Sutatauza (Sarmiento 1992) and Colorada (Guerrero & Sarmiento 1996). The core and ditch-cuttings data come from internal reports in Ecopetrol databases (Ecopetrol Petroleum Information Bank). All sections are located in the central and northern Llanos foothills and in the Llanos basin near the foothills (Fig. 1, Table 1).

The palynological data were then analyzed using the technique of graphic correlation (Edwards 1984, 1989; Shaw 1964). This technique does not assume *a priori* that any taxon is an index guide or represents a time line. Instead, the technique relies on looking at the whole assemblage to find the true stratigraphic range for each taxon (Edwards 1989), which is done by comparing taxon ranges in different sections and finding lines of correlation.

A detailed taxonomic filtering was done before using the data for the graphic correlation. For example, informal taxa (used by consultants) or some generic assignments were not used in the analysis (e.g. *Psilatricolpites* 'inciertus', *Psilatricolporites* sp.). First occurrence datums (FAD) were not used in wells that only had ditch cutting samples in order to minimize the error introduced by cavings into the wellbore during drilling.

Table 1. *Geographic location of the studied sections*

Section	N	W
ara.3	7.0	–71.8
ara.4	7.0	–71.8
bah.15	5.0	72.7
ba.3	5.0	–72.7
cu.5	5.1	–72.7
guac	4.3	–73.4
rc.1	4.9	–72.7
sanmar	6.0	–71.6
tame.2	6.4	–71.8
toco	6.3	–71.8
Colorada	5.3	–73.5
Piñalerita	4.5	–73.1
Regadera	7.4	–72.4
Sutatauza	5.1	–73.5

A spreading parameter was developed to evaluate the degree of confidence in the position of a datum in the Composite Section. Each datum in the Composite Section represents its oldest FAD or youngest last occurrence datum (LAD) after comparing all sections with the Reference Section. For example, the position of the FAD in the Composite Section of species A is 65 composite units (Fig. 5). This value means that when comparing all the sections with the Composite, the oldest position for this datum is 65 composite units. However, there are other values for this datum that are not included. These values come from other sections that have younger FADs when projected into the Composite Section (Fig. 5). The spreading parameter evaluates this degree of dispersion. A low value of the spreading parameter indicates that the datum has a similar position in the Composite regardless of what section is being analysed, and is therefore a reliable biostratigraphic marker. A high value indicates a wide range for the datum. The spreading parameter for each datum in the Composite Section is calculated using one of the following formulas. For each FAD datum, the spreading parameter = (oldest FAD in composite units – youngest FAD in composite units)/50 (Fig. 5). For each LAD datum = (oldest LAD in composite units – youngest LAD in composite units)/50. The division by the number 50 in the formula is used to standarized the parameter to the exploratory needs in the Llanos foothills. The number 50 represents fifty stratigraphic feet in the Reference Section. This number was chosen based on the level of precision in giving a top that is expected by drilling engineers in the foothills. The value of standardization could be changed for other basins or exploratory programs.

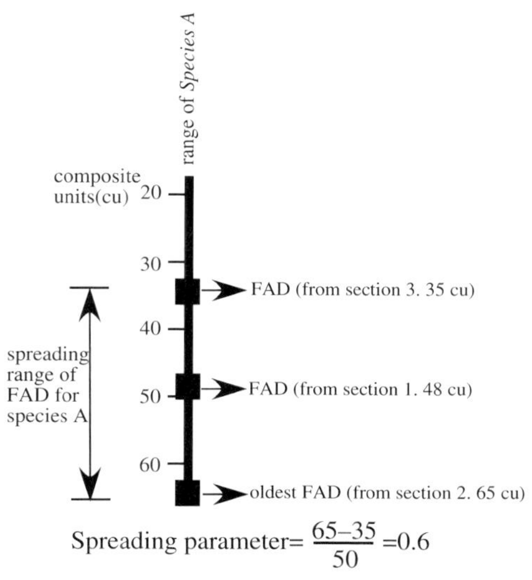

Fig. 5. Calculation of the spreading parameter for each datum (FAD and LAD) used in the analysis. For each FAD or LAD datum a spreading value was calculated using the formula: (youngest FAD or LAD in composite units – oldest FAD or LAD in composite units)/50.

Only taxa occurring in three or more wells were used to calculate this spreading parameter. Taxa occurring in less than three wells produce artificially low spreading values due to low sample size. They were considered to have an unknown spreading value. High spreading values (>10) indicate that the position of a particular FAD or LAD in the Composite Section is highly variable, within a range of 500 composite units, and therefore not very useful as a biostratigraphic marker. Using them would only introduce noise to the process of correlation.

Taxonomic nomenclature and illustration of the species used in this study can be found in diverse publications (Colmenares & Terán 1993; Doubinger 1973, 1976; Germeraad *et al.* 1968; Gonzalez 1967; Guerrero & Sarmiento 1996; Hopping 1967; Jan du Chêne *et al.* 1978*a*, 1978*b*; Jan du Chêne & Salami 1978; Jansonius & Hills 1976; Jaramillo & Dilcher 2001; Leidelmeyer 1966; Muller *et al.* 1987; Pocknall & Nichols 1996; Regali *et al.* 1974; Rull 1997, 1999; Samant & Phadtare 1997; Sarmiento 1992; Van der Hammen 1954, 1956*a*, *b*, 1957*a*, *b*, 1958; Van der Hammen & Garcia 1966; Van der Hammen & Wymstra 1964; Van Hoeken-Klinkenberg 1964, 1966; Venkatachala *et al.* 1988). Major nomenclatural changes follow those in Jaramillo & Dilcher (2001).

The Composite Section

The Composite Section was built using information from more than 105 000 palynomorph occurrences from 700 palynological samples. The BA3 section was used as the Reference Section because it has the best available information, highest density of samples, the most reliable fossil data set, and is stratigraphically the most complete. Using the technique of graphic correlation, three rounds of correlation were performed. A line of correlation was determined for each section (Table 2). Based on these correlation lines, a Composite Section was built (Table 3, Fig. 6). The Composite also contains the top of the Cuervos and Barco Formations as they were defined in the Reference Section. Fifty taxa with spreading values <10 and twelve with unknown spreading values but with a good potential for biostratigraphy (see Table 3) were plotted in Figure 6.

Biozonation for the Cuervos Formation

The Composite Section (CS) produced here allows dividing the Cuervos Formation into five informal biozones that occur along the Llanos and Llanos foothills (Fig. 6).

Zone Cu-01

Base: FAD of *Bombacacidites annae*. Position in CS: 16114 cu (composite units).

Top: FAD of *Foveotricolpites perforatus*. Position in CS: 16112 cu.

Assemblage: Bombacacidites annae, Ephedripites vanegensis, Foveotriletes margaritae, Gemmamonocolpites gemmatus, Retidiporites magdalenensis, Spinizonocolpites baculatus. Within this zone the last occurrence datum (LAD) of *Palaeocystodinium golzowense* occurs.

Comments: This zone is characterized by high abundances of *Proxapertites operculatus*.

Stratigraphic position: Uppermost Barco Formation to lower Cuervos Formation.

Zone Cu-02

Base: FAD of *Foveotricolpites perforatus*. Position in CS: 16112 cu.

Top: LAD of *Foveotricolpites perforatus*. Position in CS: 15995 cu.

Assemblage: Bombacacidites annae, Retidiporites magdalenensis. Within this zone there

Table 2. *Lines of correlation for each of the sections studied*

Section							
Piñalerita	X	16115	16110	16110	16003	15558	15350
	Y	6485	6296	6301	5698	4999	4392
Regadera	X	15440	15086				
	Y	6398	5601				
Colorada	X	16285	16273	16228	16107		
	Y	2646	2396	1970	1804		
Sutatauza	X	16285	16285				
	Y	6563	3846				
bah.15	X	16165	16164	16115	16109	15875	
	Y	15143	15143	15050	15009	14726	
rc.1	X	16432	16320	16283			
	Y	14822	14682	14677			
cu.5	X	16285	16114	15594	15509	15359	
	Y	17776	17527	17126	17067	16960	
ara.3	X	16285	16285	15786	15545		
	Y	19527	17968	17870	17870		
ara.4	X	16285	16285	15767	15342		
	Y	19231	18566	18010	18010		
guac	X	16269	15434	15099			
	Y	13320	13320	12764			
sanmar	X	16350	16283	15506	15109		
	Y	12068	11387	11387	11227		
tame.2	X	16285	15869	15550	15119		
	Y	14823	14649	14445	14282		
toco	X	16286	15579	15237	15100		
	Y	17732	17339	17166	17145		

X, values from reference section.
Y, values from correlated section.
Each X and Y couplet corresponds to a point along the line of correlation.

are the LADs of *Ephedripites vanegensis, Foveotriletes margaritae, Gemmamonocolpites gemmatus and Spinizonocolpites baculatus*.

Comments: This zone is characterized by acmes of *Proxapertites operculatus*. The top of this zone coincides with the LAD of acmes of *P. operculatus*.

Stratigraphic position: Lower Cuervos Formation.

Zone Cu-03

Base: LAD of *Foveotricolpites perforatus*. Position in CS: 15995 cu.

Top: Base of barren interval. Position in CS: 15892 cu.

Comments: Within this interval occurs the LAD of *Bombacacidites annae* and *Retidiporites magdalenensis*.

Stratigraphic position: Lower part of middle Cuervos Formation.

Zone Cu-04

Base: Base of barren interval. Position in CS: 15892 cu.

Top: FAD of *Tetracolporopollenites maculosus*. Position in CS: 15588 cu.

Assemblage: This is a barren interval with very poor recovery of organic matter and palynomorphs.

Stratigraphic position: Middle to upper Cuervos Formation.

Zone Cu-05

Base: FAD of *Tetracolporopollenites maculosus*. Position in CS: 15588 cu.

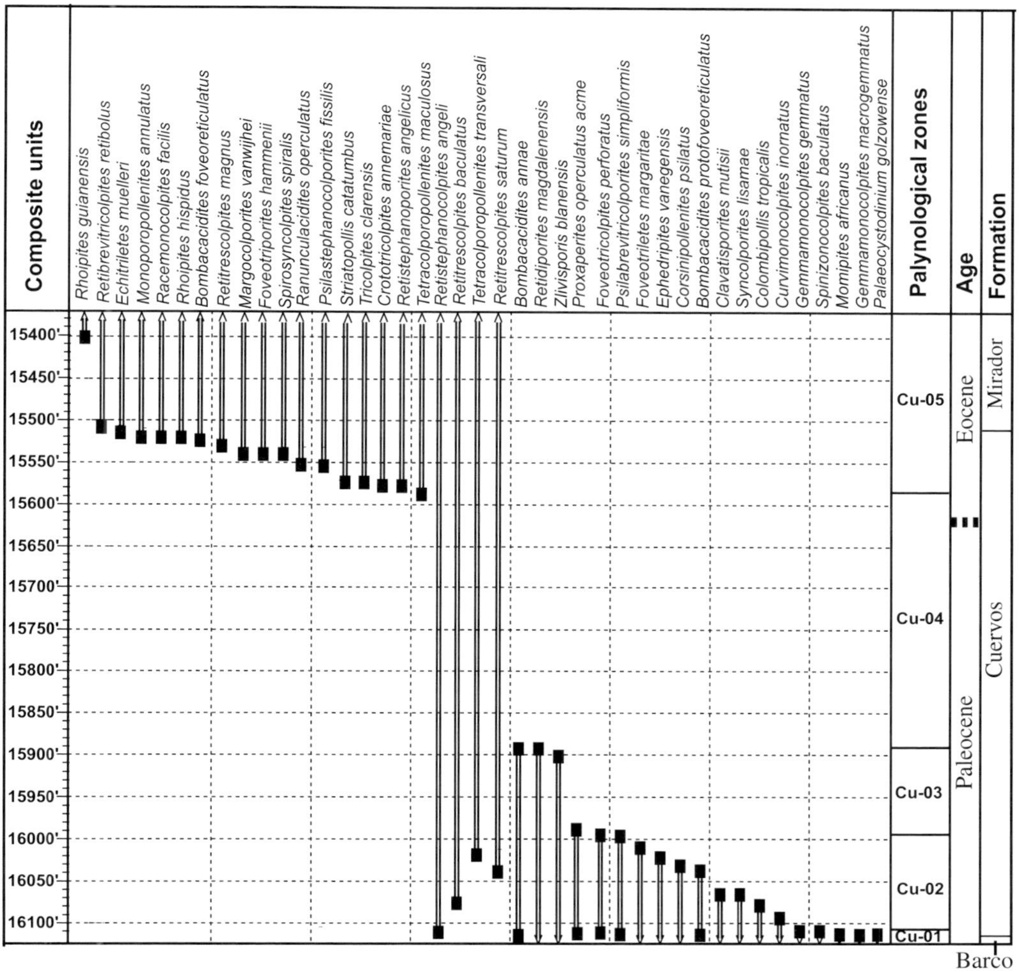

Fig. 6. Biostratigraphic range chart of the key taxa for the Cuervos Formation. Vertical units are composite units that correspond to feet in the Reference Section BA3. Position of the formational boundaries follow those defined in the Reference Section.

Top: FAD of *Rhoipites guianensis*. Position in CS: 15386 cu.

Assemblage: Crototricolpites annemariae, Tricolpites clarensis, Striatopollis catatumbus. Within this zone occurs the FAD of *Striatopollis catatumbus, Echitriletes muelleri, Foveotriporites hammenii, Margocolporites vanwijhei, Monoporopollenites annulatus, Psilastephanocolpites fissilis, Racemonocolpites facilis, Ranunculacidites operculatus, Retibrevitricolpites retibolus, Retitrescolpites magnus, Rhoipites hispidus* and *Spirosyncolpites spiralis*.

Comments: FAD of many species commonly associated to Eocene strata occurs in this zone.

Stratigraphic position: Upper Cuervos to lower Mirador Formation.

Calibration of Composite Section

The calibration of the Composite Section developed here with the geological timescale is still uncertain. There are very few published studies in northern South America that have both palynological information and planktonic foraminifera or nannoplankton that could be used for a first order calibration of the palynological data. The palynological zonation of Germeraad *et al.* (1968) is the only one for northern South America that shows

Table 3. *Composite section for the Cuervos Formation and equivalent units in the Llanos and Llanos foothills*

	Species	Author	Event	CU	S	N
1	*Rhoipites guianensis*	(Van der Hammen & Wymstra 1964) Jaramillo & Dilcher 2001	FAD	15386.1	0.7	9
2	*Cuervos*		top	15510		
3	*Retibrevitricolpites retibolus*	Leidelmeyer 1966	FAD	15512.6	7.0	3
4	*Echitriletes muelleri*	Regali *et al.* 1974	FAD	15516.2	5.6	7
5	*Bombacacidites foveoreticulatus*	Muller *et al.* 1987	FAD	15523.0	12.1	6
6	*Monoporopollenites annulatus*	(Van der Hammen 1954) Jaramillo & Dilcher 2001	FAD	15523.0	3.3	8
7	*Racemonocolpites facilis*	Gonzalez 1967	FAD	15523.0	3.4	3
8	*Rhoipites hispidus*	(Van der Hammen & Wymstra 1964) Jaramillo & Dilcher 2001	FAD	15523.0	7.0	8
9	*Retitrescolpites magnus*	(Gonzalez 1967) Jaramillo & Dilcher 2001	FAD	15530.3	4.1	3
10	*Spirosyncolpites spiralis*	Gonzalez 1967	FAD	15537.6	3.6	10
11	*Foveotriporites hammenii*	Gonzalez 1967	FAD	15537.7	4.5	5
12	*Margocolporites vanwijhei*	Germeraad *et al.* 1968	FAD	15538.0	3.6	6
13	*Ranunculacidites operculatus*	(Van der Hammen & Wymstra 1964) Jaramillo & Dilcher 2001	FAD	15553.8	7.6	8
14	*Psilastephanocolporites fissilis*	Leidelmeyer 1966	FAD	15556.2	5.0	5
15	*Striatopollis catatumbus*	(Gonzalez 1967) Takahashi & Jux 1989	FAD	15574.5	4.3	10
16	*Tricolpites clarensis*	(Gonzalez 1967) Jaramillo & Dilcher 2001	FAD	15574.6	7.5	4
17	*Crototricolpites annemariae*	Leidelmeyer 1966	FAD	15578.4	6.4	3
18	*Retistephanoporites angelicus*	Gonzalez 1967	FAD	15579.1	10.1	9
19	*Tetracolporopollenites maculosus*	(Regali *et al.* 1974) Jaramillo & Dilcher 2001	FAD	15588.3	7.3	5
20	*Bombacacidites annae*	(Van der Hammen 1954) Leidelmeyer 1966	LAD	15892.7	5.0	5
21	*Retidiporites magdalenensis*	Van der Hammen & Garcia 1966	LAD	15892.7	7.8	6
22	*Zlivisporis blanensis*	Pacltova 1961	LAD	15902.1	11.6	4
23	*Proxaperites operculatus acme*	(Van der Hammen 1954) Van der Hammen 1956	LAD	15988.5	2.1	3
24	*Foveotricolpites perforatus*	Van der Hammen & Garcia 1966	LAD	15995.2	2.9	4
25	*Psilabrevitricolporites simpliformis*	Van der Kaars 1983	LAD	15996.2	unknown	1
26	*Foveotriletes margaritae*	(Van der Hammen 1954) Germeraad *et al.* 1968	LAD	16010.2	5.9	6
27	*Tetracolporopollenites transversali*	(Dueñas 1980) Jaramillo & Dilcher 2001	FAD	16020.2	13.4	8
28	*Ephedripites vanegensis*	Van der Hammen & Garcia 1966	LAD	16022.3	5.3	3
29	*Corsinipollenites psilatus*	Jaramillo & Dilcher 2001	LAD	16031.6	unknown	1
30	*Bombacacidites protofoveoreticulatus*	Jaramillo & Dilcher 2001	LAD	16037.8	unknown	1
31	*Retitrescolpites saturum*	(Gonzalez 1967) Jaramillo & Dilcher 2001	FAD	16041.9	17.6	3
32	*Clavatisporites mutisii*	(Van der Hammen 1954) Jaramillo & Dilcher 2001	LAD	16065.8	unknown	2
33	*Syncolporites lisamae*	Van der Hammen 1954	LAD	16065.8	unknown	1
34	*Retitrescolpites baculatus*	Jaramillo & Dilcher 2001	FAD	16079.3	unknown	1
35	*Colombipollis tropicalis*	Sarmiento 1992	LAD	16079.3	unknown	2
36	*Curvimonocolpites inornatus*	Leidelmeyer 1966	LAD	16094.3	unknown	1
37	*Gemmamonocolpites gemmatus*	(Van der Hammen 1954) Van der Hammen & Garcia 1966	LAD	16109.6	3.5	5
38	*Spinizonocolpites baculatus*	Muller 1968	LAD	16110.1	3.5	5

Table 3. *(continued)*

	Species	Author	Event	CU	S	N
39	*Foveotricolpites perforatus*	Van der Hammen & Garcia 1966	FAD	16112.2	0.5	6
40	*Retistephanocolpites angeli*	Leidelmeyer 1966	FAD	16112.7	16.8	4
41	*Psilabrevitricolporites simpliformis*	Van der Kaars 1983	FAD	16113.7	unknown	1
42	*Momipites africanus*	Van Hoeken Klinkenberg 1966	LAD	16113.7	unknown	1
43	*Palaeocystodinium golzowense*	Alberti 1961	LAD	16113.9	2.2	3
44	*Bombacacidites protofoveoreticulatus*	Jaramillo & Dilcher 2001	FAD	16114.3	unknown	1
45	*Momipites africanus*	Van Hoeken Klinkenberg 1966	FAD	16114.3	unknown	1
46	*Gemmamonocolpites macrogemmatus*	Muller *et al.* 1987	LAD	16114.3	2.2	3
47	*Proxaperites operculatus acme*	(Van der Hammen 1954) Van der Hammen 1956	FAD	16115.0	2.1	3
48	*Bombacacidites annae*	(Van der Hammen 1954) Leidelmeyer 1966	FAD	16115.0	0.1	7
49	*Barco*		top	16115		

CU, composite units; FAD, first appearance datum; LAD, Last appearance datum; S, spreading parameter; N, number of wells where datum occurs, Cuervos and Barco Formation tops are also included (as they were defined in reference section).

independent data for a calibration with the geological timescale. These authors used planktonic foraminifera in several sections from South America and western Africa. However, the foraminiferal data presented for the Paleocene of South America were scarce and their stratigraphic and geographic positions were not reported. Without this information, the use of pollen and spores for chronological dating must be carefully considered in the region. The other zonation commonly used in the area is that of Muller *et al.* (1987) which was based on the Germeraad *et al.* (1968) zonation. Muller's scheme subdivided the Paleocene of NW South America into three palynological zones similar to those of Germeraad *et al.* (1968). Comparison of the Germeraad *et al.* (1968) and Muller *et al.* (1987) zonations with the Composite Section allows a preliminary calibration of the Composite Section. We expect to generate data in the coming years to obtain first-order calibrations of the palynological data.

According to the Germeraad *et al.* (1968) and Muller *et al.* (1987) zonations, the Cu-05 zone would correspond to the early Eocene zones of *Retibrevitricolpites triangulatus* (Germeraad *et al.* 1968) or *Rugotricolpites felix* (Muller *et al.* 1987) because of the FAD of *Striatopollis catatumbus*. Neither zonation recognizes the barren interval that is present along the Llanos foothills, zone Cu-04. Facies characteristics of this interval are mottled shales, which seem related to intense oxidation in well drained alluvial soils. This phenomenon is regional and could be linked to the abrupt global heating at the Paleocene–Eocene boundary (Jaramillo 2002). The zone Cu-03 cannot be identified in the Germeraad *et al.* (1968) and Muller *et al.* (1987) zonations. They place the late Paleocene *Foveotricolpites perforatus* zone underneath the Eocene *Retibrevitricolpites triangulatus* (Germeraad *et al.* 1968) or *Rugotricolpites felix* (Muller *et al.* 1987) zones. The zone Cu-03 can be found between these two zones. It has the LAD of *Bombacacidites annae* and *Retidiporites magdalenensis*, two taxa that are restricted to the Paleocene zones of both Germeraad *et al.* (1968) and Muller *et al.* (1987). The zone Cu-02 could correspond to the late Paleocene zone of *Foveotricolpites perforatus* of Germeraad *et al.* (1968) and Muller *et al.* (1987). Lastly, the zone Cu-01 could correspond to the Paleocene *Ctenolophonidites lisamae* zone of Germeraad *et al.* (1968) or *Gemmastephanocolpites gemmatus* of Muller *et al.* (1987) based on the presence of *Bombacacidites annae* and absence of *Foveotricolpites perforatus*. However, *Ctenolophonidites lisamae* and *Gemmastephanocolpites gemmatus*, two key elements of this zone, have not been reported in the studied sections used here.

The base of the Paleocene does not occur within the Composite Section, therefore the Composite Section can be considered as Paleocene in the interval from15 892 to 16 115 composite units, zones Cu-01 to Cu-03 (Fig. 6). A more detailed calibration of the Composite for the Paleocene in the Llanos and Llanos foothills is still uncertain.

Case study

The biostratigraphic framework produced for the Cuervos Formation was used for the interpretation of a well recently drilled in the northern Llanos foothills. During the original drilling, the stratigraphic position of the final depth of the well was difficult to establish. Incompatible biostratigraphic interpretations were produced by four different biostratigraphic consultants, ranging from Paleocene to upper Cretaceous for the deepest 4000 feet of the well. The final interpretation was that the Upper Cretaceous, Guadalupe Formation had been reached and that the strata were dipping almost horizontally (Fig. 7, see previous interpretation). Given the complexity of the problem, Ecopetrol, the state-owned Colombian petroleum company, formed a team that included structural geologists, petrographers, sedimentologists, stratigraphers and biostratigraphers to study the well.

The Composite Section produced in this study was used to re-interpret the well. Four palynological zones for the Cuervos Formation were identified and it was interpreted that the last 4289 feet drilled were in the Cuervos Formation, that the well was moving stratigraphically toward older rocks, and that no faults or repetitions were evident (Fig. 7). The top of the Cuervos Formation was determined to be at 7711 feet and the well ended at 12 000 feet in the zone Cu-02, Lower Cuervos (Fig. 7).

Several other sources of data, such as a Formation Micro-Imaging reinterpretation, petrographic variations and log interpretation, led to a similar conclusion (Meza *et al.* 2002), that the well was drilled in the Cuervos Formation for almost 4000 feet in beds that were steeply dipping (~60 degrees, Fig. 7).

Based on this interpretation, a drilling program was put together in order to reach the Barco Formation by drilling a sidetrack well. The kickoff point was at 10600′ in the original

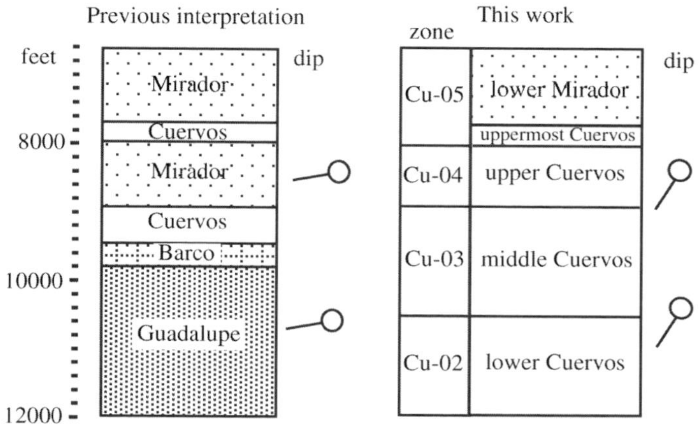

Fig. 7. Biostratigraphic interpretations of the well in the case study. Note that in the previous interpretation the Guadalupe Formation was drilled and strata are almost horizontal. The interpretation presented in this work based on the Composite Section suggests that the Cuervos Formation was drilled for the last 4289 feet and strata are steeply dipping (~60 degrees). Dip measurements coming from interpretation of a Formation Micro-Imaging log .

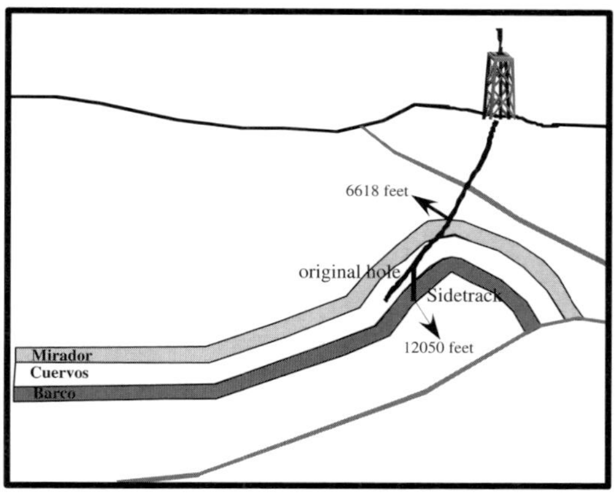

Fig. 8. Schematic geological cross-section of the case study well. The Barco Formation was reached at 11550 feet (measured depth) in the sidetrack and drilled until final depth (12 050 feet). Note that the original hole drilled in the Cuervos Formation is almost parallel to the bedding.

abandoned well (Meza *et al.* 2002). The sidetrack successfully reached the Barco Formation after drilling 1500 feet as the geological model had predicted (Fig. 8). The model was confirmed not only by palynology but also petrographic, petrophysic data, and electric logs. This test confirmed that the biostratigraphic framework developed for the Cuervos Formation works and can be successfully applied to exploration in the Llanos foothills.

We thank the Instituto Colombiano del Petróleo for permission to publish this paper. We also thank: two anonymous reviewers, M. Rueda and all the palynological consultants working in Colombia for providing raw data and palynological input for this work; H. Bartels, E. Cadena, H. Antolines, F. Hernandez for helping in the biostratigraphic analysis; all the people who helped during the field season in the towns of Sabanalarga, and Cucuta, and E. Kowalski for reading the manuscript. Special thanks go to M. I. Barreto for her continuous support and sources of ideas.

Appendix 1. *FAD and LAD's for the taxa used in the analysis. Tops (in feet) of the Cuervos and Barco Formations are also given.*

	ba.3		Piñalerita		Colorada		rc.1		Sutatuza		tame.2	
Species	FAD	LAD	FAD	LAD	FAD	LAD	FAD	LAD	FAD	LAD	FAD	LAD
Sample type	c	c	o	o	o	o	c	c	o	o	d	d
Aglaoreidia? foveolata			6300	6300								
Bombacacidites annae	16114	16000	6460	5539	1804						14664	14664
Bombacacidites foveoreticulatus	15523	15214	3986				13922	13711				
Bombacacidites protofoveoreticulatus			6460	5893								
Brevitricolpites macroexinatus			4072									
Brevitricolpites microechinatus			5857	5857								
Clavatisporites mutisii			6052	6052					6055	5307		
Colombipollis tropicalis				6129		2068						
Corsinipollenites psilatus				5857								
Crassiectoapertites columbianus	15575	15230					14137.3	13741.3			14285	13565
Cricotriporites guianensis			5001	5001			13724.3	13724.3				
Crototricolpites annemariae	15578	9450					13819	8170				
Crototricolpites protoannemariae			6460	5686								
Curvimonocolpites inornatus				6215								
Echiperiporites akanthos	15538	15516					13763.3	13746.7			14225	11355
Echitriletes muelleri	15516						13804				12705	11355
Ephedripites vanegensis	16057	16022	6300	6129					5403	3949		
Foveotricolpites perforatus	16112	16000	6294	5686	1660							
Foveotriletes margaritae	16730	16010			2412	2373	15087	14604	5622			
Foveotriporites hammenii	15538	15214	4393	4016			13925	13747				
Gemmamonocolpites gemmatus	16248	16242		6300			14671	14633	6101	3949		
Gemmamonocolpites macrogemmatus	16264	16223					14633	14633				
Heterocolpites palaeocenica	16267	16041										
Ischyosporites problematicus			5916	5001								
L. proxapertitoides reticuloides			5001	4016								
Longapertites microfoveolatus			6129	5124								
Longapertites vaneendenburgi	16231	15575			1974.6				5398			
Margocolporites vanwijhei	15526	9450					13925	13925			13785	12905
Momipites africanus			6460	6439								
Monocolpopollenites ovatus			6300									
Monoporopollenites annulatus	15523		4472	4016			13955				13725	
Palaeocystodinium golzowense	16114	16114					14671	14642				
Paleosantalaceaepites distinctus			5916	5857								
Polypodiisporites echinatus			6129	6129								
Polypodiisporites pachyexinatus			5539	5539								
Proxapertites operculatus acme	16114	16092	6300	6215								
Psilabrevitricolporites simpliformis			6439	5686								
Psilastephanocolporites fissilis	15362	15227	4350	4016			14137	13826				
Racemonocolpites facilis	15523	15214	4472				13922	13763				
Ranunculacidites operculatus	15554	9450	4229	4229			13925				14255	12705
Retibrevitricolpites retibolus	15513	15214	4229	4229			13738	13738				
Retidiporites elongatus			5916	5916					5619	4185		
Retidiporites magdalenensis	16267	16000		5686	2458		14671	14577		4185		
Retidiporites poricostatus			6300	6300								
Retistephanocolpites angeli	15579	15259	6406	5539			13829	13734				
Retistephanocolpites williamsi		15588										
Retistephanoporites angelicus	15579	9750	4138				13826	9550			14285	13565
Retitrescolpites baculatus			6129									
Retitrescolpites magnus	15530	15230	4393	4072								
Retitrescolpites peculiaris			6294	5686								
Retitrescolpites saturum	15383	15204	5916	5686			13738	13711				
Retitriporites amplireticulatus			5916	5916								
Rhoipites guianensis	15386	15217	4393				13931				14095	12395
Rhoipites hispidus	15523	15214	4229				13885	13711			14375	12505
Spinizonocolpites baculatus	16267	16233			3189	2373	14671	14633				
Spinizonocolpites breviechinatus			6460	5686								
Spinizonocolpites pachyexinatus			5539	5539								
Spirosyncolpites spiralis	15538	10890	4472				13955	7990			14335	13115
Striatopollis catatumbus	15575	10080	4472				13955				14315	10845
Striatricolporites digitatus			6439	6439								
Syncolporites lisamae			6052	6052					4257			
Tetracolporopollenites maculosus	15588	10720	4275				13829	10928				
Tetracolporopollenites spongiosus			6439	5539								
Tetracolporopollenites transversalis	15579	15214	5792	4153			13955	13747				
Tricolpites clarensis	15575	15214	5001				13783	13763				
Tricolpites protoclarensis				5686								
Verrustephanocolpites rugulatus			5001	4275								
Verrutricolporites reticulatus			4500									
Zlivisporis blanensis	16233	16010	5857	5539					5622	4215		
Cuervos		15510		4995				14087				14420
Barco		16115		6560		1804		14478				

FAD, first appearance datum. LAD, last appearance datum. All datums are in feet, in subsurface mode (even the outcrops). o, outcrop, c, core, d, ditch-cutting.

toco		sanmar		Regadera		bah.15		cu.5		ara.3		ara.4		guac	
FAD d	LAD d	FAD d	LAD d	FAD o	LAD o	FAD c	LAD c	FAD c	LAD c	FAD d	LAD d	FAD d	LAD d	FAD d	LAD d
				5696	5696	15045				17940	17940	18145	18145		
								16809	16809					12765	12765
				6284	6245										
					6458										
17075	13265	11225	10825												
					6230										
								17092	16913						
17307	14545	10675	8835									17945	17945		
16855	13445	9755	8715					16957		16055	15255				
						15039									
						14982				17940	17940	18255	18255		
						15292		18251	17782	18585	18345		18695		
				6235	6084			16945	16809						
							15141								
						15287	15045								
						15145	14957								
						14904									
16505	16445	10535	10535		6458			17087	16942						
					5696										
17185	14545	11225	8835	6284	5604			16960	16863						
						15292	15251								
					6142										
						14869	14863								
16905	16905							17091	17091						
15505	13855	11225	9465					17085	16939	15555	14905				
						15287				17940	17940	18695	18145		
								17098	17098						
17325	16105	11325	11325	5604								17945	17805	13035	12855
								16948	16942						
17185	15305	11325	9285							16055	16055	17905	17805	12765	
17307	14945	11325	10155	6234	6230									13095	13095
							15015						18695		
17307	15795	11325	9465		5604			16960	16809	17515	17515	17695		13155	12645
17325	14345	10015	8415	6235	5856			16960	16809			18005	18005	12975	12975
				6284				16960							
15305	15305	9465	8925	6235	6142			16957	16957	17235	17235				
				6284	6234										
				6284	5604										
								18386	18036						
					6560		14162		17067		17870		18010		13015
	17350						15050		17526						

References

CAZIER, E.C., HAYWARD, A.B, ESPINOSA, G., VELANDIA, J., MUGNIOT, J.F. & LEEL, W.G., JR. 1995. Petroleum geology of the Cusiana Field, Llanos Basin foothills, Colombia. *AAPG Bulletin*, **79**, 1444–1463.

COLMENARES, O.A. & TERÁN, L. 1993. A biostratigraphic study of Paleogene sequences in southwestern Venezuela. *Palynology*, **17**, 67–89.

COOPER, M.A., ADDISON, F.T. *ET AL.* 1995. Basin development and tectonic history of the Llanos basin, Eastern Cordillera and Middle Magdalena Valley, Colombia. *AAPG Bulletin*, **79**, 1421–1443.

DENGO, C.A. & COVEY, M.C. 1993. Structure of the Eastern Cordillera of Colombia: implications for trap styles and regional tectonics. *AAPG Bulletin*, **77**, 1315–1337.

DOUBINGER, J. 1973. Pollen and spores from the Paleocene coal basin of Cerrejon (Guajira Province, Colombia). *Proceedings, 96 Congres National Des Sociétés Savantes (Toulouse, 1971)*, **5**, 253–262.

DOUBINGER, J. 1976. Spore-pollen assemblages from the Tertiary of Colombia. *Acts, 97 Congres National Des Sociétés Savantes (Nantes)*, **4**, 9–18.

DZOU, L., HOLBA, A., RAMÓN, J., MOLDOWAN, M. & ZINNIKER, D. 1999. Application of new diterpane biomarkers to source, biodegradation and mixing effects on the Central Llanos Basin oils, Colombia. *Organic Geochemistry*, **30**, 5515–5534.

ECOPETROL 1996. *Opportunities for Hydrocarbon Exploration & Production in Colombia.* Ecopetrol, Bogotá.

EDWARDS, L.E. 1984. Insights on why graphic correlation (Shaw's method) works. *Journal of Geology*, **92**, 583–597.

EDWARDS, L.E. 1989. Supplemented graphic correlation: A powerful tool for paleontologists and nonpaleontologists. *Palaios*, **4**, 127–143.

FAJARDO, A., ROJAS, L. & CRISTANCHO, J. 2000. *Definición del modelo estratigráfico en el intervalo cretáceo tardío a mioceno medio en la cuenca llanos orientales y piedemonte llanero.* ICP Internal Report, Bucaramanga.

GERMERAAD, J.H., HOPPING, C.A. & MULLER, J. 1968. Palynology of Tertiary sediments from tropical areas. *Review of Palaeobotany and Palynology*, **6**, 189–348.

GONZALEZ, A.E. 1967. *A palynologic study on the upper Los Cuervos and Mirador formations (lower and middle Eocene), Tibú Area, Colombia.* E.J. Brill, Leiden.

GUERRERO, J. & SARMIENTO, G. 1996. Estratigrafía física, Palinológica, Sedimentológica y Secuencial del Cretácico Superior y Paleoceno del Piedemonte Llanero. Implicaciones en Exploración Petrolera. *Geología Colombiana*, **20**, 3–66.

HOPPING, C.A. 1967. Palynology and the Oil industry. *Review of Palaeobotany and Palynology*, **2**, 23–48.

JAN DU CHÊNE, R.E. & SALAMI, M.B. 1978. Palynology and Micropaleontology of the Upper Eocene of the well Nsukwa 1 (Niger Delta, Nigeria). *Archives des Sciences*, **13(1)**, 5–9.

JAN DU CHÊNE, R.E., ONYIKE, M.S. & SOWUNMI, M.A. 1978*a*. Some new Eocene pollen of the Ogwashi-Asaba Formation, South-Eastern Nigeria. *Revista Española de Micropaleontología*, **10**, 285–322.

JAN DU CHÊNE, R.E., PERCH-NIELSEN, K. & PETTERS, S.W. 1978*b*. New stratigraphic data on the Paleogene Ewekoro and Akinbo Formations (Southwestern Nigeria). *Archives des Sciences*, **31(3)**, 205–212.

JANSONIUS, J. & HILLS, L.V. 1976. *Genera File of Fossil Spores.* Special Publication, Dept. of Geology, University of Calgary, Calgary, Canada, cards 1–3287.

JARAMILLO, C. 2002. Response of tropical vegetation to Paleogene warming. *Paleobiology*, **28**, 222–243.

JARAMILLO, C. & DILCHER, D.L. 2001. Middle Paleogene Palynology of Central Colombia, South America: A Study of Pollen and Spores from Tropical Latitudes. *Palaeontographica B*, **285**, 87–213.

LEIDELMEYER, P. 1966. The Paleogene and Lower Eocene pollen flora of Guyana. *Leidse Geologische Mededelingen*, **38**, 49–70.

LINARES, R. 1999. *Modelamiento estructural de la Cordillera Oriental de Colombia.* ICP Internal Report, Bucaramanga.

MEZA, N., MUÑOZ, J., FAJARDO, A. & ARISTIZABAL, J. 2002. Evaluación geológica y de yacimientos pozo Gibraltar-Bloque Siriri informe final. Ecopetrol-ICP internal report, Bogotá.

MULLER, J., DI GIACOMO, E. & VAN ERVE, A. 1987. A palynologic zonation for the Cretaceous, Tertiary and Quaternary of Northern South America. *American Association of Stratigraphic Palynologist Contribution Series*, **19**, 7–76.

NOTESTEIN, F.B., HUBMAN, C.W. & BOWLER, J.W. 1944. Geology of the Barco Concession, Republic of Colombia, South America. *Bulletin of the Geological Society of America*, **55**, 1165–1216.

POCKNALL, D.T. & NICHOLS, D.J. 1996. *Palynology of coal zones of the Tongue River Member (upper Paleocene) of the Fort Union Formation, Powder River Basin, Montana and Wyoming.* American Association of Stratigraphic Palynologists Contributions Series 32. American Association of Stratigraphic Palynologists, Tulsa, Oklahoma, 58 pp.

REGALI, M., UESUGUI, N. & SANTOS, A. 1974. Palinologia dos sedimentos Meso-Cenozoicos do Brasil. *Boletin Tecnico da Petrobras*, **17**, 177–191.

RUBIO, R. 1997. *Esquemas bioestratigráficos Cuencas de Colombia.* Ecopetrol Internal Report, Bogotá.

RULL, V. 1997. Sequence analysis of western Venezuelan Cretaceous to Eocene sediments using palynology: chrono-paleoenvironmental and paleovegetational approaches. *Palynology*, **21**, 79–90.

RULL, V. 1999. Palaeofloristic and palaeovegetational changes across the Paleocene/Eocene boundary in northern South America. *Review of Palaeobotany and Palynology*, **107**, 83–95.

SAMANT, B. & PHADTARE, N.R. 1997. Stratigraphic palynoflora of the Early Eocene Rajpardi lignite,

Gujarat and the lower age limit of the Tarkeswar Formation of South Cambay Basin, India. *Palaeontographica Abt. B*, **245**, 1–108.

SARMIENTO, G. 1992. Palinología de la Formación Guaduas – estratigrafía y sistemática. Boletín *Geológico Ingeominas*, **32**, 45–126.

SHAW, A.B. 1964. *Time in Stratigraphy*. McGraw-Hill, New York.

VAN DER HAMMEN, T. 1954. The development of Colombian flora throughout geologic periods: I, Maestrichtian to Lower Tertiary. *Boletín Geológico (Bogotá)*, **2**, 49–106.

VAN DER HAMMEN, T. 1956*a*. Description of some genera and species of fossil pollen and spores. *Boletín Geológico (Bogotá)*, **4**, 103–109.

VAN DER HAMMEN, T. 1956*b*. A Palynological Systematic Nomenclature. *Boletín Geológico (Bogotá)*, **4(2–3)**, 63–101.

VAN DER HAMMEN, T. 1957*a*. Climatic periodicity and evolution of South American Maestrichtian and Tertiary Floras: a study based on pollen analysis in Colombia. *Boletín Geológico (Bogotá)*, **5**, 49–91.

VAN DER HAMMEN, T. 1957*b*. Palynologic stratigraphy of the Sabana de Bogota (East Cordillera of Colombia). *Boletín Geológico (Bogotá)*, **5**, 187–203.

VAN DER HAMMEN, T. 1958. Estratigrafía del Terciario y Maestrichtiano continentales y Tectonogénesis de los Andes Colombianos. *Boletín Geológico (Bogotá)*, **6(1–3)**, 67–128.

VAN DER HAMMEN, T. & GARCIA, C. 1966. The Paleocene pollen flora of Colombia. *Leidse Geologische Mededelingen*, **35**, 105–114.

VAN DER HAMMEN, T. & WYMSTRA, T.A. 1964. A palynological study on the Tertiary and Upper Cretaceous of British Guayana. *Leidse Geologische Mededelingen*, **30**, 183–241.

VAN HOEKEN-KLINKENBERG, P.M.J. 1964. A Palynological Investigation of some Upper-Cretaceous sediments in Nigeria. *Pollen et Spores*, **6(1)**, 209–231.

VAN HOEKEN-KLINKENBERG, P.M.J. 1966. Maastrichtian Paleocene and Eocene pollen and spores from Nigeria. *Leidse Geologische Mededelingen*, **38**, 37–48.

VENKATACHALA, B.S., CARATINI, C., TISSOT, C. & KAR, R.K. 1988. Paleocene–Eocene marker pollen from India and tropical Africa. *Palaeobotanist*, **37(1)**, 1–25.

Palynological zonation of Cenozoic non-marine sediments, Muglad Basin, Sudan

D. T. STEAD[1] & M. Z. AWAD[2]

[1]*Wellstrat Services Ltd., 1 Castle Grange, Wrexham Road, Caergwrle, Wrexham LL12 9HL, UK (e-mail: dstead@wellstrat.com)*

[2]*Greater Nile Petroleum Operating Company, Khartoum Tower, Gamhoria Street, P.O. Box 12527, Khartoum, Sudan*

Abstract: The Muglad Basin is part of the Central and West African Rift System that started to develop during late Jurassic/early Cretaceous time. The system evolved through three rift cycles spanning Berriasian to Cenomanian, Coniacian to Maastrichtian and Paleocene to Pliocene. In this work the results of a palynological study of selected wells from the Muglad Basin, Sudan are presented. The combination of samples from these wells has enabled a palynological analysis of a nearly complete Cenozoic section (Quaternary to Paleocene) which was developed during the last rift cycle. Five Zones have been recognized, most of which are further divided into two sub-zones. The Zones are (i) Gramineae – *Chenopodipollis microporatus* (Quaternary; Post-Adok Formation); (ii) *Verrucatosporites* spp. (Pliocene–Miocene; Adok Formation); (iii) *Cyathidites minor–Perfotricolpites digitatus* (Miocene–Oligocene; Tendi Formation); (iv) *Striatopollis* spp. (Oligocene–Eocene; Nayil Formation); (v) *Mauritiidites crassiexinus–Echimonocolpites rarispinosus* (Paleocene; Amal Formation). The characteristics of each zone and corresponding sub-zones are discussed.

A palynological analysis of over 1000 samples of Cenozoic and Mesozoic age, from the Muglad Basin, Sudan (Fig. 1), has been undertaken, as part of a multi-well correlation project by Wellstrat Services Limited and the Greater Nile Petroleum Operating Company. The results of this study reveal the presence of thick sequences of Cenozoic non-marine sediments in a number of exploration wells. Wells from which Cenozoic sediments have been discovered are Amal-1, Shelanugo-1, Jidian-1, Umm Batutu East-1, Kaikang-1, El Toor-6, Bamboo East-1, Kharief-1, Timsah-1, Tammor-1, Khariat-1, Heglig-1, Tafab-1 and Balome-1. By combining the data from these wells it has been possible to study a more or less complete Cenozoic section, from Paleocene to Quaternary. Of these wells Amal-1 alone provides a nearly complete Cenozoic section, of nearly 3500 m thickness, and it is the designated type well for the Cenozoic of the Muglad Basin. Palynological analysis of the samples has allowed the recognition of several distinct zones and events, allowing the construction of a formal zonation scheme. Samples studied were mostly ditch cuttings, although a small number of side wall cores were also made available. Consequently, the key events in the zonation presented are a combination of first downhole occurrences and abundance events.

Processing techniques

The majority of samples for analysis were supplied as rock samples (ditch cuttings). Processing of these samples follows a reasonably standard procedure. The precise method followed depends on the nature of the original sample and also on the microscopic appearance at various stages of processing. All samples are sieved using a 10 micron nylon mesh.

The process is as follows:

(1) original sample material pre-washed in nest of metal sieves;
(2) hydrochloric acid (HCl) to remove carbonates;
(3) hydrofluoric acid (HF) to remove silicates;
(4) hydrochloric acid (HCl) to remove silicofluoride precipitates;
(5) zinc bromide solution to float organics;
(6) preparation of unoxidized mount (kerogen);
(7) nitric acid or Schulze solution to clean the kerogen residues;
(8) potassium hydroxide or other alkali to remove humics;
(9) ultrasonic treatment;
(10) preparation of palynology mount.

From: POWELL, A. J. & RIDING, J. B. (eds) 2005. *Recent Developments in Applied Biostratigraphy*. The Micropalaeontological Society, Special Publications, 161–178. 1747-602X/$15.00

Fig. 1. Location of the Muglad Basin and other rifted basins in southern Sudan.

When samples are ready for oxidation a decision is made for each sample. That is to say bulk oxidation is not employed. This ensures that each sample receives the required amount of further treatment in order to maximize palynomorph recovery. In addition, gravity flotation methods may be used, in between steps as necessary.

Analytical procedures

(1) 200 specimen count where possible. Unfortunately many of the samples failed to yield sufficient palynomorphs to enable a 200-specimen count. In these cases an absolute count is made.
(2) Record number of traverses made. This enables an idea to be gained of palynomorph recovery.
(3) Scan preparation. Following a 200-specimen count the remainder of the preparation is scanned in order to log any forms that have not been recorded during the initial count.
(4) Additional forms recorded during the scan logged as 1.
(5) Note 'under-represented' forms. This is particularly necessary where one form dominates the assemblage.
(6) England Finder Reference taken for key/new forms. This enables the forms to be easily found at a later point for photographic purposes.

Geological setting

A full discussion of the geological history of the Muglad Basin is beyond the scope of this paper, and only a brief summary is given. Other published papers have covered this information in detail (e.g. Schull 1988; Giuraud & Maurin 1992; Giuraud *et al.* 1992; McHargue *et al.* 1992). There are several, deep, Mesozoic–Cenozoic rifted basins in southern Sudan (Fig 1.). The Muglad Basin is the largest of these, being up to 200 km wide, over 800 km long and contains up to 13 km of Cretaceous–Cenozoic sediments. The structure is NW–SE trending, and is part of a regionally linked, intracontinental Cretaceous Rift system across central Africa. The Muglad Basin was formed from the Late Jurassic to Early Cretaceous and the structural development was marked by three major rift cycles, each characterized by coarsening-upward sequences of clastic sediments. The first and the strongest of these was The Early Rifting which took place from the Late Jurassic to Early Cretaceous, lasting until the end of the Albian. This development may have been associated with the opening of the Atlantic. The intermediate rifting took place from the Cenomanian to Turonian. This period saw the development of rift lakes and the deposition of organic-rich lacustrine and flood plain sediments, generally of a fine nature. This is the Aradeiba Formation. The Late rifting took place from the Late Eocene to Oligocene. There was also gentle subsidence during the Middle Miocene. The assigned ages for these events are approximate, especially for the third stage – the late rifting. Poor age control in the Cenozoic has been a problem in the Muglad Basin. The palynozonation presented here goes some way to alleviating this problem.

The recognized Cenozoic formations from the Muglad Basin are shown in Figure 2. The Amal Formation is sandy in nature and may be reservoir quality. The Nayil Formation is shaly with sandy stringers. The Tendi Formation is sandier, with interbedded shales. The sandy stringers encountered in the Tendi and Nayil Formations may be oil bearing and be of some commercial interest. The Adok Formation is sandy. Little sedimentological information was available for the Umm Ruwaba and Zeraf Formations. Information on the different lithological units recorded from the Cenozoic of the Muglad Basin can be found in Schull (1988).

Procedure

The zonation presented below has been developed in two main stages.

The first is simply by collating information from published sources in comparable regions, notably Sudan, Central Africa and West Africa. Key papers include Germeraad *et al.* (1968), Medus (1975), Jan du Chêne *et al.* (1978), Legoux (1978), Salard-Cheboldaeff (1978, 1990), Kaska (1989), Oboh & Salami (1989), Takahashi & Jux (1989*a*, *b*), Oboh *et al.* (1992) and Awad (1994). Literature describing palynomorph distributions in South America have also been consulted. The assemblages from South America compared well with those published from West Africa sequences. However, comparison with the assemblages recovered from the Sudanese wells has been of limited use.

The second stage of development is through integrating the palynomorph assemblages with the wireline logs provided for the project. Formation tops can frequently be identified from the wireline log signatures. Where possible, distinctive palynomorph assemblages have been tied to these signatures. These assemblages may then be used as a correlation tool for those wells where the wireline log signature is less clear.

Published information concerning either palynological zones or assemblages from Sudan is very limited. Only two papers have provided information for the Cenozoic of Sudan. In 1989 Kaska produced a five-part spore/pollen zonation for the Muglad Basin. However, only two of these are from the Cenozoic, Zone D (Early Paleocene) and Zone E (Late Eocene to Oligocene), the remainder being Mesozoic. In 1994 Awad presented a palynozonation for the Kosti and Khartoum basins. Of these a single zone applies to the Cenozoic: Zone S-8 for the Paleocene. A further interval was described, but not zoned. This provides some useful information for the Oligocene to Miocene. Both of these studies provided more detailed information for the Mesozoic.

Palynozonation of the Muglad Basin

The zonation presented here ranges from Quaternary to Paleocene.

Zone 1: Gramineae – Chenopodipollis microporatus Zone

Suggested age: Quaternary

Formation: Post Adok. This includes the Umm Ruwaba Formation and the Zeraf Formation.

Note: Although the Post Adok sequence can be divided lithologically into the Umm Ruwaba and Zeraf formations, that information was not made available (i.e. wireline data) and insufficient samples were studied to enable a division of the formations. The problem is compounded by very poor recovery. The reader is referred to Schull (1988) for lithological information.

Definition: The occurrence of Gramineae and *Chenopodipollis microporatus* are constant features in this zone, which is also recognized by its stratigraphical position above zone 2.

Description: The key zonal forms are monoporate pollen grains assigned to the family

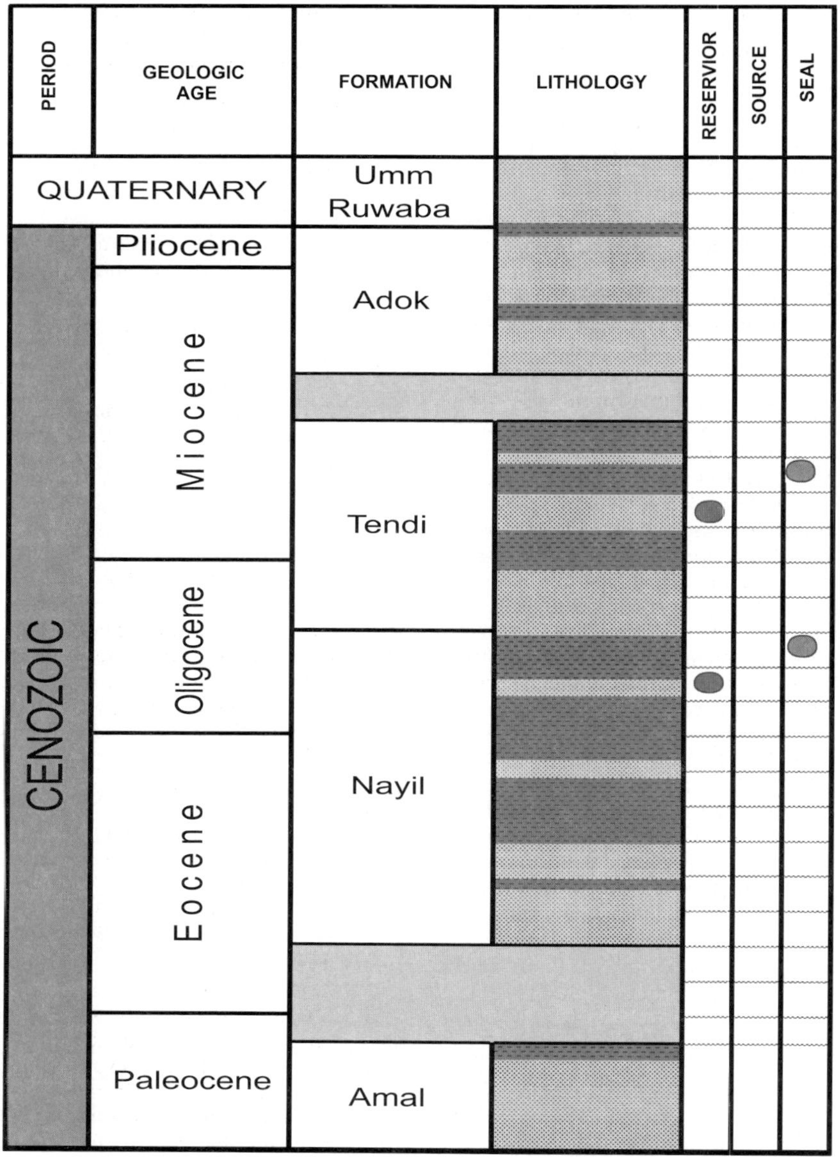

Fig. 2. Cenozoic Stratigraphy of the Muglad Basin.

Gramineae and relatively common *Chenopodipollis microporatus*. Recovery of sporomorphs from this zone is generally poor. Other forms recorded include *Alnipollenites* spp., *Betulaepollenites* spp., *Corylipollenites* spp., *Salixipollenites* spp. and also small tricolpate and tricolporate pollen. *Laevigatosporites haardtii*, *L. gracilis* and inaperturate pollen grains may also be recorded. Freshwater algae are most common, in some cases dominating the assemblages. *Botryococcus* spp. are the most common forms. *Pediastrum* spp. are also noted. None of the forms listed here are restricted to the Quaternary of the Muglad Basin. However, the assemblage is typical of samples analysed from above Zone 2; that is above the first downhole occurrence of Pliocene forms.

Zone 2: Verrucatosporites spp. Zone

Suggested age: Pliocene–Miocene

Formation: Adok

Definition: This zone characterizes the Adok Formation. It is defined by the first downhole occurrences of *Verrucatosporites usmensis, V. favus* and *V. afavus*.

Discussion: Salard-Cheboldaeff (1990) indicates the first downhole occurrence of *V. usmensis* in the Pliocene. Germaraad *et al.* (1968) indicates a range of 'throughout the Tertiary' in northern South America, although they also indicate the presence of *V. usmensis* as high as Pleistocene in Africa. In the samples studied the first downhole occurrences of *V. favus* and *V. afavus* coincide with that of *V. usmensis*, as does that of *Peregrinipollis nigericus* (Pliocene, sub-zone 2a – see below). One of the authors also records the first downhole occurrence of *V. favus* in the Pliocene in Sudan (Awad *pers obs.*). Penetration of Pliocene age sediments is suggested. Forms of *Verrucatosporites* spp. persist throughout the Cenozoic of the Muglad Basin.

Zone 2 can be sub-divided in to two sub-zones. This division is based on the published ranges of certain species.

Sub-zone 2a: Peregrinipollis nigericus–Verrucatosporites usmensis sub-zone

Suggested age: Pliocene

Formation: Adok

Definition: Defined by the first downhole occurrences of *Verrucatosporites usmensis* and *Peregrinipollis nigericus*.

Discussion: The published range of *Verrucatosporites usmensis* is discussed above. The first downhole occurrence of *Peregrinipollis nigericus* is recorded in the Pliocene by Legoux (1978), Takahashi & Jux (1989*a*) and Salard-Cheboldeaff (1990). A Pliocene age is suggested for sub-zone 2a.

Description: Sub-zone 2a characterizes the Pliocene part of the Adok Formation. The top of the sub-zone is recognized by the first downhole occurrences of the species *Verrucatosporites usmensis* and *Peregrinipollis nigericus.* Other species may also be recorded for the first time. These include *Magnastriatites howardi*, *Praedapollis flexibilis*, *V. favus* and *Echiperiporites estelae.* The assemblages recorded from this sub-zone may typically include *Botryococcus* spp, *Cyathidites minor, Cyathidites australis, Laevigatosporites haardtii, L. gracilis* and *Pachydermites diederixi.* Other less frequent forms recorded include *Quercoidites henrici, Retibrevitricolporites ibadenensis* and *Chenopodipollis microporatus.*

Sub-zone 2b: Loranthacites nataliae–Perfotricolpites digitatus sub-zone

Suggested age: Miocene

Formation: Adok

Definition: Defined by the first downhole occurrences of *Loranthacites nataliae* and *Perfotricolpites digitatus*.

Discussion: The first downhole occurrences of the palynomorphs *Loranthacites nataliae* and *Perfotricolpites digitatus* are recorded in the Miocene in Cameroon, Gabon, the Congo Basin and Angola (Salard-Cheboldaeff 1990). In West Africa the first downhole occurrence of *Perfotricolpites digitatus* is recorded in the Pliocene. A Miocene age is suggested for sub-zone 2b.

Description: Sub-zone 2b characterizes the Miocene part of the Adok Formation. The top of the sub-zone is recognized by the first downhole occurrences of *Loranthacites nataliae* and *Perfotricolpites digitatus.* Other notable features include the first downhole occurrences of *Striatopollis catatumbus, Triorites festatus, Psilastephanocolpites boureaui, Retimonocolpites irregularis* and *Praedapollis africanus*. Of these forms Salard-Cheboldeaff (1990) records the first downhole occurrence of *P. boureaui, R. irregularis* and *P. africanus* in the Miocene. Jan du Chêne *et al.* (1978) also record the first downhole occurrence of *P. boureaui* in the Miocene. Awad (*pers obs.*) has recorded the first downhole occurrence of *P. boureaui* in the Miocene of Sudan. An increase in numbers of *Verrucatosporites usmensis* may be recorded. Other forms typically recorded in this sub-zone include abundant *Botryococcus* spp., common *Cyathidites minor*, common *Laevigatosporites haardtii* and *L. gracilis*, *Peregrinipollis nigericus*, *Praedapollis flexibilis* and *Verrucatosporites favus.* Less frequent forms that may be recorded from this sub-zone include *Striatopollis bellus, Sapotaceoidaepollenites communis, S. rotundus, Zonacostites ramonae, Inaperturopollenites hiatus, Quercoidites henrici, Chenopodipollis microporatus, Retibrevitricolporites obodoensis, R. protrudens* and *Verrutricolporites rotundiporis.*

Zone 3: Cyathidites minor–Perfotricolpites digitatus zone

Suggested age: Miocene–Oligocene

Formation: Tendi

Definition: This zone characterizes the Tendi Formation. It is characterized by an influx of *Cyathidites minor*, becoming abundant in the assemblages, and by an increase in numbers of *Perfotricolpites digitatus* recorded, the form becoming a common component of the assemblages.

Discussion: The top of the Tendi Formation has been picked through the wireline log signature and sedimentological analysis (GNPOC – in house). There is no palynological evidence to indicate an age older than Miocene. A characteristic assemblage for the Miocene Tendi Formation is described below, under sub-zone 3a. The Oligocene, sub-zone 3b, is recognized by comparing published ranges of certain palynomorph species (see below).

Sub-zone 3a: Verrucatosporites usmensis sub-zone

Suggested age: Miocene

Discussion: The top of the Tendi Formation has been recognized from wireline log interpretation and sedimentological analysis. Sub-zone 3a is stratigraphically placed between the base of Zone 2, sub-zone 2b and the top of sub-zone 3b, taken at the top of the Oligocene. A Miocene age is, therefore, suggested.

Formation: Tendi

Definition: The principal feature of sub-zone 3a is an increase in the numbers of *Verrucatosporites usmensis*.

Description: This sub-zone characterizes the Miocene part of the Tendi Formation. The key zonal feature is an increase in the numbers of *Verrucatosporites usmensis* recorded. *Botryococcus* spp. remains abundant in this sub-zone. Recovery of sporomorphs is generally poor. Typical components of assemblages recorded from this sub-zone include small smooth tricolpate pollen, *Cyathidites minor, Laevigatosporites gracilis, L. haardtii, Peregrinipollis nigericus, Praedapollis flexibilis, P. africanus, Loranthacites nataliae, Retibrevitricolporites ibadensis, R. obodoensis, Verrucatosporites favus, V. afavus, Polypodiaceoisporites specious, Triporotetradites campylostemonoides, Sapotaceoidaepollenites communis, Inaperturopollenites hiatus, Pachydermites diederixi, Perfotricolpites digitatus, Magnastriatites howardi, Striatopollis nigericus, S. catatumbus* and *Deltoidospora africana.*

Sub-zone 3b: Cicatricosisporites dorogensis–Magnastriatites howardi sub-zone

Suggested age: Oligocene

Formation: Tendi

Definition: This sub-zone is characterized by the first downhole occurrence of *Cicatricosisporites dorogensis* and the occurrence of common *Magnastriatites howardi.*

Discussion: The first downhole occurrence of *Cicatricosisporites dorogensis* has been the subject of some debate. Salard-Cheboldaeff (1978, 1990) reports a first downhole occurrence at the top of the Oligocene from a number of African basins, as do Germeraad *et al.* (1968). However, Oboh & Salami (1989) and Oboh *et al.* (1992) record the form in the Miocene of the Niger Delta. One of the authors has also recorded the first downhole occurrence of the form from the Late Oligocene of North Africa (Stead *pers obs.*). Salard-Cheboldaeff (1978) describes a *Magnastriatites howardi–Cicatricosisporites dorogensis* Zone, characterizing Oligocene sediments from Cameroon, while Germeraad *et al.* (1968) indicates a *Magnastriatites howardi* Pantropical Zone and a *Cicatricosisporites dorogensis* Atlantic Zone for the Oligocene.

An Oligocene age is suggested for sub-zone 3b.

Description: Sub-zone 3b characterizes the Oligocene part of the Tendi Formation. The key zonal features are the first downhole occurrence of *Cicatricosisporites dorogensis* in association with an increase in the numbers of *Magnastriatites howardi.* Both forms are common in the recorded assemblages. Other features of this sub-zone include the first downhole occurrences of *Magnaperiporites spinosus, Racemonocolpites racematus* and *Retimonocolpites asabaensis.* The first downhole occurrence of persistent *Striatopollis bellus* may also be recorded in this sub-zone. Also, *Cricotriporites camerounensis* may be recorded for the first time. The first downhole occurrence of this form is difficult to confirm since it has not been recorded from many of the wells and, in addition, when it has been recorded it is a rare component of the assemblages. Assemblages from this sub-zone may feature common to abundant *Cyathidites minor*, common *Verrucatosporites usmensis* and common *Laevigatosporites haardtii* and *L. gracilis.* Typical components of the assemblages include *Praedapollis flexibilis, P. africanus, Verrucatosporites favus, Perfotricolpites digitatus, Magnastriatites howardi, Retibrevitricolporites ibadensis, R. obodoensis, R. triangulatus, R. protrudens, Striatocolporites ovuliformis, Striatopollis nigericus, Polypodiaceoisporites specious, Ailanthipites* spp., *Deltoidospora africana* and *Triorites festatus.* Infrequent to rare forms recorded include *Cinctiporipollis mulleri*, *Echiperiporites estelae* and *Ctenolophonidites costatus*.

The base downhole occurrence of *Peregrinipollis nigericus* may be recorded in this sub-zone. However, since this analysis is based mainly on ditch cuttings this is a difficult event to recognize.

Zone 4: *Striatopollis* spp. Zone

Suggested age: Oligocene–Eocene

Formation: Nayil

Definition: The *Striatopollis* spp. Zone characterizes the Nayil Formation. The key zonal event is an increase in numbers of *Striatricolpites* spp. Forms recorded include *Striatopollis catatumbus*, *S. bellus*, *Striatocolporites ovuliformis* and various other striatricolpate forms grouped together as *Striatopollis* spp.

Discussion: The top of the Nayil Formation has been recognized from wireline log interpretation and sedimentological analysis (GNPOC – in house). There is no palynological evidence to indicate the top of the formation is older than Oligocene. The top of the Eocene, within the Nayil Formation, is recognized by palynological evidence. The Zone is divided into two sub-zones, based on recognition of the top of the Eocene within Zone 4.

Sub-zone 4a: Racemonocolpites spp.*–Praedapollis flexibilis* sub-zone

Suggested age: Oligocene

Formation: Nayil

Definition: The *Racemonocolpites* spp. – *Praedapollis flexibilis* sub-zone characterizes the Oligocene part of the Nayil Formation. The top of the sub-zone is recognized by an increase in numbers of *Racemonocolpites* spp., including *R. facilis*, *R. hians* and *R. racematus*, and also by an increase in the numbers of *Praedapollis flexibilis* recorded.

Discussion: Sub-zone 4a is stratigraphically placed between the base of Zone 3, sub-zone 3b and the top of the underlying sub-zone 4b. The top of sub-zone 4b has initially been recognized through wireline log interpretation. The base is placed at the top of sub-zone 4b, indicated by palynological evidence for penetration of Eocene age sediments. A characteristic palynomorph assemblage is described below.

Description: A noted increase in numbers of *Racemonocolpites* spp., frequently in association with an increase in numbers of *Praedapollis flexibilis*, indicates penetration of sub-zone 4a. Other features include continued common to abundant *Cyathidites minor*, common *Laevigatosporites haardtii* and *L. gracile* and an increase in the numbers of *Foveomonocolpites* spp. Forms that are typically recorded from this sub-zone include *Cicatricosisporites dorogensis*, *Loranthacites nataliae*, *Perfotricolpites digitatus*, *Retibrevitricolporites* spp., *Praedapollis africanus*, *Magnastriatites howardi*, *Retimonocolpites irregularis*, *Striatopollis* spp., including *S. catatumbus*, *S. undulastriatus*, *S. variabilis*, *S. bellus*, *S. nigericus*, *Polypodiaceoisporites specious*, *Verrucatosporites usmensis* and *V. favus*. Less frequent forms that have been recorded include *Magnaperiporites spinosus*, *Adenantherites intermedius*, *Verrustephanocolporites complanatus*, *Echitriporites kwaense*, *Heterocolpites laevigatus*, *Psilastephanocolpites boureaui*, *Deltoidospora africana*, *Cricotriporites camerounensis*, *Polyadopollenites vancampoi* and *Ctenolophonidites costatus*. Freshwater alga are also common, particularly *Botryococcus* spp., but also *Pediastrum* spp.

Sub-zone 4b: Retimonocolpites asabaensis – Corsinipollenites jussiaeensis sub-zone

Suggested age: Eocene

Formation: Nayil

Definition: The key events that define this sub-zone are the occurrence of common *Retimonocolpites asabaensis* and the first downhole occurrence of the distinctive form *Corsinipollenites jussiaeensis*.

Discussion: Jan du Chêne *et al.* (1978) first described *Retimonocolpites asabaensis* and *Corsinipollenites jussiaeensis* from the Eocene of Nigeria. Salard-Cheboldaeff (1990) indicates that the total range of *Corsinipollenites jussiaeensis* is restricted to the Eocene in West Africa. An increase in numbers of *Retimonocolpites asbaensis* recorded appears to coincide with the first downhole occurrence of *Corsinipollenites jussiaeensis* in the Muglad Basin. An Eocene age is suggested for sub-zone 4b.

Description: An increase in the numbers of *Retimonocolpites asabaensis*, together with the first downhole occurrence of *Corsinipollenites jussiaeensis* indicate penetration of sub-zone 4b. Other events that may be recorded in this sub-zone include the first downhole occurrences of *Mauritiidites crassiexinus*, *Echimonocolpites rarispinosus* and *Gemmatricolpites pergemmatus*. Other first downhole occurrences that may be recorded are *Auriculopollenites simplex*, *Echitriporites trianguliformis*, *E. irregularis* and

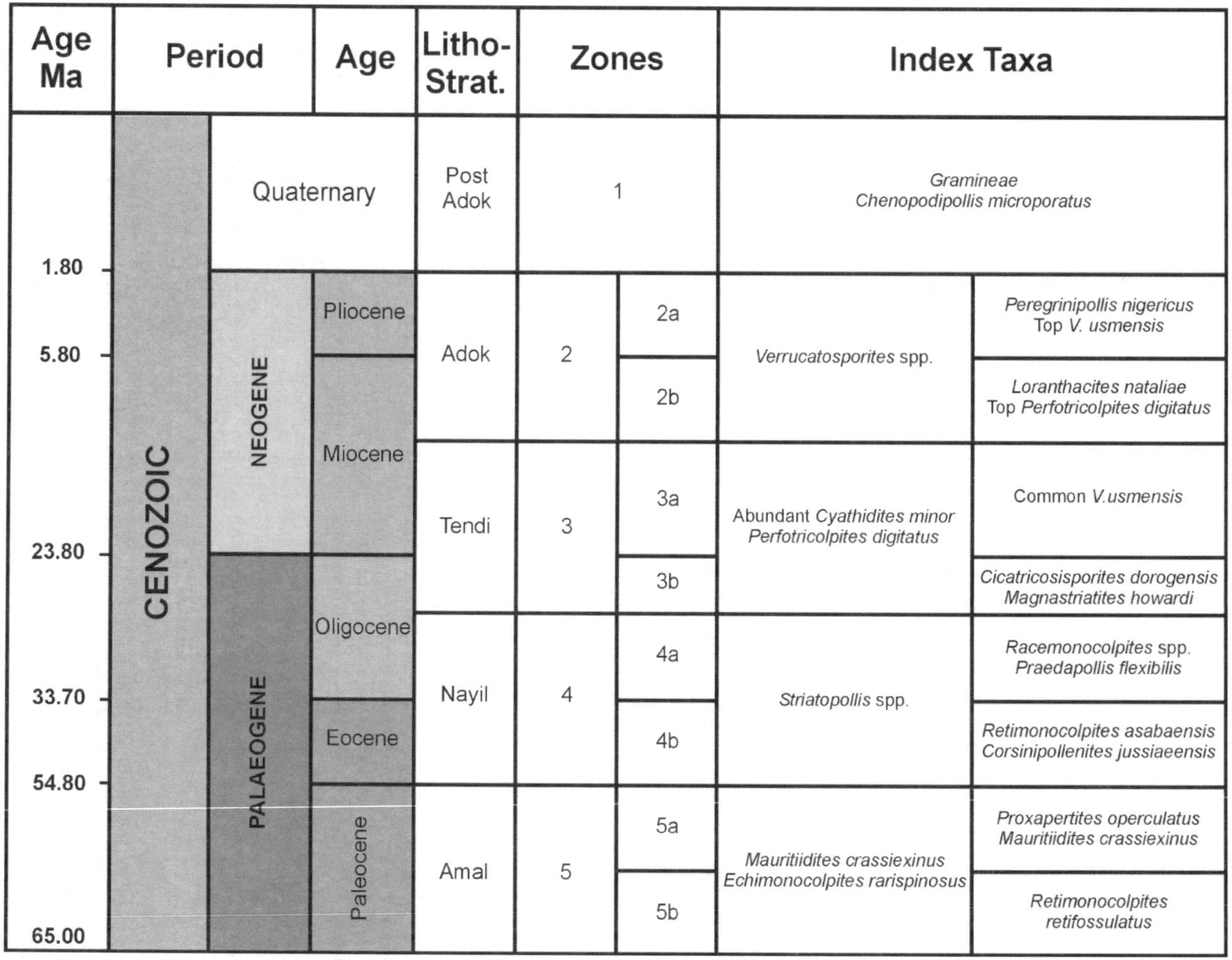

Fig. 3. Palyno-zonation of the Cenozoic of the Muglad Basin, Sudan; Age Ma column after Gradstein & Ogg (1996, 1999).

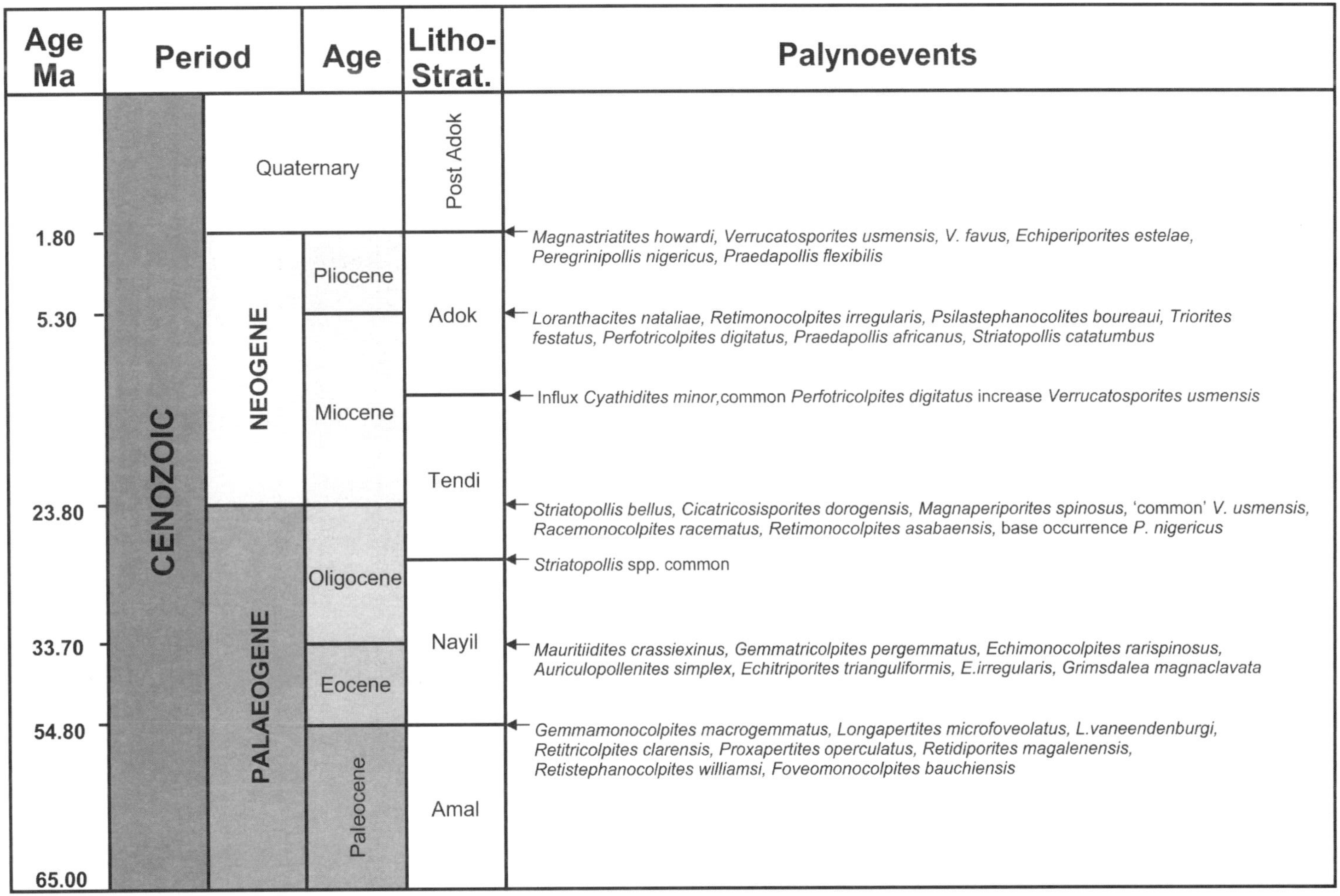

Fig. 4. Palynoevents recorded from the Cenozoic of the Muglad Basin, Sudan; Age Ma column after Gradstein & Ogg (1996, 1999).

Grimsdalea magnaclavata. Other forms typical of this sub-zone include *Cyathidites minor, Ailanthipites* spp., *Striatopollis* spp., including *S. catatumbus, S. bellus, S. nigericus, Laevigatosporites haardtii, Perfotricolpites digitatus, Racemonocolpites hians, R. racematus, Foveomonocolpites* spp., *Verrucatosporites usmensis, Echimonocolpites* spp., *Echiperiporites estelae, Spirosyncolpites spiralis, Crototricolpites densus, Retitricolporites guianensis, Psilatricolporites operculatus* and *Psilastephanocolpites boureaui.*

Zone 5: Mauritiidites crassiexinus – Echimonocolpites rarispinosus Zone

Suggested age: Paleocene

Formation: Amal

Definition: Zone 5 is recognized by the occurrence of common to abundant *Mauritiidites crassiexinus* and common *Echimonocolpites rarispinosus.*

Discussion: Jan du Chêne *et al.* (1978) first recorded *Mauritiidites crassiexinus* from the Eocene of Nigeria. Salard-Cheboldaeff (1978) recorded the first downhole occurrence of *Echimonocolpites rarispinosus* from the basal Eocene of Cameroon, while Srivistava & Binda (1991) recorded the form in the Eocene of Saudi Arabia. The first downhole occurrences of these palynomorphs, therefore, would appear to be in the Eocene. The occurrence records of these forms within the Muglad Basin supports this assessment.

At the top of the Amal Formation, however, both of these forms become common to abundant, and this is the key zonal event. The top of Amal Formation can easily be recognized from the wireline log signature. The Paleocene age assessment is based on the published records of some of the other forms recorded for the first time. These include *Gemmamonocolpites macrogemmatus, Longapertites microfoveolatus, L. vaneendenburgi, L. marginatus* and *Retistephanocolpites williamsii.* Awad (1994) and Muller *et al.* (1987) record the first downhole occurrence of *G. macrogemmatus* in the Paleocene of Sudan and Venezuela respectively. The first downhole occurrence of *Longapertites microfoveolatus* has been recorded in the Paleocene of Sudan (Awad 1994) and the Paleocene of Nigeria (Adegoke *et al.* 1978). Salard-Cheboldaeff records the first downhole occurrence of *L. vaneendenburgi* and *L. marginatus* in the Paleocene of a number of African basins. Salard-Cheboldaeff (1990) further records the first consistent downhole occurrence of *Retistephanocolpites williamsii* in the Paleocene of Africa.

Zone 5 can be subdivided into two sub-zones. The division is based on the first downhole occurrence of *Retimonocolpites retifossulatus.*

Sub-zone 5a: Proxapertites operculatus–Mauritiidites crassiexinus sub-zone

Age: Paleocene

Formation: Amal

Definition: The first downhole occurrence of *Proxapertites operculatus,* together with the occurrence of common to abundant *Mauritiidites crassiexinus.*

Discussion: The attribution of a Paleocene age to Zone 5 and its sub-zones is discussed above.

Description: Sub-zone 5b characterizes the upper part of the Amal Formation. The key zonal events are the first downhole occurrence of *Proxapertites operculatus* and the occurrence of common to abundant *Mauritiidites crassiexinus.* Other first downhole occurrences that may be recorded in this sub-zone include *Gemmamonocolpites macrogemmatus, Longapertites microfoveolatus, L. vaneendenburgi, Retitricolpites clarensis, Retidiporites magdalenensis, Retistephanocolpites williamsii* and *Foveomonocolpites bauchiensis.* An increase in the numbers of *Foveomonocolpites* spp. is also noted. Forms that are typically recorded in this sub-zone include *Cyathidites minor, Echimonocolpites rarispinosus, Retimonocolpites asabaensis, Cricotriporites camerounensis, Grimsdalea magnaclavata, Proxapertites cursus, Striatricolpites catatumbus, Racemonocolpites hians, R. racematus, Spinizonocolpites echinatus, S. baculatus, Longapertites marginatus, Spirosyncolpites spiralis, Echitriporites trianguliformis* and *Monocolpites marginatus.* Freshwater alga, notably *Botryococcus* spp. and *Pediastrum* spp., may also be common.

Sub-zone 5b: Retimonocolpites retifossulatus sub-zone

Suggested age: Paleocene

Formation: Amal

Definition: The first downhole occurrence of *Retimonocolpites retifossulatus.*

Discussion: The top of sub-zone 5b is indicated by the first downhole occurrence of *Retimonocolpites retifossulatus.* The base of the sub-zone, and of Zone 5 is at the base of the Paleocene, and is indicated by the first downhole occurrence of Cretaceous forms. The Paleocene age attribution for Zone 5 is discussed above. Published records indicate a first downhole occurrence of *Retimonocolpites retifossulatus* in the Maastrichtian (e.g. Awad 1994), but the form has consistently been noted in the Muglad Basin in the Amal Formation, that is, from Paleocene sediments.

Description: Sub-zone 5b is defined by the first downhole occurrence of *Retimonocolpites retifossulatus*. The assemblages recorded are very similar to those recorded from sub-zone 5a. Forms that are typically recorded include common *Retidiporites magdalenensis, Foveomonocolpites bauchiensis* and *Longapertites vaneendenbergii*, as well as the continued presence of common *Mauritiidites crassiexinus* and *Echimonocolpites rarispinosus.* At the base of the *Retimonocolpites retifossulatus* sub-zone there appears to be a fairly consistent occurrence of reworked Maastrichtian elements. This feature may be sufficiently consistent across the basin to enable establishment of an additional zone, characterized by a mixed assemblage containing both Paleocene and Maastrichtian forms. More samples from the Cretaceous/Cenozoic boundary would need to be studied to confirm this.

Discussion

The palynological analysis of the samples available for study and the subsequent integration of the results with available wireline log data have resulted in the development of a palynozonation scheme for the Cenozoic of the Muglad Basin. This zonation includes five zones and eight sub-zones allowing recognition of the formations shown and their further subdivision (Fig. 3).

Within the zonation a number of palynoevents have been recognized. These indicate the potential for further subdivision of the Cenozoic sequences, given more samples and a closer-sampling interval (Fig. 4). The development of this zonation scheme demonstrates an improved understanding of palynomorph distribution both within the Muglad Basin in particular, and in Sudan in general. A comparison with previously published schemes and information (Kaska 1989; Awad 1994) demonstrates a marked improvement in recognizing palynological zones. The zonation presented here provides much-needed age control for the interpretation and modelling of the petroleum geology of the Cenozoic of the Muglad Basin.

The authors wish to thank the Greater Nile Petroleum Operating Company for permission to publish this paper.

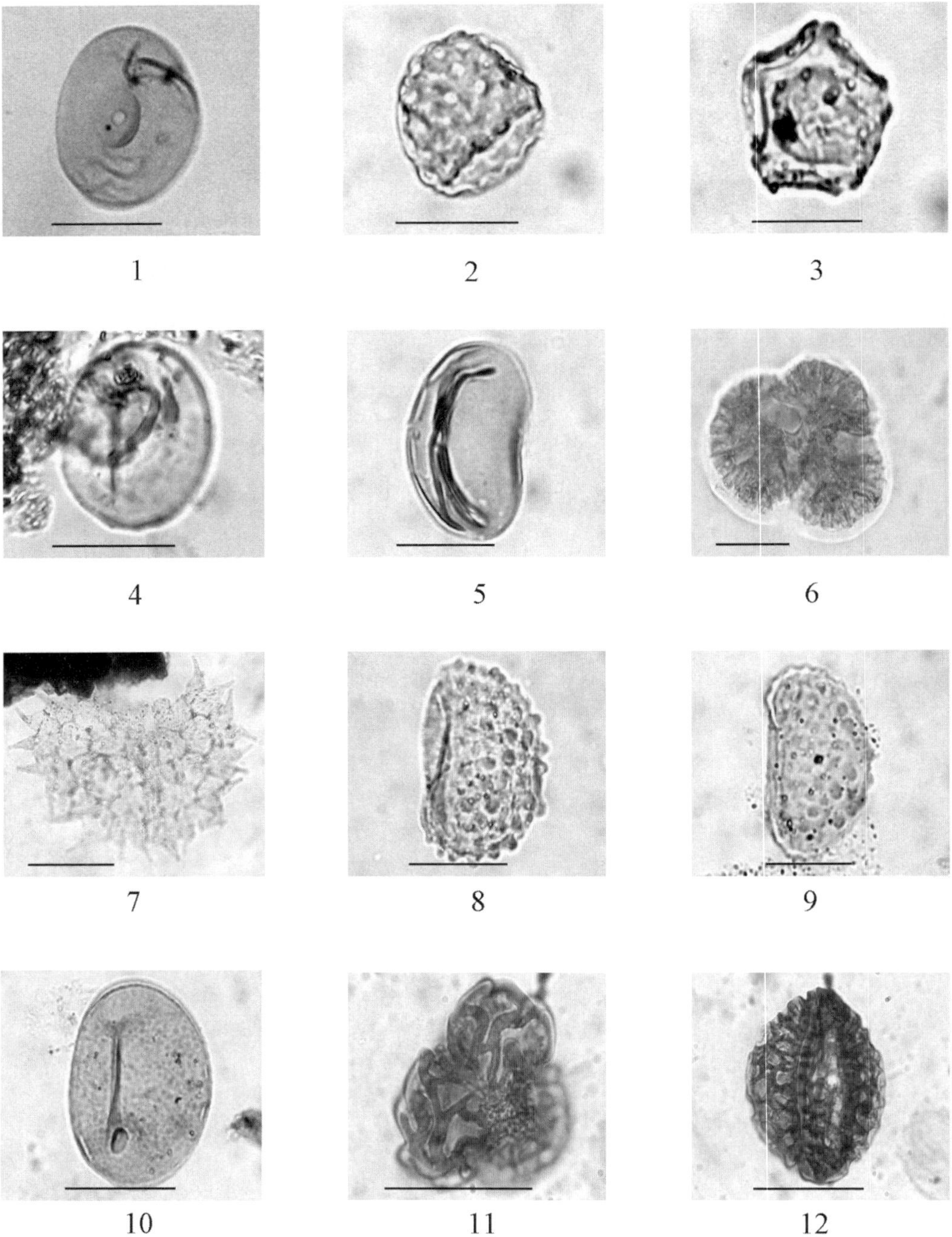

Plate 1. Range bar on plates = 20 μm. **1**, Gramineae; Timsah-1, 800 m; EFR J232; **2**, *Chenopodipollis microporatus*; Amal-1, 18.79 m (60'md); EFR F271; **3**, *Alnipollenites* sp.; Amal-1, 18.79 m (60'md); EFR T27; **4**, *Laevigatosporites gracilis*; Amal-1, 36.58 m (120'md); EFR Y321; **5**, *Laevigatosporites haardti*; Amal-1, 1780.03 m (5840'md); EFR P164; **6**, *Botryococcus* spp.; Amal-1, 1033.27 m (3390'md); EFR V304; **7**, *Pediastrum* spp.; Amal-1, 2505.46 m (8220'md); EFR N424; **8**, *Verrucatosporites usmensis*; Amal-1, 1133.86 m (3720'md); EFR T21; **9**, *Verrucatosporites afavus*; Amal-1, 996.70 m (3270'md); EFR L11; **10**, *Verrucatosporites favus*; Amal-1, 1069.85 m (3510'md); EFR O331; **11**, *Peregrinipollis nigericus,* Balome-1, 1000 m; EFR F373; **12**, *Peregrinipollis nigericus,* Balome-1, 1000 m; EFR P302.

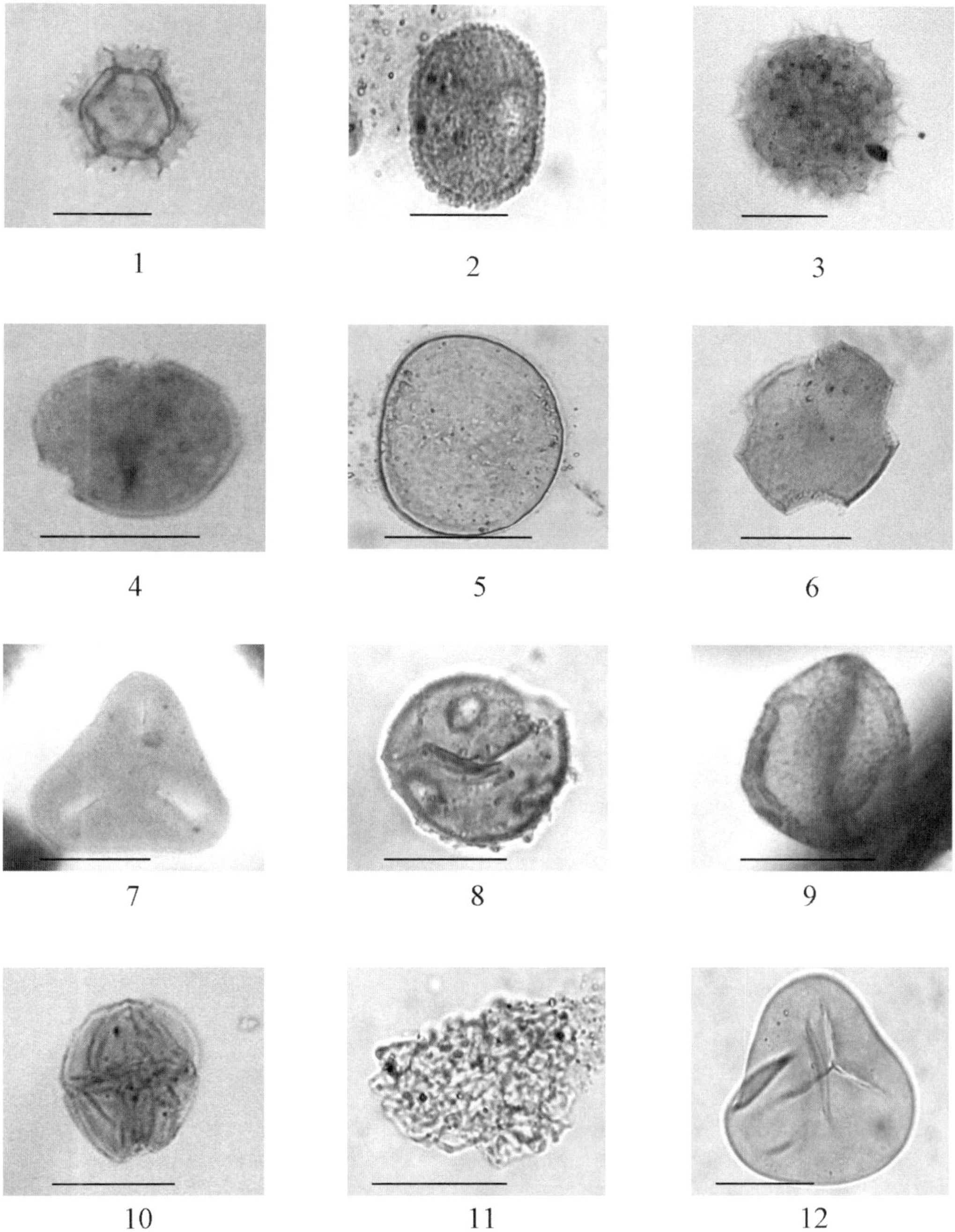

Plate 2. Range bar on plates = 20 μm. **1**, Compositae group; Umm Batutu East, 350 m, EFR G353; **2**, *Verrutricolporites rotundiporis*; Amal-1, 1060.70 m (3480'md); EFR F43; **3**, *Echiperiporites estelae*; Tammor-1, 860 m; EFR J313; **4**, *Echitriporites kwaense*; Amal-1, 2304.29 m (7560'md); EFR U331; **5**, *Verrucatosporites* sp.; Balome-1, 900 m; EFR X393; **6**, *Pachydermites diederixi*; Kaikang-1, 984.50 m (3230'md); EFR J36; **7**, *Loranthacites nataliae*; Amal-1, 1033.27 m (3390'md); EFR R33; **8**, *Cricotriporites camerounensis*; Amal-1, 1780.03 m (5840'md); EFR D17; **9**, *Retimonocolpites irregularis*; Tammor-1, 1015 m; EFR 413; **10**, *Adenantherites intermedius*; Amal-1, 2170.18 (7120'md); EFR N252; **11**, *Praedapollis flexibilis*; Amal-1, 996.70 m (3270'md); EFR M141; **12**, *Cyathidites minor*; Amal-1, 996.70 m (3270'md); EFR O294.

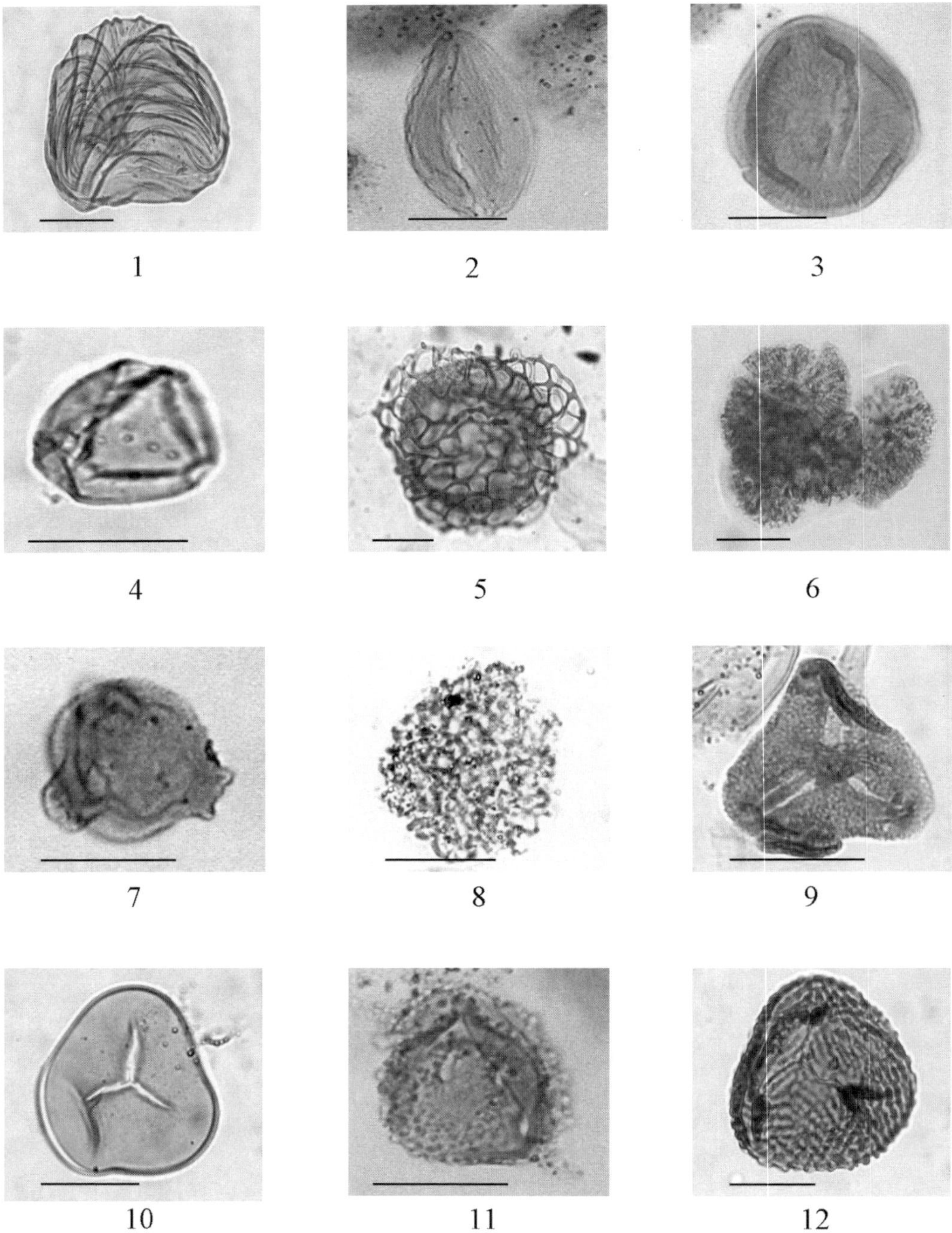

Plate 3. Range bar on plates = 20 µm. **1**, *Magnastriatites howardi*; Amal-1, 1749.55 m (5740'md); EFR P394; **2**, *Striatopollis catatumbus*; Timsah-1, 800 m; EFR K373; **3**, *Perfotricolpites digitatus*; Timsah-1, 800 m; EFR M402; **4**, *Triorites festatus*; Amal-1, 1780.03 m (5840'md); EFR Q124; **5**, *Praedapollis africanus*; Balome-1, 1000 m; EFR O463; **6**, *Botryococcus* sp.; Balome-1, 900 m; EFR T27; **7**, *Retibrevitricolporites protrudens*; Amal-1, 1780.03 m (5840'md); EFR H184; **8**, *Praedopollis flexibilis*; Amal-1, 996.70m (3720'md); EFR R40; **9**, *Loranthacites nataliae*; Amal-1, 1033.27 m (3390'md); EFR R33; **10**, *Cyathidites minor*; Amal-1, 1161.29 m (3810'md); EFR T332; **11**, *Magnaperiporites spinosus*; Amal-1, 1749.55 m (5740'md); EFR N162; **12**, *Cicatricosisporites dorogensis*; Amal-1, 1780.03 (5840'md); EFR S143.

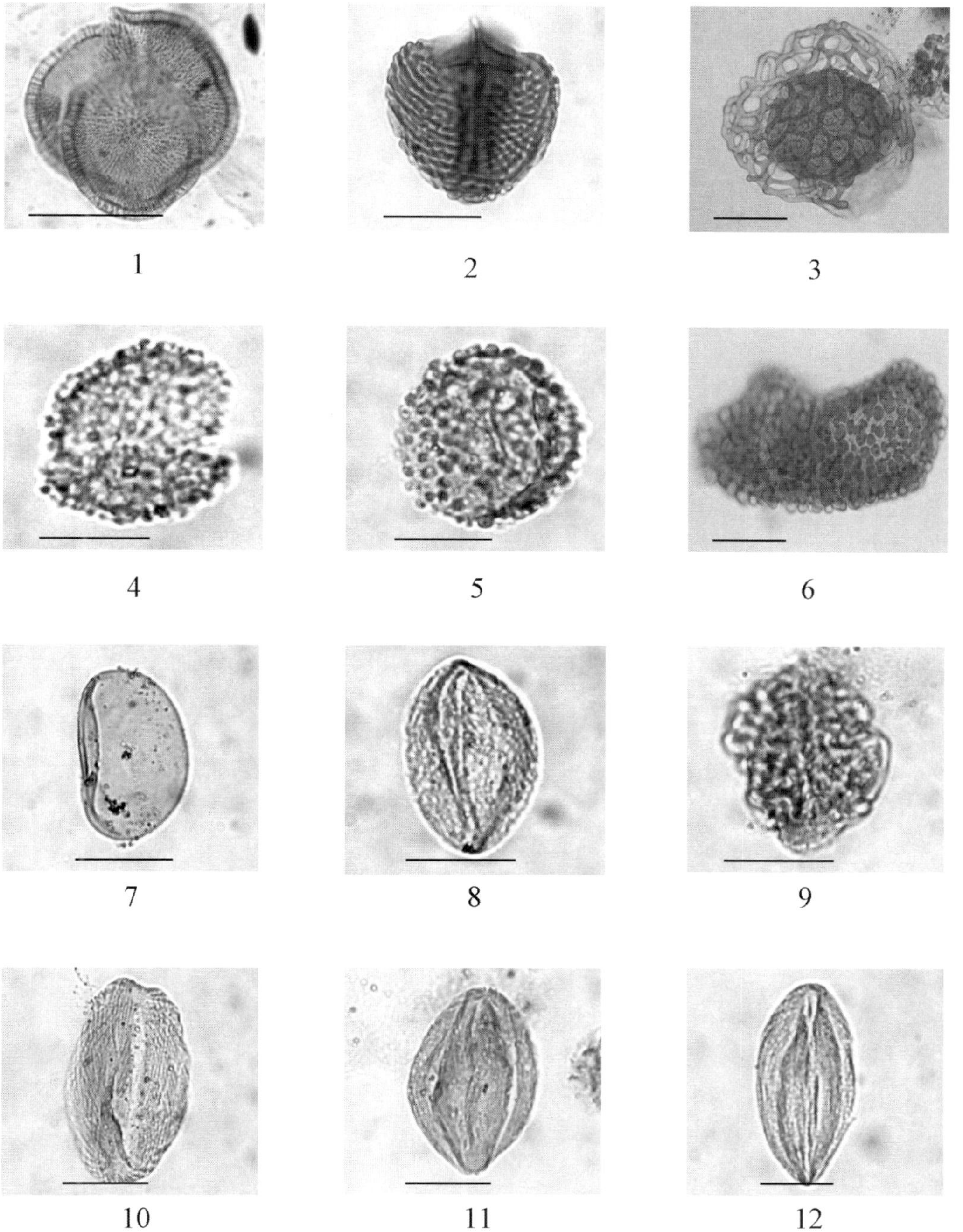

Plate 4. Range bar on plates = 20 μm. **1**, *Perfotricolpites digitatus*; Balome-1, 1000 m; EFR N391; **2**, *Cicatricosisporites dorogensis*; Amal-1, 1627.63 (5340'md); EFR M144; **3**, *Praedopollis africanus*; Timsah-1, 1195 m; EFR F323; **4**, *Racemonocolpites facilis*; Amal-1, 3087.62 m (10130'md); EFR L234; **5**, *Racemonocolpites hians*; Amal-1, 2743.20 m (9000'md); EFR Q17; **6**, *Racemonocolpites racematus*; Balome-1, 1620 m; EFR W303; **7**, *Laevigatosporites haardtii*; Amal-1, 2935.22 m (9630'md); EFR S433; **8**, *Ailanthipites* spp.; Amal-1, 1554.48 m (5100'md); EFR T72; **9**, *Peregrinipollis nigericus*; Amal-1, 1033.27 m (3390'md) EFR J274; **10**, *Striatopollis bellus*; Amal-1, 996.70 m (3270'md); EFR O292; **11**, *Striatocolporites ovuliformis*; Amal-1, 1627.63 m (5340'md); EFR O151; **12**, *Striatricolpites* sp.; Amal-1, 1069.85 m (3510'md); EFR K424.

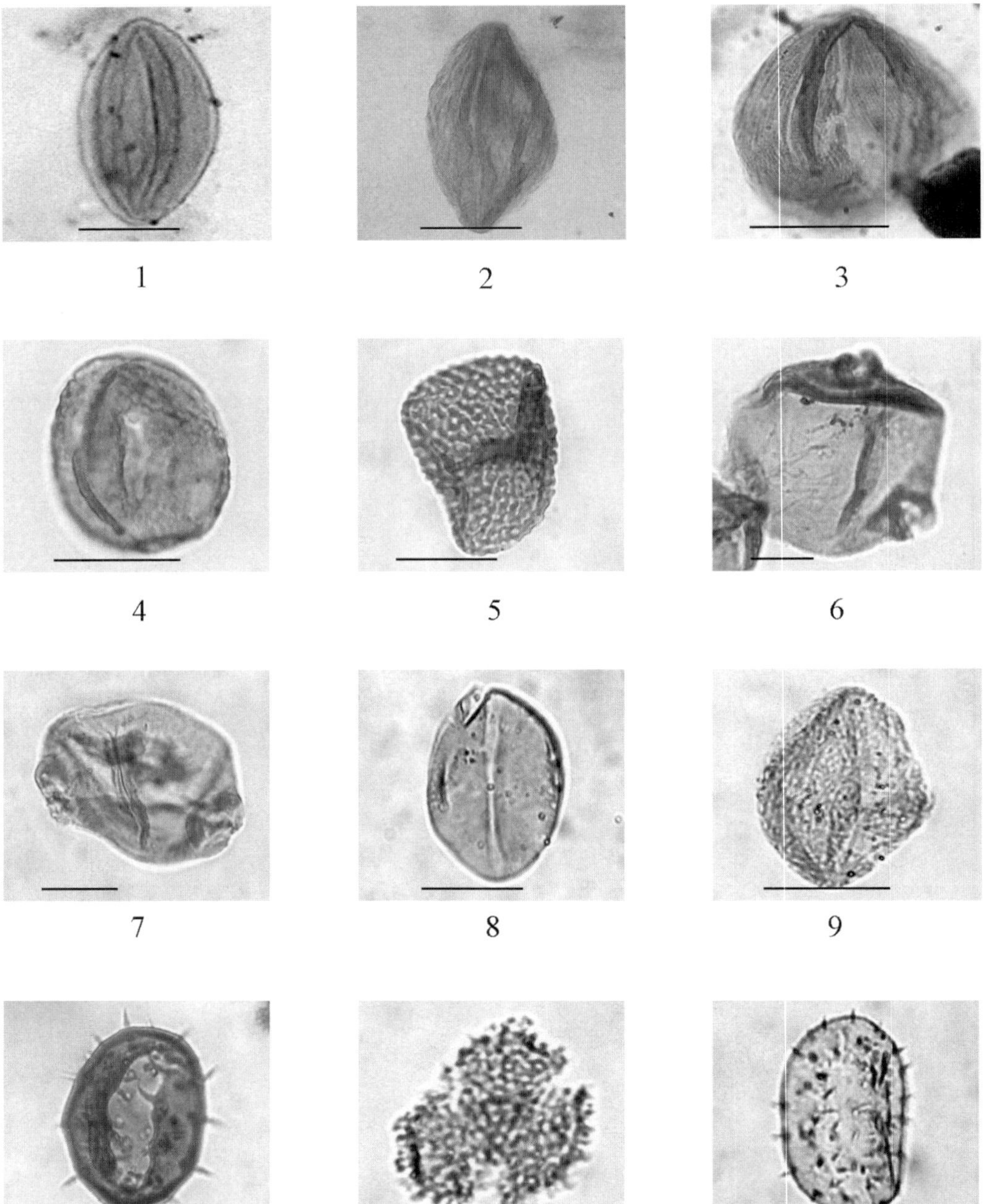

Plate 5. Range bar on plates = 20 µm. **1**, *Striatopollis nigericus*; Kaikang-1, 1706.88 m (5600'md); EFR P16; **2**, *Striamonocolpites undulastriatus*; Heglig-1, 518.16 m (1700'md); EFR R37; **3**, *Striatopollis bellus*: Balome-1, 1000 m; EFR U544; **4**, *Verrustephanocolporites complanatus*; Balome-1, 900 m; EFR U313; **5**, *Cicatricosisporites dorogensis*; Amal-1, 1554.48 m (5100'md); EFR S242; **6**, *Corsinipollenites jussiaeensis*; Amal-1, 1780.03 (5840'md); EFR L342; **7**, *Corsinipollenites jussiaeensis*; Amal-1, 1554.48 m (5100'md); EFR G27; **8**, *Retimonocolpites asabaensis*; Amal-1, 3087.62 (10130'md); EFR V264; **9**, *Ailanthipites* sp.; Amal-1, 1719.07 (5640'md); EFR H23; **10**, *Mauritiidites crassiexinus*; Amal-1, 2840.74 (9320'md); EFR J224; **11**, *Gemmatricolpites pergemmatus*; Amal-1, 3584.45 m (11760'md); EFR V412; **12**, *Echimonocolpites rarispinosus*; Amal-1, 3154.68 m (10350'md); EFR V35.

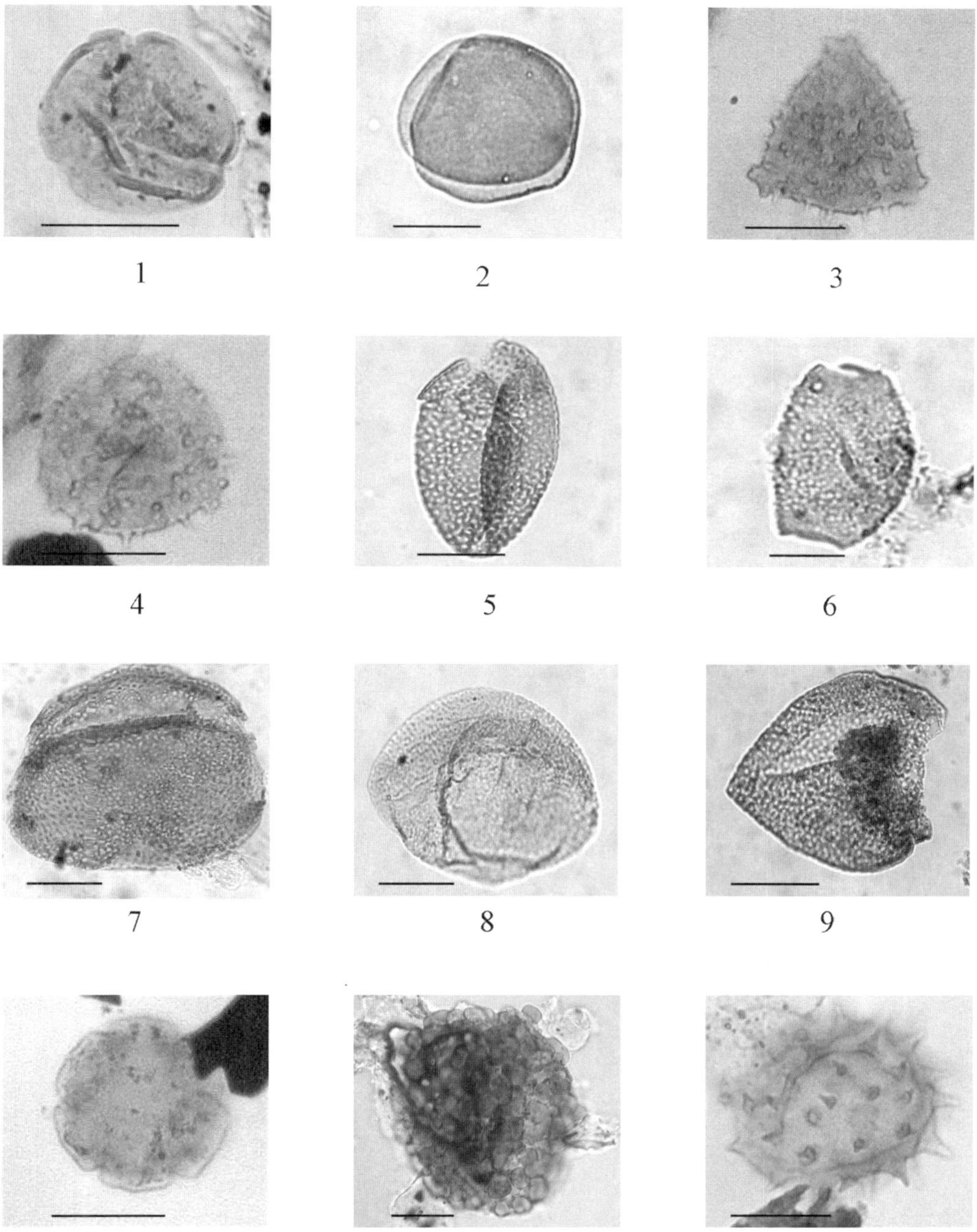

Plate 6. Range bar on plates = 20 μm. **1**, *Retitricolpites clarensis*; Tafab-1, 1445 m; EFR M302; **2**, *Proxapertites operculatus*; Jidian-1, 582.17 m (1910'md); EFR G16; **3**, *Echitriporites trianguliformis*; Shelanugo-1, 605 m, EFR F284; **4**, *Echitriporites irregularis*; Shelanugo-1, 610 m; EFR D304; **5**, *Foveomonocolpites bauchiensis*; Amal-1, 2868.17 m (9410'md); EFR T274; **6**, *Retidiporites magdalenensis*; Amal-1, 3087.62 m (10130'md); EFR P384; **7**, *Longapertites microfoveolatus*; Balome-1, 2390 m; EFR U47; **8**, *Longapertites vaneendenburgi*; Amal-1, 3608.83 m (11840'md); EFR P184; **9**, *Retimonocolpites retifossulatus*; Amal-1, 3956.30 m (12980'md); EFR Q273; **10**, *Retistephanocolpites williamsi*; Munga-4, 685 m; EFR L273; **11**, *Gemmamonocolpites macrogemmatus*; Tafab-1, 1635 m; EFR L47; **12**, *Spinizonocolpites echinatus*; Shelanugo-1, 935 m; EFR L342.

References

ADEGOKE, O.S., JAN DU CHÊNE, R.E., AGUMANU, A.E. & AJAYI, P.O. 1978. Palynology and age of the Kerri Kerri Formation, Nigeria. *Revista Española de Micropaleontología,* **10,** 267–283.

AWAD, M.Z. 1994. Stratigraphic, palynological, and paleoecological studies in the east-central Sudan (Khartoum and Kosti Basins), Late Jurassic to Mid-Cenozoic. *Berliner Geowissenschaftliche Abhandlungen* (A), **161**.

GERMERAAD, J.H., HOPPING, C.A. & MULLER, J. 1968. Palynology of Cenozoic sediments from tropical areas. *Review of Palaeobotany and Palynology*, **6**, 189–348.

GRADSTEIN, F.M. & OGG, J.G. 1996. A Phanerozoic Time Scale. *Episodes*, **19**, 3–4.

GRADSTEIN, F.M. & OGG, J.G. 1999. A Phanerozoic Time Scale. *http://www.iamg.org*

GUIRAUD, M. & MAURIN, J-C. 1992. Early Cretaceous rifts of Western and Central Africa: an overview. *Tectonophysics*, **213**, 153–168.

GUIRAUD, M., BIRKS, R.M., FAIRHEAD, J.D. & WILSON, M. 1992. Chronology and geodynamic setting of Cretaceous-Cenozoic rifting in West and Central Africa. *Tectonophysics*, **213**, 227–234.

JAN DU CHÊNE, R.E., ONYIKE, M.S. & SOWUNMI, M.A. 1978. Some new Eocene pollen of the Ogwashi-Asaba Formation, South-eastern Nigeria. *Revista Española de Micropaleontología*, **10**, 285–322.

KASKA, H.V. 1989. A spore and pollen zonation of Early Cretaceous to Cenozoic nonmarine sediments in central Sudan. *Palynology*, **13**, 79–90.

LEGOUX, O. 1978. Quelques espèces de pollen caractéristiques du Néogène du Nigéria. *Bulletin des Centres de Recherche Exploration-Production Elf-Aquitaine*, **2**, 265–317.

MCHARGUE, T.R., HEIDRICK, T.L. & LIVINGSTON, J.E. 1992. Tectonostratigraphic development of the Interior Sudan rifts, Central Africa. *Tectonophysics*, **213**, 187–202.

MEDUS, J. 1975. Palynologie de sédiments tertiaires du Sénégal méridional. *Pollen et Spores*, **17(4),** 545–608.

MULLER, J. DE DI GIACOMO & VAN ERVE, A.W. 1987. A palynological zonation for the Cretaceous, Cenozoic and Quaternary of northern South America. *American Association of Stratigraphic Palynologists Contribution Series*, **19**.

OBOH, F.E. & SALAMI, M.B. 1989. Lithostratigraphical and palynostratigraphical studies of Igbomotoru Well, Niger Delta. *Journal of African Earth Sciences*, **9(3/4)**, 531–540.

OBOH, F.E., SALAMI, M.B. & CHAPMAN, J.L. 1992. Palynological interpretation of the palaeo-environments of Miocene strata of the well Igbomotoru-1, Niger delta. *Journal of Micropalaeontology*, **11(1)**, 1–6.

SALARD-CHEBOLDAEFF, M. 1978. Palynologie Maestrichtienne et Tertiaire du Cameroun. Etude qualitative et repartition verticale des principales espèces. *Review of Palaeobotany and Palynology*, **28**, 365–388.

SALARD-CHEBOLDAEFF, M. 1990. Intertropical African palynostratigraphy from Cretaceous to Late Quaternary times. *Journal of African Earth Science*, **11**, 1–24.

SCHULL, T.J. 1988. Rift basins of interior Sudan: petroleum exploration and discovery. *AAPG Bulletin*, **72**, 1128–1142.

SRIVASTAVA, S.K. & BINDI, P.L. 1991. Depositional history of the Early Eocene Shumaysi Formation, Saudi Arabia. *Palynology*, **15**, 47–61.

TAKAHASHI, J. & JUX, U. 1989*a*. Palynology of Middle Cenozoic lacustrine deposits from the Jos Plateau, Nigeria. *Bulletin of the Faculty of Liberal Arts, Nagasaki University, Natural Science*, **29(2)**, 181–367.

TAKAHASHI, J. & JUX, U. 1989*b*. Palynologic investigation of Late Eocene to Early Oligocene lignites from Fayum Oasis, Egypt. *Bulletin of the Faculty of Liberal Arts, Nagasaki University, Natural Science*, **29(2)**, 369–463.

Statistical modelling of ecological signals: a new method for biostratigraphy

B. DALE[1], A. L. DALE[2], & I. PRINCE[3]
[1]*Geosciences Department, University of Oslo, PB 1047 Blindern, N-0316 Oslo, Norway*
(e-mail: Barrie.dale@geo.uio.no)
[2]*GeoResearch Consulting, 2100 Skarnes, Norway*
[3]*Statoil, Forus 50, N-4035 Stavanger, Norway*

Abstract: Research over the past few decades has shown that recent marine dinoflagellate cyst assemblages are strongly influenced by environmental factors, and statistical modelling of ecological signals (SMES) therefore has potential application for interpretation of fossil dinocysts in biostratigraphy. Towards this end a global database of recent cyst distributions from known environments has been developed using statistical methods that most accurately reflect and quantify the ecological signals expressed by the cysts. The first test is reported here of the application of SMES to industrial biostratigraphy using a palynology dataset from four wells along an onshore–offshore transect from the Norwegian North Sea provided by Statoil. The wells were part of a larger study on palaeoenvironments previously completed by Statoil on approximately 48 wells, where some of the traditional multivariate statistical methods had been tried but without success. The palaeoenvironment was thus partly understood, allowing the SMES method to be tested against a control.

Correspondence analysis on the dinocyst data from each well allowed the identification of the statistically most important species, and the relative positions of these species along the two most important axes suggested ecological trends (the coastal/oceanic being most dominant). Tracking these down-hole showed two major shifts in each well that could be correlated between all wells; the shifts corresponded to major flooding surfaces previously identified, and suggested orientation of the well sites relative to a palaeo-shoreline. This first test strongly suggests that these methods offer a much quicker, and more robust application for palaeoenvironmental interpretations from the dinocysts in biostratigraphic datasets.

The industrial data set used

Statoil provided frequency count data for palynology from the Paleocene, generated by the same contractor in order to ensure maximum taxonomical consistency, in four wells from the Norwegian North Sea. These data were chosen by Statoil to represent a profile to illustrate the palaeoenvironmental changes progressing from shelf to basin floor, based on a previous regional study by the company that integrated the results from the sedimentology, seismic and biostratigraphy of 48 wells. The results from this larger palaeoenvironmental study were reported by Laursen & Prince (2002). Some of the traditional multivariate statistical methods had been tried but without success (in particular various types of cluster analyses, both on individual wells and on interpreted time slice data for all wells). The palaeoenvironment was therefore at least partly understood, allowing the SMES method to be tested against a control.

The analytical methods used

The SMES test was purposely run 'blind', i.e. without any previous knowledge of the age or locality of the well data used, or the results of the larger palaeoenvironmental study, in order to ensure independent results that could be tested against the other study. The statistical approach used is explained in Jongman *et al.* (1995) and ter Braak (1996) and references therein, and its application to palynology is discussed by A. L. Dale and B. Dale (2002). A brief summary of the main elements of correspondence analysis (CA) is given below, to aid the reader in interpreting the ordination results and diagrams.

Brief description of CA

CA is a unimodal ordination method for displaying ecological information along perpendicular axes, representing directions of greatest variation in the dataset. It is based on the

From: POWELL, A. J. & RIDING, J. B. (eds) 2005. *Recent Developments in Applied Biostratigraphy*.
The Micropalaeontological Society, Special Publications, 179–203. 1747-602X/$15.00

Table 1. *Species names and codes for the statistically important taxa from each well discussed in the text*

Code	Indicator Species	Code	Indicator Species
Well A		**Well C**	
Apec spp.	*Apectodinium* spp.	Areo Cfm	*Areoligera cf. medusettiformis* sensu *Williams & Downie*
Areo mic	*Areosphaeridium michoudii*	Areo spp.	*Areoligera* spp.
Areo spp.	*Areoligera* spp.	Cary siG	*Caryapollenites simplex* group
Cary siG	*Caryapollenites simplex* group	Defl oeb	*Deflandrea oebisfeldensis*
Cero spp.	*Cerodinium* spp.	Glap ord	*Glaphyrocysta ordinata*
Come sp.	*Cometodinium* spp.	Homo ten	*Homotryblium tenuispinosum*
Glap ord	*Glaphyrocysta ordinata*	Hyst tub	*Hystrichosphaeridium tubiferum*
Hyst tub	*Hystrichosphaeridium tubiferum*	Impl spp.	*Impletosphaeridium* spp.
Impa spp.	*Impagidinium* spp.	Inap spp.	*Inaperturopollenites* spp.
Impl lig	*Impletosphaeridium ligospinosum*	Pala pyr	*Palaeoperidinium pyrophorum*
Impl spp.	*Impletosphaeridium* spp.	Phth sto	*Phthanoperidinium stockmansii*
Inap spp.	*Inaperturopollenites* spp.	Reti hia	*Retitricolpites hians*
Pala pyr	*Palaeoperidinium pyrophorum*	Syst pla	*Systematophora placacantha*
Spin raG	*Spiniferites ramosus* group	**Well D**	
Subt spp.	*Subtilisphaera* spp.	Areo spp.	*Areoligera* spp.
Syst pla	*Systematophora placacantha*	Areo mic	*Areosphaeridium michoudii*
Tria tria	*Triatriopollenites triangulus/subtriangulus*	Cary siG	*Caryapollenites simplex* group
Well B		Eato urs	*Eatonicysta ursulae*
Apec spp.	*Apectodinium* spp.	Glap div	*Glaphyrocysta divaricata*
Areo mic	*Areosphaeridium michoudii*	Glap ord	*Glaphyrocysta ordinata*
Areo spp.	*Areoligera* spp.	Glap spp.	*Glaphyrocysta* spp.
Cary siG	*Caryapollenites simplex* group	Homo ten	*Homotryblium tenuispinosum*
Fung spo	Fungal spores and hyphae	Hyst tub	*Hystrichosphaeridium tubiferum*
Homo ten	*Homotryblium tenuispinosum*	Impl spp.	*Impletosphaeridium* spp.
Impa sp1	*Impagidinium* sp. 1 Heilman-Clausen 1985	Inap spp.	*Inaperturopollenites* spp.
Impa spp.	*Impagidinium* spp.	Ling mac	*Lingulodinium machaerophorum*
Impl spp.	*Impletosphaeridium* spp.	Pala pyr	*Palaeoperidinium pyrophorum*
Inap spp.	*Inaperturopollenites* spp.	Pedi ast	*Pediastrum* spp.
Lent wet	*Lentinia wetzelii*	Spin raG	*Spiniferites ramosus* group
Oper spp.	*Operculodinium* spp.	Syst pla	*Systematophora placacantha*
Pala pyr	*Palaeoperidinium pyrophorum*	Syst spp.	*Systematophora* spp.
Phth com	*Phthanoperidinium comatum*		
Spin raG	*Spiniferites ramosus* group		

assumption that the species composition at a site is a result of the environmental conditions at that site, and that their abundance is highest when conditions are optimal, and tails off gradually to either side on any given gradient. The input consists of a matrix of species (Table 1) by sample data, from which species and sample scores are derived by way of a reciprocal averaging algorithm, also called the two-way weighted averaging algorithm (Hill 1973). The species and samples are dispersed along geometric axes on the basis of their species composition, such that species and samples that are similar to each other plot near to each other, whereas species and samples that are dissimilar plot far apart on the axes. Thus, an environmental gradient will contain a number of different species, whose varying amounts can be assumed to reflect ecological differences along that gradient. The axes are presumed to represent the main environmental gradients influencing the species distribution, such that the ecological requirements of species or samples can be inferred from their position (score) in the ordination diagram (Jongman *et al.* 1995; ter Braak 1996).

The first axis always represents the gradient of greatest species variation, the second axis the greatest source of the variation *remaining* after the first axis is accounted for; the third axis the majority of variation *remaining* after the second axis is accounted for, and so on, until all of the species variation is accounted for. A species ordination shows the spread of species along the chosen axes, a sample ordination shows the spread of samples, whereas a biplot shows the spread of both simultaneously.

Each axis is associated with a coefficient reflecting the degree of species dispersion along that axis, which is its '*eigenvalue*'. The *eigenvalue* thus gives a measure of the relative importance of the various axes in explaining the overall species variation. The resulting placement of the species and samples along the axes reveals the inherent ecological structure in the data. In this study, the majority of variation is accounted for by Axes 1 and 2, and these are shown in the ordination diagrams presented. Important environmental effects also may be reflected by the subsequent axes (Axes 3,4 and so on), but these tend to be more local, and a fuller treatment of these was beyond the scope of this study.

The CA was carried out using the CANOCO™ program (CANOnical Community Ordination; ter Braak & Smilauer 1998). For all the analyses, the Hill scaling option was used (Hill & Gauch 1980), such that the environmental optimum for each species is given in standard deviations. In this way, a species can be interpreted to appear, rise to its maximum, and disappear again within four standard deviations. Thus, species that plot more than four standard deviations from each other are not expected to appear in the same samples, and samples that plot more than four standard deviations from each other are not expected to have any species in common (Jongman *et al.* 1995).

The SMES approach

The approach used consisted of the following steps:

(a) Correspondence analysis (CA) of the data sets created for all four wells, both for all groups provided and for dinoflagellate cysts only. Wells C and D contained outliers such that several runs of these wells were necessary in order to arrive at results which best accounted for the outliers. Outliers were generally caused by single species that appeared in very high abundance in only one sample, exceptionally low species counts, or anomalous occurrences of species interpreted by company geologists to suggest sample contamination.

(b) CA for all groups included a classified site ordination showing pie diagrams of the relative proportions of all groups present in each sample, relative to the CA axes. This enables an assessment of the relative inputs from marine versus terrestrial environments for each site. The site and species ordinations are shown for each well (Figs 1–14, with corresponding scores listed in Tables 2–17). The samples in the site ordination are represented as proportional pie diagrams in the run including all groups.

(c) Results of the first two or three axes were plotted downcore for each of the eight runs, with the lengths of the axes being made proportional to the depths of the wells. Preliminary analysis of these showed the analyses including only dinoflagellate cysts to be slightly preferable, on the basis of their higher eigenvalues (relative to total species/samples) and defined boundaries relative to raw data for the identified indicator species. Steps d–e were thus carried out for dinoflagellate cysts only.

(d) The percentages of the main indicator species identified from step a) above were also plotted downcore to aid in the interpretation of the sample score plots.

(e) The sample plots were aligned in suspected position relative to the palaeo-shoreline (interpreted by comparison with patterns of modern cyst distribution), with main environmental boundaries correlated from well to well (Fig. 15). A number of additional analyses were carried out in this initial attempt to analyse a series of industrial wells, but were not found to be of significant value. Mainly, these included: composite runs of a master database comprising species information from all four wells; grouping the main indicator species identified by the CA analyses into four main groups corresponding to quadrants in the CA ordinations; and plotting changes in these species groupings downcore. The composite run of all four wells was unduly dominated by the wells having higher numbers of samples, and samples having higher counts, whereas wells with fewer samples, and samples with lower counts, had less influence on the overall results. Both this and combining the main indicator species into groups had the effect of 'smoothing over' important details that were deemed to be of value in each well.

It was important to try these analyses in order to discover their shortcomings, but these will not be repeated in future. Results are reported here using the letter code format required by the programs, with the first 3–4 letters from the genus name followed by a space and then the first 3–4 letters from the species name. A list of the full names of the statistically important species discussed in the text is included in Table 1.

Statistical weight

Because the species scores in the CA are based on a weighted average of their values in the samples, the *statistical weight* of a species is an important criterion for assessing its significance in the SMES method. A species' weight is a measure of its total abundance, and this influences the relative species' positions in the resulting ordination plot. Species with both large weights and large scores (on axes having the largest eigenvalues), for example, will be indicator species: they are both influential in defining the main directions of variation (= axes, representing ecological gradients), and for representing end-member positions along the important gradients. If the end-member positions were occupied by known coastal and oceanic species plotting highly positively and negatively, respectively, then the axis could be interpreted to reflect a nearshore–offshore gradient, for example. A species with a high weight and a low score on an important axis is likely to be a cosmopolitan species, or a species with a high tolerance for the prevailing conditions in the sample area, whereas a species with both a low weight and a low score is considered to be on the outer fringe of its ecological tolerance, or a random occurrence. The statistical weight of a species (wt.) is a fundamental element used in the results presented here.

The statistical weight of a sample is merely a measure of the total number counted, and is not used for ecological interpretation. Relative species weights are reflected in the figures by the font type used: higher-weight species were plotted using larger, bold print, whereas lesser-weight species were plotted using smaller, plain print. The size categories varied on the basis of the species distibutions for each well and are described individually in the corresponding figure text. All species with weights below a certain cut-off (5–20, depending on the abundance distribution for the well) were omitted from the plots, and species with weights below 5 were omitted from the tables. In cases where a number of species plotted very close to one another, space conflict necessitated that one or several were omitted from the plot, but these are listed in the tables.

Results, Well A

Well A, all groups included

The CA ordinations for Well A, all groups, are shown in Figures 1 (samples) and 2 (species) and are listed in Tables 2 and 3. The samples are represented as pie diagrams showing the total proportions of different microfossil groups present. The eigenvalues for Axes 1 and 2 were both high: 0.59 and 0.51.

Well A, all groups included: samples. The sample ordination shows three main divisions: between depths 2800–2880 m and all others on Axis 1, and between 2900–3020 m and 3040–3220 m on Axis 2 (Axis 3 suggested depths 2940 m, 2980 m, and 3000 m to be significant, but without knowing initially that these were in fact identified as flooding surfaces using other company criteria, efforts were concentrated on the first two axes which had the markedly highest eigenvalues). The site composition plot shown in Figure 1 indicates that the largest proportion of pollen occurs in the samples from 2900–3020 m, suggesting this interval to have existed nearest shore during deposition, probably representing the lowest sea-level stand.

Well A, all groups included: species. The main indicator species for Axis 1 are *Cary sig* (wt. 117) and *Syst pla* (wt. 120), and on Axis 2 are *Cary sig* and *Paly pyr* (wt. 333). By their association in the same quadrant, the following species can be interpreted to reflect nearshore deposition: *Apec.* spp., *Come* spp., *Subt.* spp, and to a lesser extent, *Glap ord.* Since species of *Impagidinium* in our recent database indicate oceanic environments, the association of *Impagidinium* spp. with *Paly pyr* suggests that the samples from 2900–3020 m indicate oceanic conditions. *Spin raG* is identified as an abundant, cosmopolitan species, as it occurs near the centre of both gradients.

Well A, dinoflagellate cysts only

The CA ordination results for Well A, cysts only, are shown in Figures 3 (samples) and 4 (species) and are listed respectively in Tables 4 and 5.

Well A, cysts only: samples. The sample ordination shows the same main divisions on Axis 1 as for all-groups-included, that is, between depths 2800–2880 m and all others lower in the core. On Axis 2, the depth grouping that included most pollen (i.e., presumably nearest shore) in the run with all-groups-included extended from 2900 to 3000 m, with sample 3020 m grouping in the lower left quadrant with the lowest samples in the core (3020–3220 m).

Cluster analysis and 'eyeballing' of the raw data supported the inclusion of sample 3020 with the lowest grouping in the core, that is, following the breakdown given by the ordination using cysts only instead of that including all-groups. Since the data for all-groups-included comprises input from pollen transport from land, whereas the cyst signal reflects conditions in the water only, the cyst data were used for more detailed study, including subsequent percentage calculations and downcore plots of indicator species identified from the CA analyses.

Well A, cysts only: species. The main indicator species for Axis 1 are *Pala pyr* (wt. 333), *Hyst tub* (wt. 51), *Impa spp.* (wt. 50), *Impl lig* (wt. 32) and *Apec spp.* (wt. 32) in the negative domain, and *Syst pla* (wt. 120), and to a lesser extent *Areo mic* (wt. 24) and *Impl. spp.* (wt. 19) in the positive domain. The main variation on Axis 2 is represented by end members *Apec* spp. (wt. 32) and *Paly pyr* (wt. 333) (Fig. 3, Table 4).

All of the raw counts for Well A were converted to percent of total cyst assemblage to test the relative treatment by CANOCO of both types of data input. The results using percentages were essentially the same as those based on raw counts, so subsequent analyses were run on the raw count data only since this reflects the actual cyst concentrations.

Well B

Well B, all groups included

The CA results for Well B, all groups included, are shown in Figures 5 (samples) and 6 (species) and listed in Tables 6 and 7. Again, the eigenvalues for the first two axes were high: 0.57 and 0.49.

Well B, all groups included: samples: The main breakdown suggested by Axis 1 is between 1758.70 m and 1776.98 m, whereas the main breakdown on Axis 2 is between 1633.73 m and 1642.87 m. The site composition plot in Figure 5 shows that the largest concentration of pollen, suggesting more nearshore conditions, occurs in the middle grouping, consistent with the results from Well A. The analysis using all-groups gives a larger distinction between the bottom section of the well and the remainder, whereas Well A had indicated the largest environmental shift to be between the top two sections. Since the ordination using cysts only (below) also supported the largest shift between the top two sections, this suggests that the analyses of cysts alone is probably the most informative.

Well B, all groups included: species. The species ordination shown in Figure 6 indicates the main indicator species for each axis: *Pala pyr* and *Areo spp.* (wts. 157 and 577, respectively) on the positive side of Axis 1 in contrast to spore and pollen groups *Cary sig*, *Fung spo* and *Inap pol* (wts. 156, 137 and 125, respetively) on the negative side. On Axis 2, the main variation is indicated by end-members *Homo ten* (wt. 46) in the positive domain and in the negative domain. *Lent wet* (wt. 42) and the spore/pollen group from Axis 1.

Well B, dinoflagellate cysts only

The CA results for well B using cysts only are shown in Figures 7 (samples) and 8 (species) and listed in Tables 8 and 9. The eigenvalues were again high for the first two axes: 0.57 and 0.49, respectively.

Well B, cysts only: samples. The main variation on Axis 1 was between depth intervals 1511.81–1642.87 m and 1652.02–1938.53 m, that is, between the top and middle sample groupings as in Well A. On Axis 2, the main division was between 1652.02–1758.70 m and 1776.98–1939.53 m, that is, between the middle (nearest shore) and bottom sample groupings.

Well B, cysts only: species. The main species responsible for the variation on Axis 1 can be seen on the ordination plot in Figure 8. *Pala pyr* and *Areo spp.* plot most negatively (wts. 157 and 577, respectively) and *Homo ten* and *Phth com* (wts. 46 and 26, respectively) most positively. On Axis 2, *Apec spp.* (wt. 123) and *Pala pyr* make up the positive and negative end-members, respectively. It is notable that so many species of *Apec* plot positively on Axis 2, in the same quadrant that contained proportionally most pollen, suggesting *Apec* to be an indicator of nearshore conditions at this location. *Lent wet.* may also be interpreted as a relatively nearshore species by virtue of its highly positive position on Axis 2.

Well C

The first run identified two outliers, samples at 1426.46 m and 1444.75 m. The first of these contained anomalously high proportions of one species: *Areo cfse* (wt. of 38, as compared with the next highest in any sample of 4), and the latter of which contained only seven palynomorphs. These were removed in the results reported below.

Well C, all groups included

The CA results for Well C, all groups included, are shown in Figures 9 (samples) and 10 (species) and listed in Tables 10 and 11. The

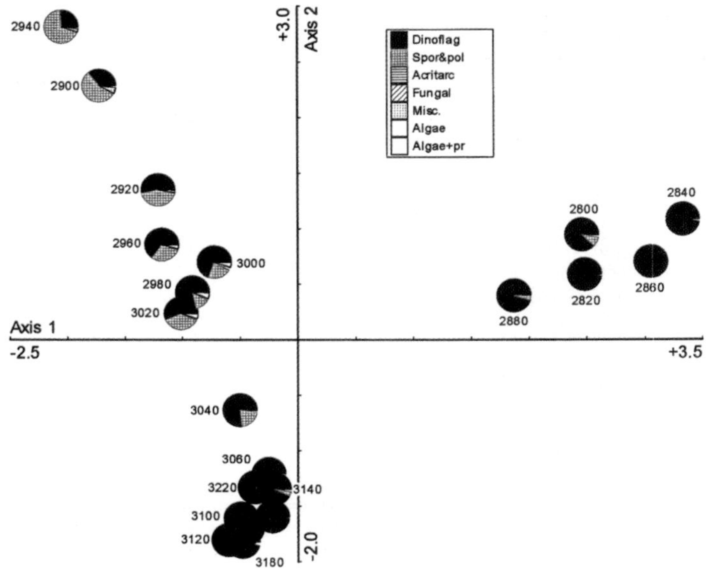

Fig. 1. Well A, CA analysis of all groups included, samples plotted as pie diagrams showing proportions of groups present. Note abundance of spores and pollen in upper left. Sample scores listed in Table 2.

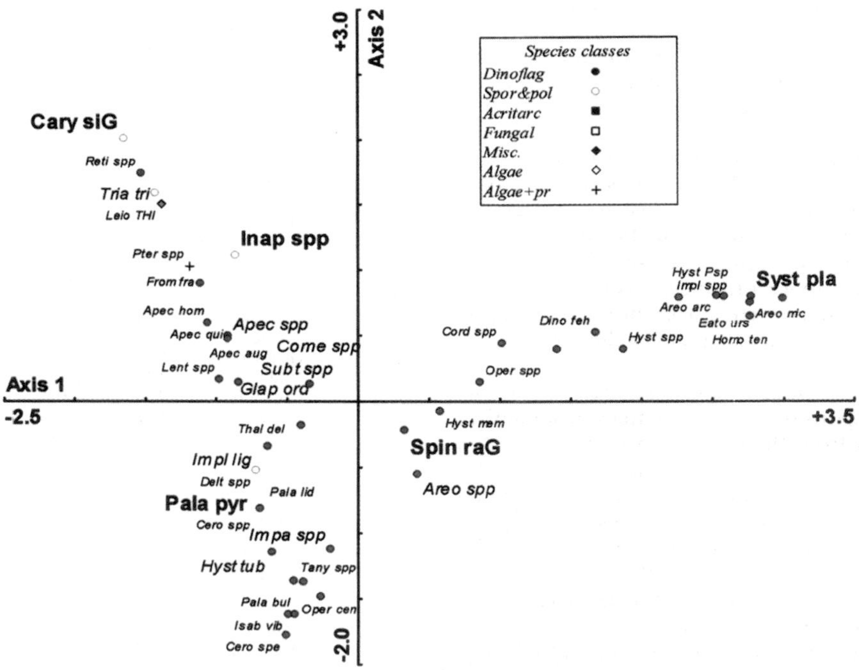

Fig. 2. Well A, CA analysis of all groups included, species classified according to group. Font size and style reflects relative statistical weights. Largest font reflects weights > 100, next largest 25–100, and the smallest 9–25. Species scores listed in Table 3.

Table 2. *Well A, CA of all groups, sample scores, statistical weights and eigenvalues. N = 22*

Sample	Depth (m) eigenvalue	Axis 1 0.58	Axis 2 0.35	Stat. wt.
1	2800	2.62	–0.16	102
2	2820	2.20	–0.11	100
3	2840	3.03	–0.18	101
4	2860	2.58	–0.32	109
5	2880	1.62	–0.07	101
6	2900	–0.41	0.87	42
7	2920	–0.57	2.05	72
9	2940	–0.29	1.09	28
10	2960	–0.85	1.97	79
11	2980	–0.83	3.85	87
12	3000	–0.39	3.00	25
13	3020	–0.82	–0.20	39
14	3040	–0.82	–0.32	92
15	3060	–0.84	–0.58	103
16	3080	–1.19	–0.64	99
18	3100	–1.25	–1.02	73
19	3120	–1.39	–1.15	112
20	3140	–0.86	–0.89	98
21	3160	–0.95	–0.86	99
22	3180	–1.23	–0.94	94
23	3200	–1.20	–0.71	55
24	3220	–0.98	–0.29	107

eigenvalues for Axes 1 and 2 were somewhat lower than were obtained for Axes 1 and 2 in Wells A and B: 0.42 and 0.33, respectively.

Well C, all groups included: samples. The main variation on Axis 1 was between sample groups 1207.01–1280.16 m and all lower samples, consistent with the largest source of variation in Wells A and B. Axis 2 showed no discernable divisions without further omission of suspect taxa (discussed later), but the site composition plot showed roughly that depths from 1292.35 to 1377.70 m contained proportionally more spores and pollen than cysts (Fig. 9), consistent with the middle section again representing the nearest shore environment in the core.

Well C, all groups included: species. The main taxa responsible for the samples groupings can be seen from the ordination plot in Figure 10. The taxa are classified according to group and plotted with a symbol designating their microfossil affinity. Axis 1 is represented by end-members *Impl spp.* (wt. 560), *Pala pyr* (wt. 302) and *Areo spp.* (wt. 191) in the negative domain and *Syst pla* (wt. 36), *Homo ten* (wt. 38) and *Phth sto* (wt. 32) in the positive domain. The main variation on Axis 2 is represented by the pollen *Cary sig* (wt. 152) and *Hyst tub* (wt. 146) in the positive domain and *Pala pyr* in the negative domain.

Table 3. *Well A, CA of all groups, species scores, statistical weights and eigenvalues. N = 190. Species with weights < 5 omitted*

Species	Spp. code eigenvalue	Axis 1 0.59	Axis 2 0.51	Stat. wt.
1	Spin raG	0.32	–0.22	512
2	Pala pyr	–0.46	–1.36	333
3	Inap spp	–0.86	1.12	175
4	Syst pla	2.77	0.79	120
5	Cary siG	–1.65	2.01	117
6	Areo spp	0.42	–0.55	63
7	Hyst tub	–0.60	–1.14	51
8	Impa spp	–2.00	–1.12	50
10	Impl lig	–0.64	–0.34	33
11	Apec spp	–0.91	0.48	32
15	Come spp	–0.09	0.50	32
17	Glap ord	–0.94	0.12	31
18	Subt spp	–0.34	0.13	26
19	Tria tri	–1.43	1.59	25
20	Areo mic	2.76	0.75	24
21	Cero spp	–0.69	–0.81	23
22	Impl spp	2.26	0.79	19
23	Apec hom	–1.06	0.60	17
26	Pala bul	–0.45	–1.62	17
27	Thal del	–0.40	–0.18	17
29	Lent spp	–0.97	0.17	16
31	Pter spp	–1.19	1.03	16
32	Reti spp	–1.52	1.74	16
33	Delt spp	–0.72	–0.52	15
35	Eato urs	2.99	0.78	15
37	Tany spp	–0.39	–1.37	15
38	Dino feh	1.67	0.52	13
39	Hyst mem	0.58	–0.08	13
41	Isab vib	–0.50	–1.62	12
44	Oper spp	0.85	0.14	12
45	Cero spe	–0.51	–1.77	11
47	Homo ten	2.76	0.65	11
49	Leio THI	–1.38	1.51	11
50	Oper cen	–0.26	–1.48	11
51	Apec aug	–0.84	0.14	10
53	Areo arc	2.52	0.80	10
56	Cord spp	1.02	0.44	10
57	Pala lid	–0.65	–0.75	10
59	Hyst pSP	2.58	0.80	10
62	Apec qui	–0.92	0.50	9
63	From fra	–1.11	0.90	9
65	Thal pel	1.40	0.40	9
69	Cary ver	–1.21	1.02	8
70	Cero dep	–0.62	–1.38	8
71	Acho alc	1.17	0.12	7
72	Apec par	–0.73	–0.16	7
73	Cero str	–0.49	–1.71	7
74	Crib ten	0.53	–0.27	7
75	Glap spp	–0.13	0.02	7
76	Para ind	–1.10	0.72	7
77	Phth gem	2.50	0.83	7
78	Reti ret	–1.57	1.96	7
80	Seno ino	–0.39	–1.37	7
81	Cere bar	2.81	0.80	6
82	Cero war	–1.35	1.61	6
83	FUNG SPO	0.39	–0.07	6
84	Glei spp	–0.69	–0.75	6
86	Leio LRG	–1.76	2.25	6
88	Ling mac	2.04	0.96	6
89	Meli pse	–0.78	–0.51	6
90	Spha spp	–1.37	1.44	6
91	Till mic	–1.32	1.37	6
92	Acan Acr	–0.68	–0.69	5
93	Areo CFm	2.35	0.51	5
94	Cord gra	0.87	0.88	5
95	Cupu cin	–1.51	1.91	5
96	Micr spp	0.10	–1.10	5
97	Pala spp	–0.53	–1.31	5

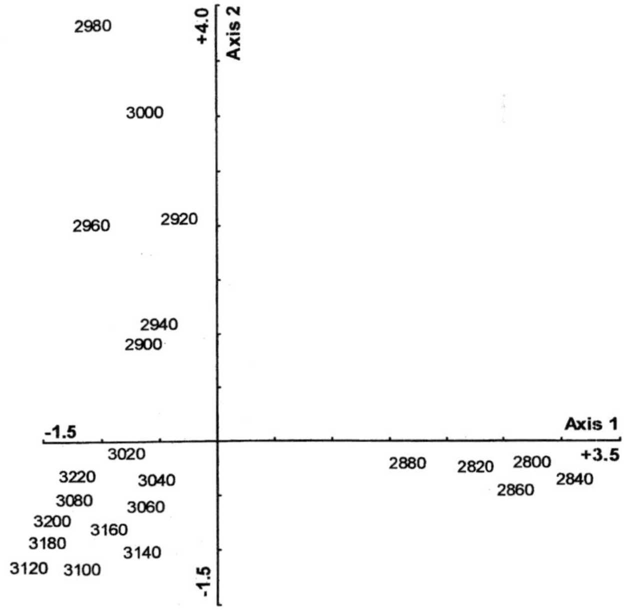

Fig. 3. Well A, CA analysis of dinoflagellate cysts only, sample ordination. Sample scores listed in Table 4.

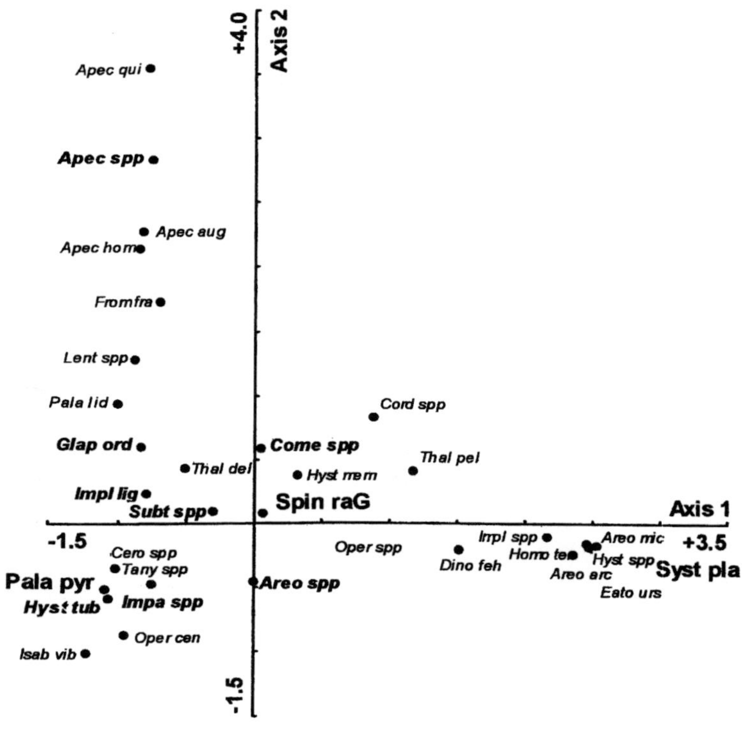

Fig. 4. Well A, CA analysis of dinoflagellate cysts only, species ordination. Font size and style reflects relative statistical weights. Largest font reflects weights > 100, next largest 25–100, and the smallest 9–25. All species with weights < 9 were omitted. Species scores listed in Table 5.

Table 4. *Well A, CA of dinoflagellate cysts only, sample scores, statistical weights and eigenvalues. N = 22*

Sample	Depth eigenvalue	Axis 1 0.59	Axis 2 0.51	Stat. wt.
1	2800	2.46	0.94	116
2	2820	2.48	0.59	103
3	2840	3.33	1.08	104
4	2860	3.06	0.69	110
5	2880	1.87	0.39	106
6	2900	−1.72	2.29	160
7	2920	−1.20	1.35	137
8	2940	−2.04	2.82	113
9	2960	−1.17	0.86	126
10	2980	−0.91	0.42	112
11	3000	−0.72	0.70	36
12	3020	−1.00	0.23	72
13	3040	−0.50	−0.64	119
14	3060	−0.25	−1.21	108
15	3080	−0.44	−1.70	99
16	3100	−0.52	−1.81	79
17	3120	−0.60	−1.81	118
18	3140	−0.21	−1.35	106
19	3160	−0.22	−1.59	101
20	3180	−0.47	−1.83	101
21	3200	−0.49	−1.61	61
22	3220	−0.37	−1.33	112

Well C, dinoflagellate cysts only

The CA results for well C using cysts only (and outliers omitted) are shown in Figures 11 (samples) and 12 (species) and listed in Tables 12 and 13. The eigenvalue for Axis 1 using cysts only was slightly higher than for all-groups-included (0.47 as compared with 0.42), and was 0.33 for Axis 2.

Well C, cysts only: samples. Axis 1 showed the same sample groupings as for all-groups-included. In order to obtain a better spread of the variation and try to identify environmental shifts along Axis 2, three taxa that plotted most highly on Axis 2 were omitted: *Glap ord*, *Defl oeb* and *Areo Cfm*. This resulted in an increased separation between two groups plotting positively, represented by samples 1292.35–1335.02 m and 1359.41–1408.18 m, and one group plotting negatively, 1405.13–1517.90 m, with slight overlap between the boundaries of the three groups.

Well C, cysts only: species. The main variation on Axis 1 was between *Impl. spp*, *Pala pyr* and *Areo spp* (wts. 560, 302 and 191, respectively) plotting in the negative domain, and *Homo tenu*, *Syst pla* and *Phth sto* (wts. 38, 36 and 32) in the positive domain (all three latter species plot so closely together that only *Homo tenu* is shown in Figure 12 due to conflicting plotting positions). *Glap ord*, *Defl oeb* and *Areo Cfm* plus *Hyst tub* (wt. 146) plotted most positively on

Table 5. *Well A, CA of dinoflagellate cysts only, species scores, statistical weights and eigenvalues. N = 116. Species with weights < 5 omitted*

Sample	Spp. code eigenvalue	Axis 1 0.58	Axis 2 0.35	Stat. wt.
1	Spin raG	0.07	0.08	512
2	Pala pyr	−1.06	−0.59	333
3	Syst pla	2.53	−0.18	120
4	Areo spp	0.00	−0.45	63
5	Hyst tub	−1.09	−0.51	51
6	Impa spp	−0.75	−0.47	50
7	Impl lig	−0.79	0.23	33
9	Come spp	0.05	0.58	32
10	Apec spp	−0.76	2.83	32
11	Glap ord	−0.83	0.60	31
12	Subt spp	−0.29	0.10	26
13	Areo mic	2.50	−0.19	24
14	Cero spp	−1.01	−0.24	23
15	Impl spp	2.17	−0.11	19
16	Thal del	−0.50	0.43	17
18	Apec hom	−0.85	2.14	17
19	Pala e bu	−1.16	−0.88	17
20	Lent spp	−0.88	1.28	16
21	Eato urs	2.60	−0.24	15
22	Tany spp	−1.01	−0.35	15
23	Hyst mem	0.33	0.38	13
24	Dino feh	1.52	−0.20	13
25	Oper spp	0.76	−0.12	12
26	Isab vib	−1.22	−1.01	12
27	Homo ten	2.36	−0.24	11
28	Cero spe	−1.26	−1.01	11
29	Oper cen	−0.95	−0.87	11
30	Areo arc	2.47	−0.18	10
31	Cord spp	0.88	0.82	10
32	Hyst spp	2.46	−0.16	10
33	Pala lid	−1.00	0.94	10
34	Apec aug	−0.83	2.28	10
35	Thal pel	1.18	0.41	9
36	From fra	−0.70	1.73	9
37	Apec qui	−0.78	3.55	9
38	Cero dep	−1.23	−0.87	8
39	Acho alc	0.89	0.00	7
40	Para ind	−0.64	0.77	7
41	Phth gem	2.48	−0.14	7
42	Glap spp	−0.20	0.25	7
43	Crib ten	0.18	0.61	7
44	Apec par	−0.83	1.59	7
45	Cero str	−1.23	−0.83	7
46	Seno ino	−1.01	−0.35	7
47	Ling mac	2.06	0.22	6
48	Cere bar	2.52	−0.14	6
49	Meli pse	−0.99	0.26	6
50	Cero war	−0.46	1.80	6
51	Cord gra	1.01	0.07	5
52	Micr spp	−0.44	−0.75	5
53	Areo CFm	2.00	−0.17	5

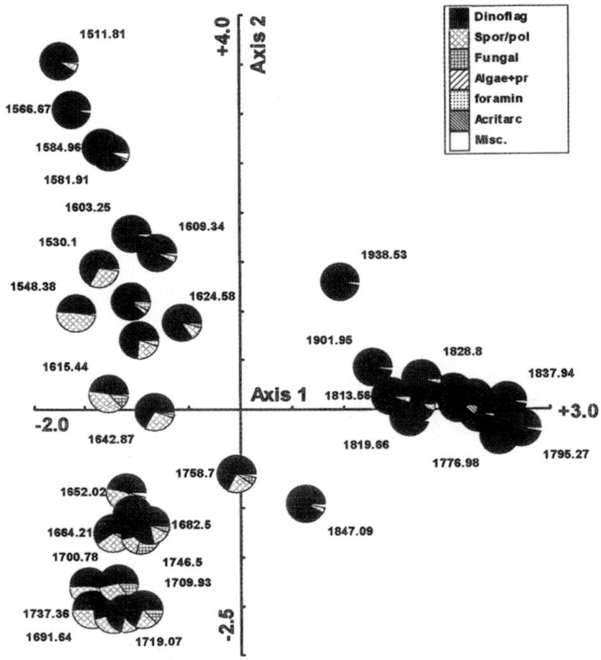

Fig. 5. Well B, CA analysis of all groups included, samples plotted as pie diagrams showing proportions of groups present. Note abundance of spores and pollen in lower-middle left. Sample scores listed in Table 6.

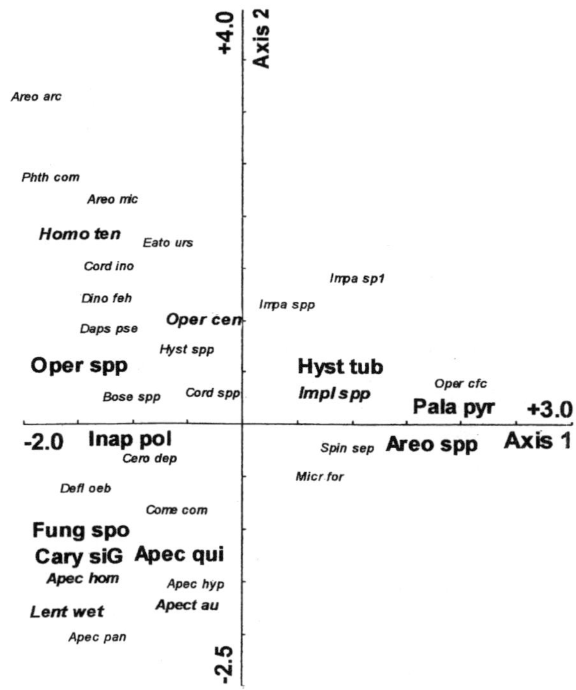

Fig. 6. Well B, CA analysis of all groups included, species ordination. Font size and style reflects relative statistical weights. Largest font reflects weights > 90, next largest 25–90, and the smallest 11–25. All species with weights < 11 were omitted. Species scores listed in Table 7.

Table 6. *Well B, CA of all groups, sample scores, statistical weights and eigenvalues. N = 39*

Sample	Depth (m) eigenvalue	Axis 1 0.57	Axis 2 0.49	Stat. wt.
1	1511.81	−1.74	3.54	55
2	1530.10	−1.36	1.43	70
3	1548.38	−1.59	0.97	59
4	1566.67	−1.63	3.04	68
5	1581.91	−1.25	2.60	78
6	1584.96	−1.33	2.66	85
7	1597.15	−1.05	1.09	69
9	1603.25	−1.04	1.78	80
10	1609.34	−0.80	1.59	74
11	1615.44	−1.27	0.16	60
12	1624.58	−0.56	0.88	95
13	1633.73	−0.97	0.70	82
14	1642.87	−0.83	−0.03	85
15	1652.02	−1.11	−0.84	75
16	1664.21	−1.23	−1.26	77
17	1673.35	−1.04	−1.07	83
18	1682.50	−0.87	−1.18	88
19	1691.64	−1.43	−2.03	95
20	1700.78	−1.18	−1.77	102
21	1709.93	−0.94	−2.03	84
22	1719.07	−1.23	−2.08	94
23	1728.22	−1.11	−2.07	81
24	1737.36	−1.47	−1.80	79
25	1746.50	−0.96	−1.27	91
26	1758.70	−0.04	−0.65	84
27	1776.98	2.51	−0.27	97
28	1795.27	2.73	−0.19	97
29	1813.56	1.47	0.13	74
30	1819.66	1.64	−0.08	73
31	1828.80	1.76	0.30	67
32	1837.94	2.59	0.08	87
33	1847.09	0.63	−0.96	63
34	1856.23	2.12	0.05	96
35	1865.38	2.26	0.11	98
36	1874.52	2.07	0.17	98
37	1883.66	2.31	−0.05	98
38	1901.95	1.29	0.43	78
39	1920.24	1.71	0.05	17
40	1938.53	0.97	1.29	89

Axis 2, with *Pala pyr* plotting most negatively. The fact that the three species plotting positively on Axis 1 plot roughly 4 standard deviations away from those plotting negatively indicates that the samples from 1207.01 to 1298.45 m represent a markedly different environment than those lower in the well (see Hill & Gauch 1980 or Ter Braak 1996 for explanation). This environmental shift appears to be more pronounced in Well C (and D, see below) than in Wells A or B.

Well D

The initial run of Well D was the first in which samples that were not chronological plotted together at the positive end of Axis 1. The species ordination indicated the presence of several taxa (notably, *Spin spp*. and *Acho ram*) that were not present at any other depth. Since *Spin raG* (*Spiniferites ramosus* group) was the most abundant cyst type in Wells A, C and D and

Table 7. *Well B, CA of all groups, species scores, statistical weights and eigenvalues. N = 154. Weights < 5 omitted*

Sample	Spp. code eigenvalue	Axis 1 0.57	Axis 2 0.49	Stat. wt.
1	Areo spp	1.67	−0.17	577
2	Spin ram	−0.03	0.54	481
3	Pala pyr	2.06	0.04	157
4	Cary siG	−1.02	−1.10	156
5	Fung spo	−0.97	−0.97	137
6	Inap spp	−1.08	−0.16	125
7	Apec spp	−0.84	−1.55	123
9	Hyst tub	0.45	0.40	104
10	Oper spp	−0.99	0.41	91
11	Impl spp	0.71	0.21	65
12	Homo ten	−1.07	1.70	46
13	Lent wet	−1.23	−1.65	42
14	Oper cen	−0.39	0.91	41
15	Apec aug	−0.85	−1.60	38
16	Apec hom	−1.08	−1.34	36
18	Cero dep	−0.55	−0.45	35
19	Apec hyp	−0.73	−1.41	34
20	Come com	−0.93	−0.83	27
21	Cord spp	−0.19	0.17	26
22	Phth com	−1.47	2.26	26
23	Momi spp	−0.95	−1.52	25
24	Bose spp	−0.70	0.14	24
25	Eato urs	−0.96	1.63	24
26	Laev spp	−0.87	−0.02	23
27	Daps pse	−0.91	0.80	22
28	Pala gol	−0.79	−0.17	22
29	Oper cfc	2.06	0.15	22
30	Cord ino	−0.94	1.50	21
31	Areo mic	−0.90	2.15	21
32	Spin sep	1.15	−0.13	21
33	Areo arc	−1.60	3.04	17
34	Impa spp	0.11	1.24	17
35	Impa sp1	0.97	1.29	17
36	Micr for	0.43	−0.38	15
37	Delt spp	−0.77	−1.03	15
38	Hyst spp	−0.18	0.71	14
39	Glap div	−0.29	−0.91	14
40	Apec qui	−0.11	−1.08	14
41	Acan Acr	1.62	0.05	11
42	Dino feh	−0.97	1.10	11
43	Defl oeb	−1.15	−0.50	11
44	Apec pan	−1.00	−1.91	11
45	Cupu cin	−0.98	−1.44	11
46	Crib giu	−1.20	2.13	10
47	Defl pho	−1.44	2.44	10
48	Ling mac	−1.04	1.07	10
49	Azol spp	−1.15	1.74	10
50	Acho alc	−1.38	2.36	9
51	Areo dik	−1.11	1.62	9
52	Phth spp	−1.02	1.45	9
53	Wetz art	−1.33	2.54	9
54	Sene oSP	−1.03	0.86	8
55	Apec par	−0.15	−1.10	8
56	Bac– Osm	−0.82	0.33	8
57	Sequ spp	−1.38	0.46	8
58	Nyss spp	−0.66	−0.82	8
59	Reti hia	−1.09	−1.78	8
60	Thal pel	−1.10	−0.22	7
61	Phth ind	−1.28	2.28	7
62	Oper sev	−0.61	−1.04	7
63	Reti spp	−0.79	−0.69	7
64	Meli pse	−1.17	1.76	6
65	Phth amo	−1.54	3.06	6
66	Cord o mi	−1.36	2.03	6
67	Cord fun	−1.31	2.65	6
68	Trip spp	−1.09	−0.61	6
69	Char col	−1.27	2.31	5
70	Cord gra	−1.07	1.73	5
72	Leje spp	−1.10	0.32	5
73	Glap ord	−0.35	−0.44	5
74	Para ind	−1.17	−1.38	5
75	Tril spp	−0.11	−0.37	5

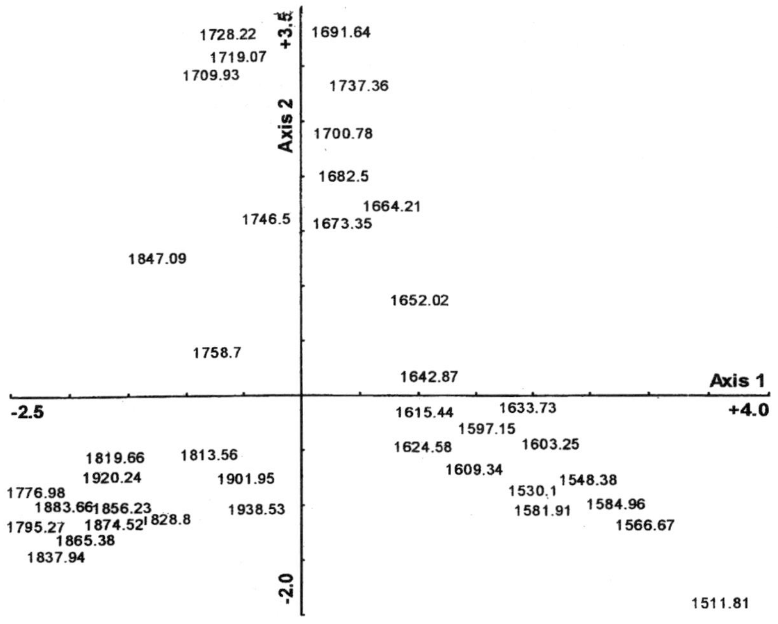

Fig. 7. Well B, CA analysis of dinoflagellate cysts only, sample ordination. Sample scores listed in Table 8.

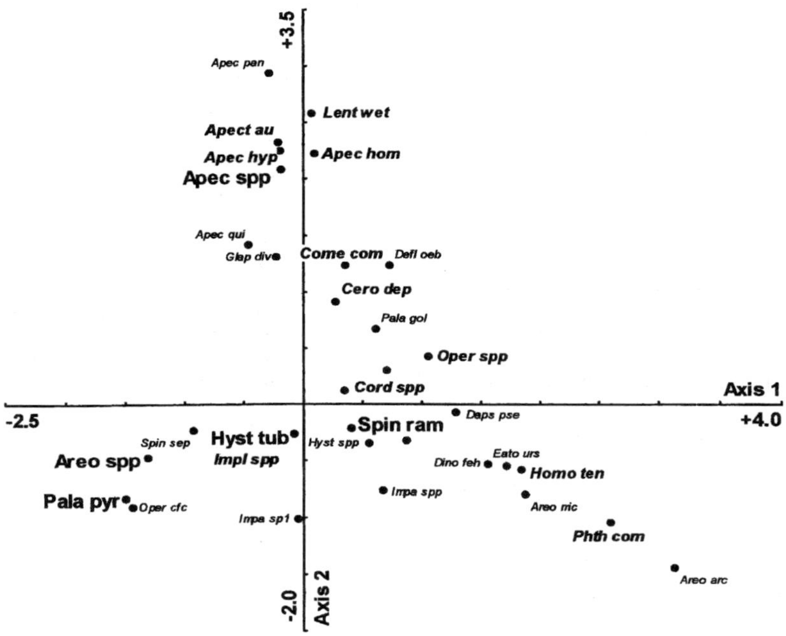

Fig. 8. Well B, CA analysis of dinoflagellate cysts only, species ordination. Font size and style reflects relative statistical weights. Largest font reflects weights > 100, next largest 25–100, and the smallest 10–25. All species with weights < 10 were omitted. Species scores listed in Table 9.

Table 8. *Well B, CA of dinoflagellate cysts only, sample scores, statistical weights and eigenvalues. N = 39*

Sample	Depth (m) eigenvalue	Axis 1 0.57	Axis 2 0.49	Stat. wt.
1	1511.81	3.57	−1.79	50
2	1530.10	2.26	−0.86	47
3	1548.38	2.18	−0.76	29
4	1566.67	2.91	−1.07	66
5	1581.91	2.39	−0.93	72
6	1584.96	2.42	−0.87	80
7	1597.15	1.63	−0.16	61
8	1603.25	1.83	−0.43	77
9	1609.34	1.57	−0.56	68
10	1615.44	1.26	−0.35	29
11	1624.58	1.03	−0.38	80
12	1633.73	1.32	−0.18	61
13	1642.87	0.80	0.05	57
14	1652.02	0.68	0.85	35
15	1664.21	0.44	1.70	46
16	1673.35	0.28	1.73	58
17	1682.50	0.08	1.96	69
18	1691.64	0.00	3.28	48
19	1700.78	0.01	2.36	55
20	1709.93	−0.44	2.90	58
21	1719.07	−0.22	3.07	51
22	1728.22	−0.32	3.29	58
23	1737.36	0.16	2.79	40
24	1746.50	−0.01	1.59	49
25	1758.70	−0.47	0.27	57
26	1776.98	−1.98	−0.86	95
27	1795.27	−2.08	−1.02	93
28	1813.56	−1.03	−0.66	70
29	1819.66	−1.25	−0.57	70
30	1828.80	−1.16	−1.00	64
31	1837.94	−1.87	−1.19	85
32	1847.09	−0.95	1.14	58
33	1856.23	−1.49	−0.87	86
34	1865.38	−1.58	−1.01	97
35	1874.52	−1.44	−0.89	96
36	1883.66	−1.70	−0.91	95
37	1901.95	−0.68	−0.77	76
38	1920.24	−1.32	−0.76	16
39	1938.53	−0.04	−1.01	87

was the second most abundant type in Well B (after *Areo spp.*), this was suspected to represent counting inconsistencies. Closer examination of the raw data supported this assumption, i.e. *Spin raG* was not counted in the samples containing *Spin spp.*, and *Ach ram*, whereas it was abundant or dominant in adjacent samples. These taxa were therefore grouped with *Spin raG* for subsequent runs (Fig. 13 and Tables 14 and 15). Taxonomic inconsistencies were also identified between *Syst spp.* and *Sys pla* and these also were combined. Additional taxonomic sources of error are suspected (e.g. sample 270 is assumed to not have been counted

Table 9. *Well B, CA of dinoflagellate cysts only, species scores, statistical weights and eigenvalues. N = 107. Weights < 5 omitted*

Sample	Depth (m) eigenvalue	Axis 1 0.57	Axis 2 0.49	Stat. wt.
1	Areo spp	−1.30	−0.48	577
2	Spin ram	0.41	−0.22	481
3	Pala pyr	−1.49	−0.83	157
4	Apec spp	−0.19	2.24	123
6	Hyst tub	−0.07	−0.26	104
7	Oper spp	1.06	0.42	91
8	Impl spp	−0.33	−0.36	65
9	Homo ten	1.84	−0.59	46
10	Lent wet	0.07	2.57	42
11	Oper cen	0.88	−0.32	41
12	Apec tau	−0.21	2.31	38
13	Apec hom	0.10	2.22	36
15	Cero dep	0.27	0.91	35
16	Apec hyp	−0.18	2.07	34
17	Come com	0.36	1.23	27
18	Cord spp	0.35	0.12	26
19	Phth com	2.60	−1.04	26
20	Bose spp	0.70	0.30	24
21	Eato urs	1.71	−0.55	24
22	Daps pse	1.29	−0.07	22
23	Oper cfc	−1.43	−0.90	22
24	Pala gol	0.62	0.66	22
25	Areo mic	1.87	−0.80	21
26	Cord ino	1.66	−0.55	21
27	Spin sep	−0.93	−0.23	21
28	Areo arc	3.13	−1.44	17
29	Impa sp1	−0.04	−1.01	17
30	Impa spp	0.67	−0.76	17
31	Apec qui	−0.46	1.41	14
32	Glap div	−0.22	1.30	14
33	Hyst spp	0.56	−0.35	14
34	Apec pan	−0.28	2.93	11
35	Defl oeb	0.73	1.23	11
36	Dino feh	1.56	−0.53	11
37	Crib giu	2.37	−1.17	10
38	Defl pho	2.47	−0.85	10
39	Ling mac	1.48	−0.01	10
40	Acho alc	2.47	−0.90	9
41	Areo dik	1.96	−0.62	9
42	Phth spp	1.73	−0.49	9
43	Wetz art	2.44	−0.92	9
44	Apec par	−0.48	1.24	8
45	Sene oSP	1.43	−0.23	8
46	Oper sev	−0.10	1.46	7
47	Phth ind	2.29	−0.78	7
48	Thal pel	0.85	1.05	7
49	Cord fun	2.41	−0.88	6
50	Cord o mi	2.26	−0.73	6
51	Meli pse	2.06	−0.71	6
52	Phth amo	2.96	−1.26	6
53	Char col	2.28	−0.81	5
54	Cord gra	1.84	−0.48	5
55	Glap ord	0.17	0.63	5
56	Leje spp	1.13	0.45	5
57	Para ind	0.26	2.02	5

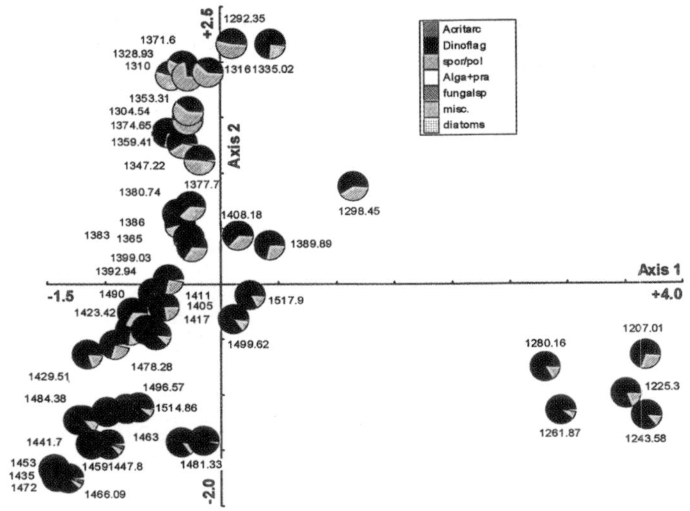

Fig. 9. Well C, CA analysis of all groups included, samples plotted as pie diagrams showing proportions of groups present. Note abundance of spores and pollen in upper left. Sample scores listed in Table 10.

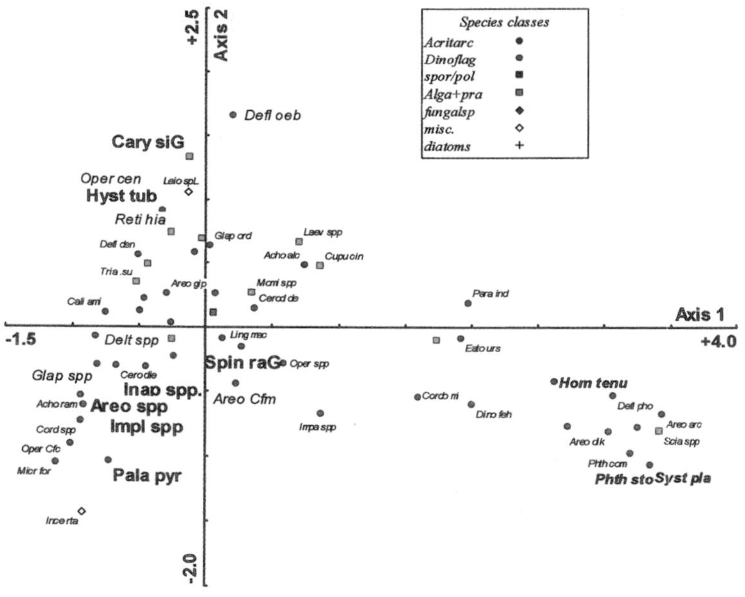

Fig. 10. Well C, CA analysis of all groups included, species classified according to group. Font size and style reflects relative statistical weights. Largest font reflects weights > 100, next largest 25–100, and the smallest 10–25. All species with weights < 10 were omitted. Species scores listed in Table 11.

by the same analyst who counted samples 258, 264, 276 and 282 and this plots out of chronological order). This indicates the potential for using this kind of statistical analysis for quality control at an early stage in data collection.

Well D, all groups included

With the suspected counting inconsistencies corrected for, the eigenvalues for Axes 1 and 2 were high: 0.45 and 0.42, respectively.

Table 10. *Well C, CA of all groups, sample scores, statistical weights and eigenvalues. N = 49*

Sample	Depth (m) eigenvalue	Axis 1 0.42	Axis 2 0.33	Stat. wt.
1	1207.01	3.67	–0.65	137
2	1225.30	3.50	–1.00	89
3	1243.58	3.68	–1.19	110
4	1261.87	2.94	–1.14	131
5	1280.16	2.80	–0.76	116
6	1292.35	–0.10	1.90	107
7	1298.45	1.15	0.88	163
8	1304.54	–0.45	1.36	121
9	1310.64	–0.32	1.96	121
10	1316.74	0.10	2.16	137
11	1328.93	0.42	1.88	87
12	1335.02	0.44	2.16	152
13	1347.22	–0.33	1.26	96
14	1353.31	–0.28	1.48	109
15	1359.41	–0.17	1.11	109
16	1365.50	–0.34	0.54	90
17	1371.60	–0.28	1.87	118
18	1374.65	–0.27	1.56	111
19	1377.70	–0.25	0.70	111
20	1380.74	–0.37	0.63	103
21	1383.79	–0.24	0.33	113
22	1386.84	–0.27	0.41	101
23	1389.89	0.42	0.35	103
24	1392.94	–0.58	–0.08	87
25	1399.03	–0.45	0.04	102
26	1405.13	–0.56	–0.47	97
27	1408.18	0.15	0.44	34
28	1411.22	–0.49	–0.21	100
29	1417.32	–0.64	–0.42	104
30	1423.42	–0.76	–0.26	99
32	1429.51	–1.15	–0.64	109
33	1435.61	–0.99	–1.16	103
34	1441.70	–1.23	–1.23	103
36	1447.80	–1.12	–1.45	88
37	1453.90	–1.44	–1.67	123
38	1459.99	–0.97	–1.45	98
39	1463.04	–0.34	–1.43	138
40	1466.09	–1.41	–1.74	106
41	1472.18	–1.32	–1.76	105
42	1478.28	–0.92	–0.55	119
43	1481.33	0.15	1.42	109
44	1484.38	–1.19	–1.24	97
45	1490.47	–0.78	–0.42	101
46	1496.57	–0.82	–1.13	95
47	1499.62	0.11	–0.33	112
48	1514.86	–0.70	–1.12	66
49	1517.90	0.26	–0.11	100

Well D, all groups included, samples. With the *Spiniferites* treated as above, the top section of the core (samples 198–252 m) accounted for the main source of variation in Well D, consistent with Wells A–C. The main variation on Axis 2 was represented by sample depths 258–282 m, with 270 m plotting lower and together with 288 m on the positive side of Axis 1, presumably reflecting the counting inconsistencies mentioned above. The site composition ordination indicates that samples 270 m and 288–324 m contain more than 25% pollen and spores, identifying this middle section of the core again as being nearest shore at the time of deposition.

Well D, all groups included, species. The species plot for Well D all groups is represented as a biplot including both samples and species (Fig. 13), due to the importance of considering which

Table 11. *Well C, CA of all groups, species scores, statistical weights and eigenvalues. N = 189. Weights < 5 omitted*

Sample	Spp. code eigenvalue	Axis 1 0.42	Axis 2 0.33	Stat. wt.
1	Spin raG	0.27	–0.16	1147
2	Inap spp	–0.03	0.69	890
3	Impl spp	–0.75	–0.67	560
5	Pala pyr	–0.72	–1.05	302
6	Areo spp	–0.92	–0.61	191
7	Cary siG	–0.12	1.33	152
8	Hyst tub	–0.32	0.91	146
10	Areo Cfm	0.23	–0.45	108
11	Defl oeb	0.21	1.66	70
12	Reti hia	–0.26	0.74	65
13	Glap ord	0.04	0.64	60
14	Glap spp	–0.82	–0.30	55
15	From fra	–0.24	–0.24	53
16	Oper Cfc	–1.01	–0.92	52
17	Oper cen	–0.45	1.06	45
18	Delt spp	–0.25	–0.10	44
19	Areo gip	–0.29	0.26	41
20	Hom tenu	2.62	–0.43	38
22	Cord gra	0.08	0.26	37
23	Syst pla	3.34	–1.07	36
24	Cary spp	–0.43	0.49	34
25	Phth sto	3.25	–0.78	32
26	Areo CFs	–0.46	0.22	32
27	Defl den	–0.51	0.56	25
28	Laev spp	0.70	0.67	25
29	Ling mac	0.13	–0.09	24
30	Glap div	–0.83	–0.08	22
34	Acan Acr	–0.91	–0.78	21
36	Oper spp	0.59	–0.29	20
37	Cero d de	0.37	0.14	18
38	Cali ami	–0.75	0.10	17
40	Acho alc	0.74	0.48	17
41	Areo mic	2.72	–0.77	16
43	Dino feh	2.01	–0.60	15
45	Cord o mi	1.59	–0.55	15
46	Hyst spp	–0.45	–0.31	15
47	Acho ram	–0.94	–0.54	15
48	Micr for	–1.12	–1.06	15
49	Ince rta	–0.93	–1.44	15
50	Tria .su	–0.52	0.34	15
51	Para ind	1.98	0.18	14
52	Defl pho	3.07	–0.54	14
54	Cord spp	–0.94	–0.74	13
55	Leio spL	–0.13	1.06	13
56	Pter spp	0.07	0.11	12
57	Areo dik	3.03	–0.81	12
58	Pala gol	–0.26	0.03	12
59	Nyss kru	1.73	–0.11	12
60	Scia spp	3.41	–0.81	12
61	Momi spp	0.35	0.26	12
62	Areo arc	3.43	–0.68	11
64	Impa spp	0.86	–0.68	11
66	Subt spl	–0.50	0.12	11
67	Phth com	3.20	–0.97	10
68	Eato urs	1.92	–0.09	10
69	Lent wet	–0.08	0.58	10
70	Cero die	–0.67	–0.31	10
71	Cupu cin	0.86	0.47	10
72	Tili miG	0.01	1.21	9
73	Apec spp	–0.96	–0.66	8
74	Cero spe	–0.18	–0.69	8
75	Areo CFc	–0.82	–0.40	8
76	Sequ spp	–0.11	0.92	8
77	Trip cor	–0.32	1.37	8
78	Crib giu	0.08	0.16	7
79	Ster spp	0.65	0.22	7
80	Reti spp	–0.65	0.14	7
81	Sele nep	2.93	–0.94	6
82	Thal del	–0.06	1.47	6
83	Glap pas	0.30	–0.02	6
84	Apec aug	–0.06	1.59	6
85	Impl lig	–0.55	–0.11	6
86	Daps sim	3.00	–0.81	5
87	Cord ino	0.82	0.27	5
88	Cero war	–0.25	1.39	5
89	Micr Cfo	–0.20	–1.06	5
90	Cord exi	–0.73	–0.54	5
91	Fung spo	–0.65	0.10	5
92	Alni ver	–0.20	1.40	5

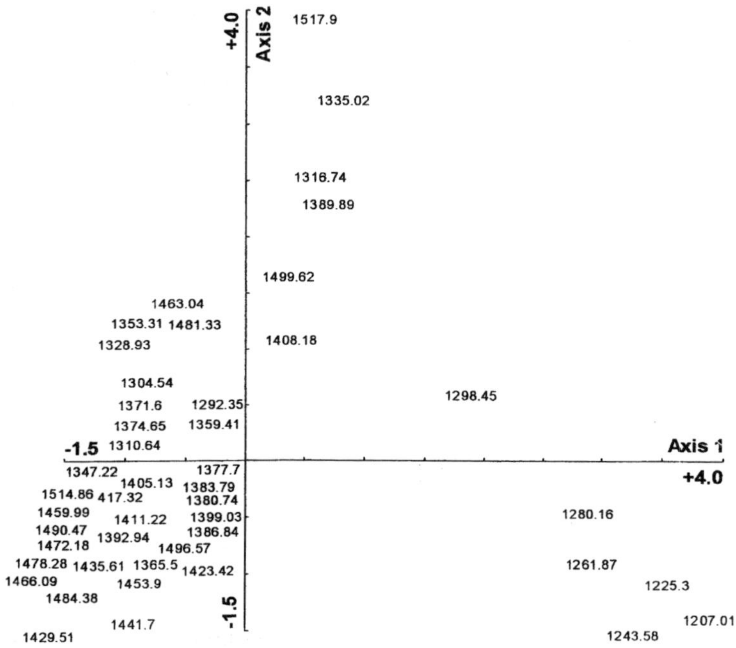

Fig. 11. Well C, CA analysis of dinoflagellate cysts only, sample ordination, two outliers removed. Sample scores listed in Table 12.

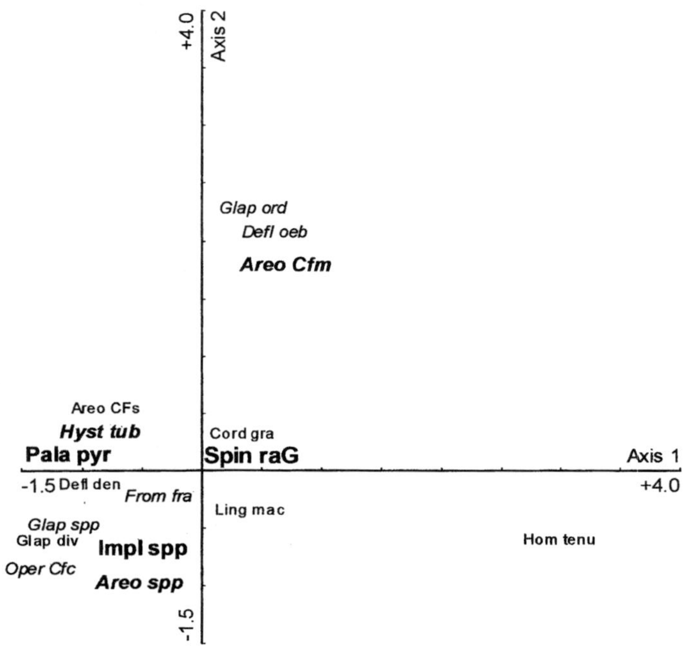

Fig. 12. Well C, CA analysis of dinoflagellate cysts only, species ordination. Font style reflects relative statistical weights. Largest font reflects weights > 300, next largest 100–200, the next largest 50–100, and the smallest 20–25. All species with weights < 20 were omitted. Species scores listed in Table 13.

Table 12. *Well C, CA of dinoflagellate cysts only, sample scores, statistical weights and eigenvalues. N = 49*

Sample	Depth (m) eigenvalue	Axis 1 0.47	Axis 2 0.33	Stat. wt.
1	1207.01	3.77	−1.29	96
2	1225.30	3.31	−0.98	71
3	1243.58	3.53	−1.42	95
4	1261.87	2.65	−0.91	118
5	1280.16	2.61	−0.35	97
6	1292.35	−0.11	0.59	41
7	1298.45	1.62	0.48	96
8	1304.54	−0.55	0.59	87
9	1310.64	−0.64	0.04	52
10	1316.74	0.37	2.41	63
11	1328.93	−0.74	0.94	40
12	1335.02	0.56	3.09	112
13	1347.22	−0.55	−0.02	57
14	1353.31	−0.40	1.12	52
15	1359.41	−0.23	0.12	53
16	1365.50	−0.49	−0.60	49
17	1371.60	−0.65	0.39	32
18	1374.65	−0.37	0.20	44
19	1377.70	−0.28	−0.06	69
20	1380.74	−0.42	−0.17	63
21	1383.79	−0.27	−0.04	75
22	1386.84	−0.34	−0.54	56
23	1389.89	0.44	2.17	74
24	1392.94	−0.63	−0.53	63
25	1399.03	−0.50	−0.32	73
26	1405.13	−0.55	−0.06	85
27	1408.18	0.12	0.97	22
28	1411.22	−0.51	−0.43	78
29	1417.32	−0.66	−0.37	74
30	1423.42	−0.91	−0.86	66
32	1429.51	−1.28	−1.49	87
33	1435.61	−0.94	−0.81	96
34	1441.70	−1.15	−1.33	96
36	1447.80	−0.97	0.80	78
37	1453.90	−1.25	−0.90	115
38	1459.99	−0.85	−0.39	84
39	1463.04	−0.27	1.18	117
40	1466.09	−1.28	−0.81	94
41	1472.18	−1.14	−0.63	88
42	1478.28	−1.08	−0.79	83
43	1481.33	−0.15	1.11	104
44	1484.38	−1.17	−1.09	82
45	1490.47	−0.86	−0.50	74
46	1496.57	−0.71	−0.53	85
47	1499.62	0.10	1.53	95
48	1514.86	−0.62	−0.36	59
49	1517.90	0.35	3.90	83

Table 13. *Well C, CA of dinoflagellate cysts only, species scores, statistical weights and eigenvalues. N = 132. Weights < 5 omitted*

Sample	Spp. code (eigenvalue)	Axis 1 0.47	Axis 2 0.33	Stat. wt.
1	Spin raG	0.27	0.07	1147
2	Impl spp	−0.73	−0.48	560
3	Pala pyr	−0.67	0.05	302
4	Areo spp	−0.93	−0.83	191
6	Hyst tub	−0.37	0.43	146
7	Areo Cfm	0.28	1.94	108
8	Defl oeb	0.30	1.95	70
9	Glap ord	0.10	2.16	60
10	Glap spp	−0.81	−0.49	55
11	From fra	−0.22	−0.01	53
12	Oper Cfc	−1.00	−0.73	52
13	Oper cen	−0.61	0.03	45
14	Areo gip	−0.32	0.17	41
15	Hom tenu	2.64	−0.48	38
16	Cord gra	0.01	0.08	37
17	Syst pla	3.16	−1.06	36
18	Areo CFs	−0.47	0.43	32
19	Phth sto	3.16	−0.88	32
20	Defl den	−0.59	−0.02	25
21	Ling mac	0.06	−0.22	24
22	Glap div	−0.93	−0.64	22
23	Oper spp	0.61	0.27	20
24	Cero dep	0.32	0.12	18
25	Acho alc	0.76	0.59	17
26	Cali ami	−0.81	−0.06	17
27	Areo mic	2.59	−0.61	16
28	Acho ram	−0.99	−0.64	15
29	Cord o mi	1.53	−0.84	15
30	Dino feh	1.95	−0.51	15
31	Hyst spp	−0.47	0.19	15
32	Defl pho	3.13	−0.90	14
33	Para ind	2.03	0.46	14
34	Cord spp	−0.88	−0.60	13
36	Areo dik	2.94	−0.92	12
37	Pala gol	−0.32	−0.12	12
38	Areo arc	3.46	−1.03	11
39	Impa spp	0.80	−0.56	11
40	Subt sp1	−0.54	−0.05	11
41	Cero die	−0.69	−0.37	10
42	Eato urs	1.80	−0.43	10
43	Lent wet	−0.08	2.04	10
44	Phth com	2.99	−0.82	10
45	Apec spp	−0.95	−0.83	8
46	Areo CFc	−0.90	−0.93	8
47	Cero spe	−0.15	1.24	8
48	Crib giu	0.02	−0.02	7
49	Apec aug	−0.14	1.35	6
50	Glap pas	0.22	0.24	6
51	Impl lig	−0.61	−0.63	6
52	Sele nep	2.78	−0.30	6
53	Thal del	−0.07	1.49	6
58	Cero war	−0.40	0.01	5
59	Cord exi	−0.78	−0.48	5
60	Cord ino	0.81	0.72	5
62	Daps sim	2.85	−0.65	5
64	Micr Cfo	−0.17	1.12	5

species groups were identified by which analyst in each sample. The analyst that is suspected of having counted samples 264 m, 258 m, 276 m and 282 m also failed to document any pollen or spores, further contributing a source of error to the ordination results. The main species variation on Axis 1 is between *Pala pyr* in the

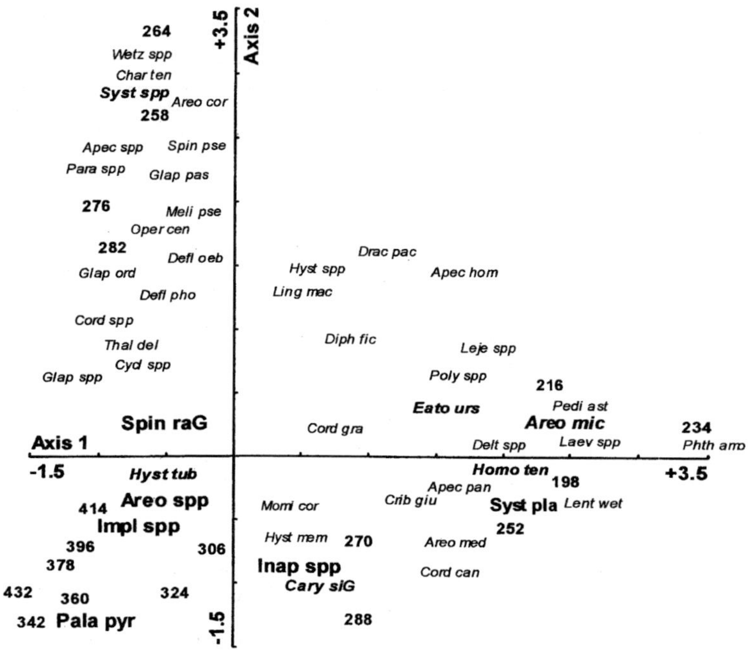

Fig. 13. Well D, CA analysis of all groups included, biplot of samples and species. Data corrected for suspected counting inconsistencies. Font style reflects statistical weights of species: largest font reflects weights > 80, next largest 20–80, and the smallest 5–20. All species with weights < 5 were omitted. Sample and species scores are listed in Tables 14 and 15, respectively.

negative domain and *Syst pla* in the positive domain (wts. 125 and 88, respectively), with *Areo mic* and *Homo ten* also plotting positively (wts. 25 and 19, respectively) and *Impl spp.*, *Hyst tub* and *Glap spp.* plotting negatively (wts. 164, 74 and 33).

On Axis 2, the main variation is between *Syst spp.* plotting highly positively and *Pala pyr* plotting highly negatively (wts. 39 and 125, respectively). The suspicion that *Syst spp.* might be the same as *Syst pla* identified by other analysts was confirmed, and these two taxa were grouped for the final compilation shown in Figure 14.

The documentation of *Pediastrum*, sometimes considered to reflect fresh water to lagoonal conditions, in this well only, suggested that Well D might be the nearest to a palaeo-shoreline of the four wells investigated. Since *Pediastrum* appears in the very top of the core (an interval interpreted from the other wells to represent a return of high sea-level stand), it might be that *Pediastrum* is being washed out from areas of lagoonal to fresh water salinities as the sea level rose and penetrated into these areas.

Well D, dinoflagellate cysts only

Axes 1 and 2 had high eigenvalues of 0.44 and 0.43, respetively. Due to the counting inconsistencies mentioned above and smaller number of samples in Well D, the ordination results are shown as a biplot, with samples and species plotted together (Fig. 14).

Well D, cysts only, samples. Consistent with all of the above ordinations, the samples at the top of the core comprised the main variation on Axis 1, including samples 198 m to 288 m plotting positively and samples 306 to 432 m plotting negatively. The main variation on Axis 2 was comprised of samples 306 m and 324 m from the middle grouping highly positively.

Well D, cysts only, species. The main variation on Axis 1 is represented by *Syst pla* plotting most positively and *Pala pyr* and to a lesser extent, *Impl spp.* most negatively (wts. 127, 125 and 164, respectively). *Areo spp.* plotted most positively on Axis 2 (wt. 125) and *Impl spp.* most negatively. Again, *Spin raG* plots as a highly abundant cosmopolitan species, near the

Table 14. *Well D, CA of all groups, corrected for counting inconsistencies, sample scores, statistical weights and eigenvalues. N = 18*

Sample	Depth (m) eigenvalue	Axis 1 0.45	Axis 2 0.42	Stat. wt.
1	198.00	2.22	−0.03	77
2	216.00	2.20	0.46	73
3	234.00	3.27	0.14	93
4	252.00	1.90	−0.45	121
5	258.00	−0.44	2.57	110
6	264.00	−0.44	3.22	94
7	270.00	0.79	−0.56	102
8	276.00	−0.87	1.87	107
9	282.00	−0.76	1.54	109
10	288.00	0.79	1.17	126
11	306.00	−0.01	−0.62	113
12	324.00	−0.28	−0.97	93
13	342.00	−1.27	−1.21	110
14	360.00	−1.31	−1.01	98
15	378.00	−1.12	−0.85	114
16	396.00	−0.97	−0.61	99
17	414.00	−0.73	−0.59	109
18	432.00	−1.43	−0.96	128

centre of the gradients represented by both axes.

Integration of Wells A–D

The information gleaned from each of the individual wells was integrated into a composite plot in order to elucidate trends and attempt an environmental interpretation. Figure 15 shows the plots of the scores on CA Axes 1–2 for all four wells. It can be seen that *Syst pla* is an important species in the top of Wells A, C and D but in Well B, *Spin raG*, *Oper cen* and *Homo ten* (wts of 481, 91 and 86) are more important than *Syst pla*, which has a wt. of only 2 in Well B. *Phth com* plotted strongly positively in the top section of Well B, where it had a weight of 26, compared with 2, 10 and 1 in Wells A, C and D, where it also plotted strongly positively. *Spin raG* was abundant in all the sections but was considered to be a cosmopolitan species whose variance contributed little to the interpretations, so it was omitted from the plots.

The middle sections of Wells A and B were dominated by *Apec spp.*, whereas *Impl. spp.* dominated this section in Well C and *Syst pla* in Well D.

The bottom sections of Wells A–D all contained high proportions of *Pala pyr.* Wells A–C also contained high proportions of *Areo spp.* Wells B–D all contained large proportions of *Impl. spp. Hyst tub* was abundant in the bottom of Wells A, B and D. Since only Well A contained large proportions of *Impagidinium* species (elsewhere considered to be oceanic), it was interpreted to be located the farthest offshore, with B, C and then D being interpreted as relatively nearer to the palaeo-shoreline.

Table 15. *Well D, CA of all groups, corrected for counting inconsistencies, species scores, statistical weights and eigenvalues. N = 142. Weights < 5 omitted*

Sample	Spp. code eigenvalue	Axis 1 0.45	Axis 2 0.42	Stat. wt.
1	Spin raG	−0.15	0.14	574
2	Impl spp	−0.64	−0.62	164
3	Pala pyr	−1.23	−0.99	125
4	Areo spp	−0.32	−0.25	103
5	Syst pla	1.74	−0.38	88
6	Inap spp	0.14	−0.84	88
7	Hyst tub	−0.65	−0.08	74
8	Cary siG	0.35	−0.91	66
9	Syst spp	−0.49	2.75	39
10	Glap spp	−0.93	0.53	33
11	Areo mic	2.02	0.29	25
12	Homo ten	1.69	−0.02	19
13	Glap div	−0.22	−0.93	18
14	Eato urs	1.29	0.40	17
15	Ling mac	0.25	1.29	17
16	Glap ord	−0.70	1.54	17
17	Pedi ast	2.33	0.31	17
18	Leio spp	−0.81	1.68	16
19	Cord gra	0.50	0.12	14
20	Oper cen	−0.29	1.79	13
21	Cord min	0.43	−0.30	11
22	Hyst mem	0.72	−0.55	11
23	Daps spp	1.68	−0.35	11
24	Acho alc	1.32	0.01	10
25	Areo med	1.38	−0.57	10
26	Areo sen	−0.47	2.59	10
27	Laev spp	2.38	0.10	10
28	Hyst spp	0.86	1.38	9
29	Para spp	−0.78	2.17	9
30	Defl oeb	−0.53	1.46	8
31	Pala gol	−0.61	−0.92	7
32	Defl pho	−0.25	1.17	6
33	Areo cor	−0.44	2.67	6
34	Cord ino	−0.90	−0.94	6
35	Diph fic	0.63	0.83	5
36	Phth amo	2.85	0.07	5
37	Poly spp	1.41	0.55	5
38	Cere bar	3.00	0.02	5
39	Leje spp	1.64	0.76	5
40	Meli pse	−0.06	1.82	5
41	Thal pel	−0.18	0.99	5
42	Glap pas	−0.68	2.21	5
43	Thal del	−0.52	0.88	5
44	Tili miG	0.92	−0.43	5

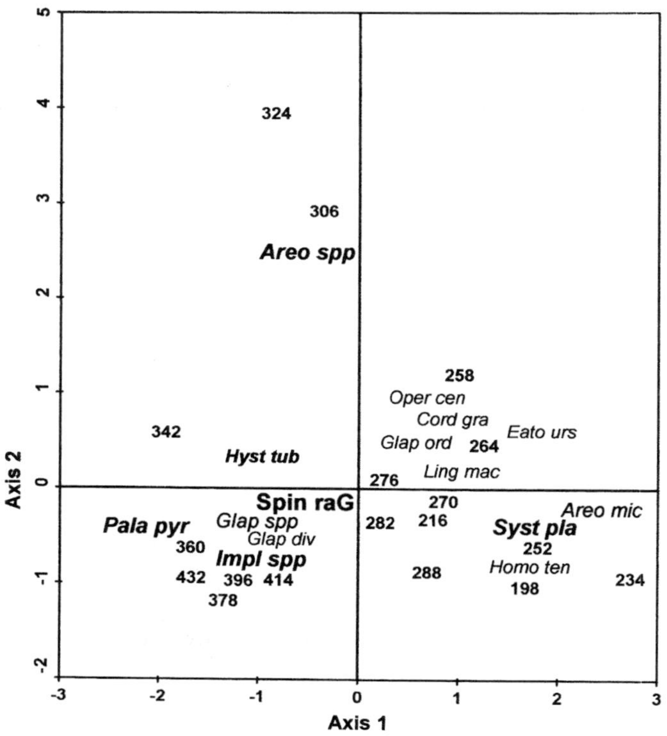

Fig. 14. Well D, CA analysis of dinoflagellate cysts only, biplot of samples and species. Data corrected for suspected counting inconsistencies. Font style reflects statistical weights of species. Largest font reflects weights > 400, next largest 100–400, and the smallest 10–100. All species with weights < 10 were omitted. Sample and species scores listed in Tables 16 and 17, respectively.

Comparison between results from SMES and the previous study

The main subdivisions identified in the SMES test matched closely with two of the flooding surfaces that had been identified in the previous Statoil study (Laursen & Prince 2002). The wells also had been positioned correctly with respect to their positions relative to the previously identified palaeo-shoreline. However, since additional flooding surfaces had been identified previously in the middle and bottom sections, attempts were made to re-treat the data to see if the SMES methods being tested here could detect these, too. For this, the taxa were re-grouped in accordance with the suspected taxonomic inconsistencies exposed by the methods (discussed above) and each subsection was rerun individually. By treating each subsection individually, it was hoped that smaller-scale changes between the larger-scale changes identified in the first round, would be more clearly represented. This work was carried out for cysts only, to the point of being able to plot the ordination results downcore for both the middle and bottom sections (Fig. 16 shows results from the middle sections).

A fuller interpretation was beyond the scope of this initial project, but preliminary results are presented here (Fig. 16). Even from this first treatment it appears likely that the 140 flooding surface is reflected in the middle section of all wells (Fig. 16) and the 60 flooding surface in the bottom of Wells A, B and D, with an alternative suggestion for its location in Well C (data not shown here).

Discussion

Two previous investigations have used forms of correspondence analysis to aid palaeoecological interpretations from pre-Quaternary dinoflagellate cyst data. Versteegh & Zonneveld (1994) applied Detrended Correspondence Analysis

Table 16. *Well D, CA of dinoflagellate cysts only, corrected for Spin spp and Syst spp grouped with Syst pla. Sample scores, statistical weights and eigenvalues. N = 18*

Sample	Depth (m) eigenvalue	Axis 1 0.44	Axis 2 0.33	Stat. wt.
1	198.00	1.71	–0.95	63
2	216.00	0.69	–0.26	48
3	234.00	2.79	–0.92	86
4	252.00	1.58	–0.44	117
5	258.00	1.02	1.18	110
6	264.00	1.20	0.45	103
7	270.00	0.81	–0.11	74
8	276.00	0.23	0.07	97
9	282.00	0.14	–0.33	98
10	288.00	0.67	–0.86	79
11	306.00	–0.05	2.78	81
12	324.00	–0.61	3.93	69
13	342.00	–1.93	0.56	94
14	360.00	–1.67	–0.58	95
15	378.00	–1.35	–1.07	106
16	396.00	–1.12	–0.82	98
17	414.00	–0.89	–0.81	107
18	432.00	–1.72	–0.68	126

Table 17. *Well D, CA of dinoflagellate cysts only, corrected for Spin spp and Syst spp grouped with Syst pla. Species scores, statistical weights and eigenvalues. N = 110. Weights < 5 omitted*

Sample	Spp. code eigenvalue	Axis 1 0.44	Axis 2 0.33	Stat. wt.
1	Spin raG	–0.03	–0.16	585
2	Impl spp	–0.86	–0.46	164
3	Syst pla	1.26	–0.24	127
4	Pala pyr	–1.62	–0.26	125
5	Areo spp	–0.31	2.20	103
6	Hyst tub	–0.51	0.15	74
7	Glap spp	–0.53	–0.19	33
8	Areo mic	1.78	–0.27	25
9	Homo ten	1.55	–0.57	19
10	Glap div	–0.42	–0.35	18
11	Eato urs	1.40	0.36	17
12	Glap ord	0.25	0.19	17
13	Ling mac	0.53	0.09	17
14	Cord gra	0.64	0.25	14
15	Oper cen	0.70	0.72	13
16	Cord min	0.19	0.27	11
17	Daps spp	1.42	0.33	11
18	Hyst mem	0.36	0.64	11
19	Acho alc	1.17	–0.52	10
20	Areo med	1.12	–0.51	10
21	Areo sen	0.96	0.89	10
22	Hyst spp	1.31	0.26	9
23	Para spp	0.45	0.15	9
24	Defl oeb	0.37	0.88	8
25	Pala gol	–0.82	–0.76	7
26	Areo cor	1.05	1.06	6
27	Cord ino	–1.13	–0.90	6
28	Defl pho	0.56	0.20	6
29	Cere bar	2.55	–0.82	5
30	Diph fic	1.12	0.53	5
31	Glap pas	0.56	0.29	5
32	Leje spp	1.74	–0.61	5
33	Meli pse	0.97	0.64	5
34	Phth amo	2.35	–0.93	5
35	Poly spp	1.11	–0.22	5
36	Thal del	0.11	0.05	5
37	Thal pel	0.42	–0.07	5

(DCA) and Canonical Correspondence Analysis (CCA) to a short interval from the Late Pliocene of southern Italy (2.8–2.2 Ma BP), and Brinkhuis *et al.* (1998) applied DCA to the Cretaceous–Tertiary boundary at El Kef in NW Tunisia. In both these cases, the DCA provided plots of the cyst data spread along axes that could be interpreted to represent ecological trends, based on comparison with modern cyst distributions or ancillary data from other sources (planktonic foraminifera and stable oxygen isotopes). However, apart from these two examples, the DCA-based methods of Versteegh & Zonneveld have not been applied elsewhere in the many studies using cysts as palaeoecological indicators in the past ten years, suggesting that the method may have only limited application.

Differences between SMES and DCA-based methods used previously

The statistical treatments used here in SMES differ substantially from those used in the previous two studies:

CA versus DCA. Both CA and DCA were tested on our recent-cyst data, and CA produced a more distinctive representation of the known ecological trends in the data (discussed in A. L. Dale & B. Dale 2002). CA is therefore used here rather than DCA.

Statistical weights. A further main difference concerns the prominent use here of statistical weights, whereas they were not utilized in the previous two studies. One of the strengths in the CA family of methods is the *quantification* of the statistical placing of species and samples along the axes, expressed as both eigenvalue coefficients (scores) and statistical weights. These weights, together with the scores on axes having high eigenvalues, allow the rapid identification of the statistically most important species

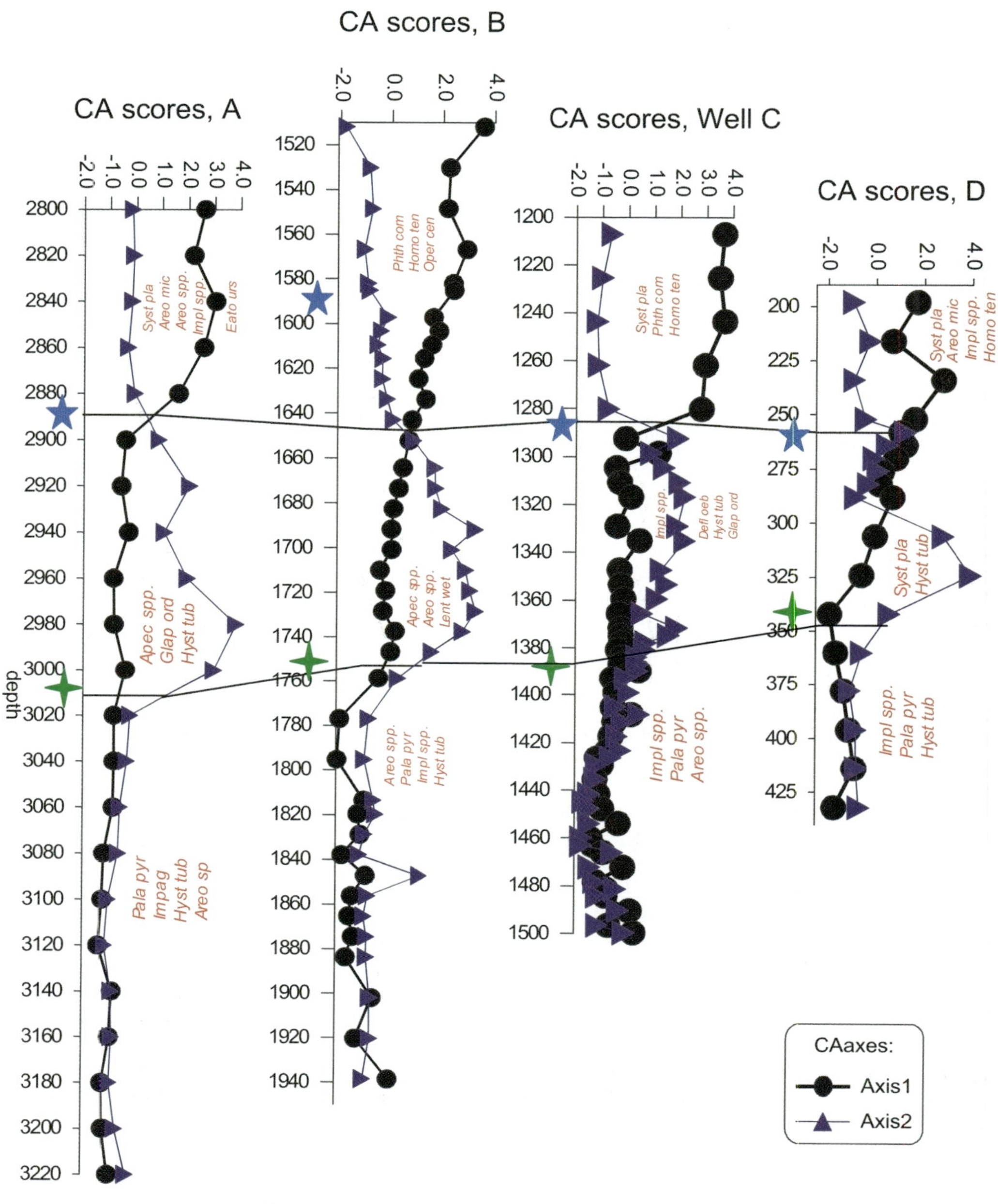

Fig. 15. Composite plots showing scores on Axes 1 and 2 of CA of Wells A–D, dinoflagellate cysts only. Main environmental shifts identified by CA (correlation lines) correspond closely to Flooding Surfaces 165 (blue stars) and 90 (green stars) previously identified by Statoil. In the one discrepancy (FS165 in Well B), Statoil biostratigraphers subsequently considered the pick based on CA of the dinoflagellate cysts to be most likely correct. Main indicator species listed in red at depths for which they are significant. Data corrected for suspected counting inconsistencies and outliers removed. Wells are aligned from furthest offshore (A) to nearest to the palaeo-shoreline (D), based on palaeoecological interpretations developed in the study.

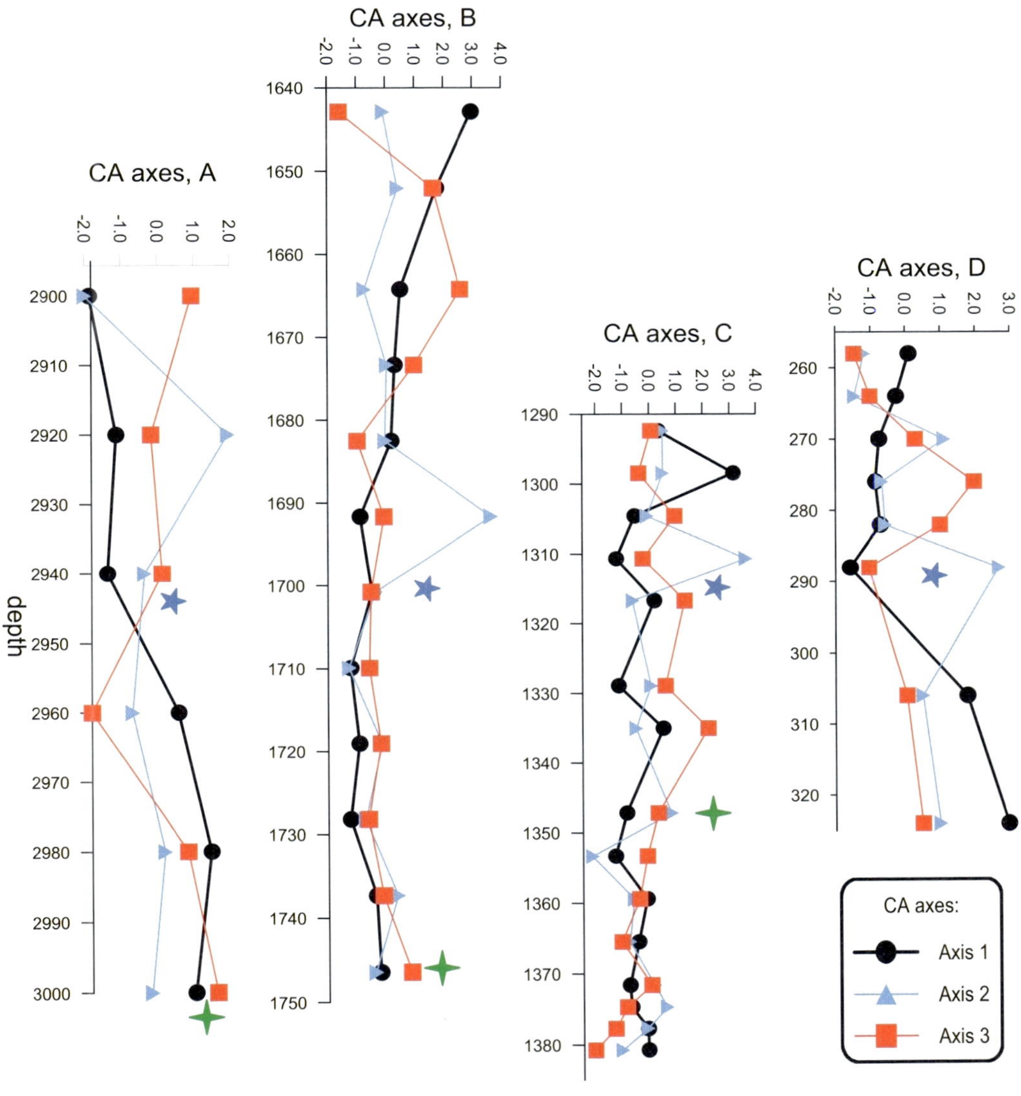

Fig. 16. Middle section as identified from the first CA for Wells A–D treated individually, dinoflagellate cysts only. Scores on CA Axes 1–3 plotted, and Flooding Surfaces 130 (green stars) and 140 (blue stars) identified previously by Statoil indicated. Note coincidence of environmental shifts on Axes 1 and 2.

contributing to the ecological signals (indicator species). This has the potential, as here, to reduce industrial datasets with up to several hundreds of species to just a relatively few important species to be further interpreted.

Quantification of ecological change with time. A distinctive feature of the SMES method is the use of plots tracking changes in the main axes with time through the well data. In the present study, these revealed the consistent, systematic shifts in all four wells that allowed correlations based on the identification of what subsequently proved to be flooding planes.

Advantages of the SMES application to biostratigraphy

The realization that recent cyst assemblages are strongly influenced by ecological factors suggested the possibility that pre-Quaternary assemblages were also similarly affected (e.g. Dale 1983). This has been confirmed to be the case in many investigations in the past twenty years (e.g. Stover *et al.* 1996), with profound implications for the use of cysts in biostratigraphy. Traditionally, as with the use of other microfossil groups, the biostratigraphical application of cysts is largely based on first and last occurrences of species, and bioevents such as species acmes, used to construct 'timelines' for correlations. This dependence on individual species has proved unreliable, since basinwide interpretations may involve different ecological settings (as here) with different species assemblages at any given time. For these methods to be effective, it is therefore necessary to understand the palaeoecological settings, and how these have changed with changing environmental factors through time (the 'ecological signals' of Dale 1996).

In some cases, extensive experience working with the same time interval has eventually suggested subjective palaeoecological models for aiding cyst-based biostratigraphy (e.g. Stover *et al.* 1996, fig. 53). It has also become increasingly apparent that such models are more robust if the data from as many different microfossil groups as possible are integrated. Furthermore, as sequence stratigraphy has become the main method for establishing timelines (flooding surfaces), it is necessary to integrate the microfossil data with seismic and sedimentary data (e.g. Laursen & Prince 2002). While these developments have improved biostratigraphical resoludion and palaeoenvironmental interpretations, they have major disadvantages. Reliance on previous experience is of limited use to exploration in new regions/basins, which may be characterized by different palaeoecological conditions. Reliance on integration of different microfossil groups may be limited by economical constraints, or absence of some groups due to dissolution or palaeoecological conditions (e.g. planktonic foraminifera, calcareous nanoplankton, and radiolaria are mainly restricted to oceanic waters).

The SMES method tested here has several advantages over the more traditional biostratigraphical methods. It provides an objective statistical quantification of ecological change expressed by one group, the dinoflagellates, which lived in the different parts of the marine basin of interest. In doing so, it takes into account the fact that different species may characterize different parts of the basin at the same time. SMES thus provides comparable information to that which is utilized otherwise in sequence stratigraphy. The latter is based on the physical changes seen in the seismic and sedimentological records, but these major changes in relative sea level also represented major ecological changes to the organisms living at that time, and SMES provides evidence of these ecological signals marking the same events.

This first test strongly suggests that SMES offers a much quicker, and more robust application for palaeoenvironmental interpretations from the dinocysts in biostratigraphical datasets. A few days' work using these methods produced main results which were confirmed to coincide with flooding surfaces and palaeoenvironments identical with those identified after many months of work integrating seismic and biostratigraphic data from a range of different microfossil groups (Laursen & Prince 2002). In the one discrepancy (flooding surface 165 in Well B, Fig. 15), Statoil biostratigraphers reviewed the data and considered the pick based on CA of the dinoflagellate cysts to be most likely correct. A future publication is planned to discuss details of the ecological signals revealed, which in several cases confirmed the previous use of certain cyst species as palaeoenvironmental indicators (e.g. by Stover *et al.* 1996; Laursen & Prince 2002). Further investigations are underway to test and develop this method on other industrial datasets.

References

BRINKHUIS, H., BUJAK, J.P., SMIT, J., VERSTEEGH, G.J.M. & VISSCHER, H. 1998. Dinoflagellate-based sea surface temperature reconstructions across the Cretaceous-Tertiary boundary. *Palaeogeography, Palaeoclimatology, Palaeoecology*, **141**, 67–83.

DALE, A.L. & DALE, B. 2002. Application of ecologically-based statistical treatments for micropaleontology. *In*: HASLETT, S. (ed.) *Quaternary Environmental Micropalaeontology*. Edward Arnold, London, 259–286.

DALE, B. 1983. Dinoflagellate resting cysts: "benthic plankton". *In*: FRYXELL, G.A. (ed.) *Survival Strategies of the Algae*. Cambridge University Press, New York, 69–135.

DALE, B. 1996. Dinoflagellate cyst ecology: modelling and geological applications. *In*: JANSONIUS, J. & MCGREGOR, D.C. (eds) *Palynology: principles and applications*. American Association of Stratigraphic Palynologists Foundation, **3**, 1249–1275.

HILL, M.O. 1973. Reciprocal averaging: an eigenvector method of ordination. *Journal of Ecology*, **61**, 237–251.

HILL, M.O. & GAUCH, H.G. 1980. Detrended correspondence analysis: an improved ordination technique. *Vegetatio*, **42**, 47–58.

JONGMAN, R.H.G., TER BRAAK, C.J.F. & VAN TONGEREN, O.F.R. (eds) 1995. *Data Analysis in*

Community and Landscape Ecology. Cambridge University Press, Cambridge.

LAURSEN, G.V. & PRINCE, I. 2002. *Palaeoenvironments: A multidisciplinary approach. Northern European Cenozoic Stratigraphy*. Proceedings of the 8th Biannual Meeting of the Regional Committee on Northern Neogene Stratigraphy/Regional Committee on Northern Paleogene Stratigraphy, - Flintbeck 2002. Landesamt für Natur und Umwelt des Landes, Flintbeck, Germany, 47–56.

STOVER, L.E., BRINKHUIS, H. ET AL. 1996. Mesozoic-Tertiary dinoflagellates, acritarchs and prasinophytes. *In*: JANSONIUS, J. & MCGREGOR, D.G. (eds) *Palynology: Principles and Applications*. American Association of Stratigraphic Palynologists Foundation, **2**, 641–750.

TER BRAAK, C.J.F. 1996. Unimodal models to relate species to environment. DLO-Agricultural Mathematics Group, Wageningen.

TER BRAAK, C.J.F. & SMILAUER, P. 1998. CANOCO Reference Manual and User's Guide to CANOCO for Windows: Software for Canonical Community Ordination (version 4). Microcomputer Power (Ithaca, N.Y., USA).

VERSTEEGH, J.M. & ZONNEVELD, K.A.F. 1994. Determination of (palaeo-)ecological preferences of dinoflagellates by applying Detrended and Canonical Correspondence analysis to Late Pliocene dinoflagellate cyst assemblages of the south Italian Singa section. *Review of Palaeobotany and Palynology*, **84**, 181–199.

A method for two-well correlation using multivariate biostratigraphical data

A. C. GARY[1], G. W. JOHNSON[1], D. D. EKART[1], E. PLATON[1] & M. I. WAKEFIELD[2]

[1]*Energy and Geoscience Institute, 423 Wakara Way, Suite 300, Salt Lake City, Utah 84108, USA*

[2]*BG Group, 100 Thames Valley Park Drive, Reading RG6 1PT, UK*

Abstract: Biostratigraphical resolution has increased markedly over the past few decades. The use of origination and extinction data has been supplemented, particularly at the reservoir scale, by more subjective local 'bioevents and biozones' in the form of successions of unique abundance fluctuations of individual taxa and distinct faunal/floral assemblages. Herein, a methodology is described that uses a sample-by-sample multivariate similarity matrix of biostratigraphical census data to define a high-resolution correlation path between two wells. Three different similarity/dissimilarity metrics are used to calculate the matrix (cosine-theta, and Euclidean and Manhattan distance) for each of three transformations of the data: raw data (counts), presence–absence, and per cent–range transformation. Recommendations are presented for use of various metrics and transformations, given different types of data sets and the objectives of the analyst. The correlation path through the matrix is monotonic and must increase in depth in one or both wells. The correlation path moves one sample-pair or matrix element at a time to an adjacent element that has the highest average, local path. After an adjacent element is selected it becomes the current state and the algorithm repeats the process until the matrix has been transited. The correlation path can be constrained using *a priori* information, if available (e.g. maximum flooding surfaces). Such events are established as tie-points, and the correlation path is required to go through these points. Additional steerage of the correlation path may be accomplished interactively, by graphical analysis of individual taxon abundance plots, followed by establishment of additional tie-points. The method also includes calculation and graphic representation of a 95% confidence interval about the calculated faunal/floral percentages. This enables the biostratigrapher to take into account the statistical confidence one has in a peak faunal abundance. The method is demonstrated on palynological data from two wells from the Blake Field, Outer Moray Firth, UKCS.

This paper summarizes the results of a research program conducted by the Technical Alliance for Computational Stratigraphy (TACS) at the University of Utah and supported by a consortium of oil companies. This program focuses on the application of multivariate and intelligent systems technologies in industrial biostratigraphy, and the development of software for the implementation of those technologies. One of the higher-level research goals is to explore and develop techniques to better apply biostratigraphical data at the field and reservoir scale. A problem in field-scale biostratigraphy is the sparsity of discrete biostratigraphical events, such as the origination (i.e. first appearance datum or FAD) or termination (i.e. last appearance datum or LAD) of a taxon range. The rarity of such unambiguous ('on' or 'off') signals necessitates use of non-discrete data (i.e. differences in the relative abundance among multiple taxa). Efficiently processing large numbers of data vectors populated by continuous or ordinal abundance values is a difficult task for the unaided human mind. Therefore, our objective was to create techniques and tools for the industrial biostatigrapher that will assist in processing and interpreting complex data in a stratigraphical context.

The basic method outlined here involves calculating a sample-by-sample similarity/dissimilarity matrix between two wells using the abundances of fossil taxa, and then determining a correlation path between the matrix elements based on a combination of their similarity/dissimilarity values and user-defined constraints, such as discrete biostratigraphical events (e.g. LAD). The correlation path is determined by traversing the matrix starting at the top of the two wells and moving sequentially to adjacent elements of highest similarity without recursion. In general, this is a sequence-slotting problem that others (Clark 1985; Gordon 1973, 1980; Gordon & Reyment 1979; Thompson & Clark

From: POWELL, A. J. & RIDING, J. B. (eds) 2005. *Recent Developments in Applied Biostratigraphy*. The Micropalaeontological Society, Special Publications, 205–217. 1747-602X/$15.00

1989, 1993) have addressed by calculating a global, optimum common path between the two wells or sequences using dynamic programming methods. Several programmes using these approaches, based on the SLOTSEQ software of Gordon (1980), have been written (e.g. *SlotDeep* (Maher 1993, 2000) and *PC Slot* (Clark 1985, 1992, 1993*a*, *b*, 1994, 1995)); see also Birks (1979). Our approach differs in that we use a series of local optimizations and a variation of steepest ascent hill-climbing or gradient search (Rich & Knight 1991) to determine the correlation path, which though analogous to the common path of Thompson & Clark (1989, 1993) is conceptually more akin to lateral tracing (see Swan & Sandilands 1995 p. 379).

The procedure described herein also differs from other approaches in that the method involves simultaneous inspection of taxon ranges and user-defined constraints of the sequence slotting solution based on inspection of taxon abundance curves. This allows the knowledge and expertise of the data analyst to be efficiently brought to bear on the problem. Individual taxa abundance distributions can be evaluated in the context of the biostratigrapher's full experience, and in turn can be easily related to the sample similarity/dissimilarity matrix. The method is designed to quickly calculate a correlation path in order to allow for frequent user interaction and the updating of correlation constraints, but it doesn't necessarily produce the globally optimal correlation path. We sacrificed a global solution for speed of computation, and to allow the system to be highly interactive, as described above.

The techniques discussed herein are illustrated using North Sea palynological data generated by Duxbury Stratigraphic Consultants from two Blake Field (Outer Moray Firth, UKCS) wells (13/24a-4 and 13/24a-6). All samples were prepared using standard palynological techniques, including initial hydrochloric acid treatment, followed by hydrofluoric digestion of silicates. Oxidation with warm nitric acid further lightened palynomorphs and removed pyrite. Residues were strew-mounted on large (32 mm × 22 mm) coverslips and all palynomorphs were counted, including dinocysts, acritarchs and miospores. Counts were typically over 1000 specimens per slide in shalier samples, and sometimes exceeded 2500 specimens. In sandier sections, palynomorph numbers were typically reduced but still numbered 200 to 300. Preservation of palynomorphs was good to excellent in all sections, particularly in sandier intervals, where specimens were often exceptionally well preserved.

Correlation-path finding algorithm

Similarity/dissimilarity indices

Similarity and dissimilarity coefficients provide a quantitative measure of the degree to which two samples have common attributes. Three different methods were used in this research, and are available options in the software. One of the three (cosine-theta coefficient) is referred to as a 'similarity' coefficient because high values indicate similar samples. The other two (Euclidean distance and Manhattan distance) are 'dissimilarity' coefficients because low values indicate similarity. Each coefficient will produce different results as they operate differently on the data, as detailed below. By using matrices derived from different metrics, various slotting routes can be investigated and a preferred route decided upon (Clark 1992). As the slotting paths derived from the different metric matrices converge, confidence in the final slotting path will increase.

The primary metric used for this research is the *cosine-theta coefficient*. This is the default option in the developed software. For each multivariate sample (where the number of variables or taxa is equal to n), a sample composition may be thought of as an n-dimensional vector and the angle between two sample vectors is a function of their faunal/floral compositions. Two samples with identical compositions will be coincident, the angle between them will be 0°, and the resultant cosine of the angle will be one. Similarly, if two samples share no common taxa, the angle defined between their vectors will be 90° and the cosine of that angle will be zero. The cosine-theta coefficient is always a positive number, bounded between zero and one, and is conceptually similar to a correlation coefficient, r (i.e. perfect correlation is assigned a r = 1.0, completely dissimilar samples have a r = 0).

When comparing two samples (samples i and j), with the abundance of n taxa measured in each sample, although $n_i = n_j$ is not a requisite, cosine-theta is calculated as (Jöreskog *et al.* 1976; Davis 1986):

$$\cos\theta = \sum_{k=1}^{n} x_{ik} x_{jk} \left(\sum_{k=1}^{n} x_{ik}^2 \sum_{k=1}^{n} x_{jk}^2 \right)^{-1/2} . \quad (1)$$

The second metric used is the *Euclidean distance*, also known as the *chord distance*. As the name implies, this metric is literally the direct line distance measured between two samples plotted in multivariate space. Euclidean distance is a 'dissimilarity' coefficient (Romesburg 1984), so as the distance between a pair of samples increases, the similarity between them decreases. Like all distance measures, Euclidean distance is affected by additive and proportional changes in the measured communities (Romesburg 1984; Krebs 1989). The calculation of Euclidean distance is performed using an equation based on the Pythagorean theorem (Davis 1986), where X and Y represent the data vectors for the two samples being compared:

$$d = \left(\sum_{i=1}^{n} (X_i - Y_i)^2 \right)^{1/2} . \qquad (2)$$

The minimum distance possible (zero) indicates two identical samples and the maximum distance possible is unbounded. In contrast, cosine-theta has an upper bound of 1.0. In order to make the user's interpretation of slotting graphics consistent regardless of the coefficient chosen, the method described here includes a final transformation bounding d maximum at one.

Another often used dissimilarity metric is *Manhattan distance* (also called the *city block* method). It is so named because it is conceptually similar to travel distance in a city with blocks defined by perpendicular streets. One cannot travel directly from point to point, but instead must travel a greater distance along the perpendicular grid of streets. Manhattan distance results are normally similar to those produced by the Euclidean-distance metric; however, the effect of single large differences (outliers) is reduced, as they are not squared. The formula for Manhattan distance is given by Kaufman & Rousseeuw (1990), where X and Y represent the sample data vectors:

$$d = \sum_{i=1}^{n} \left| X_i - Y_i \right| . \qquad (3)$$

Data metrics/transformations

Prior to calculation of similarity/dissimilarity, it is extremely important to consider whether the data should be transformed and if so, in what manner. While a transformation is performed prior to the calculation of similarity/dissimilarity, the choice of transformations must be made with knowledge of how it will affect the results, therefore, we choose to present transformations after introducing the similarity/dissimilarity metrics. The slotting software allows for several data transformation options. One choice is to use raw data (i.e. raw, untransformed abundance counts). Running the sequence-slotting algorithm using raw data will result in the dominant taxa having a large influence on the results. This metric can be used when a strict sampling methodology has been followed or if there is less confidence in the reported abundance of rare or sparse taxa. From a philosophical standpoint, it is generally wise to not stray too far, analytically, from an understanding of the raw, unaltered data.

When using the raw-data metric, it is highly recommended that the cosine-theta coefficient be used as the similarity measure. Compared to Euclidean distance and Manhattan distance, cosine-theta is less sensitive to differences in the number of specimens in each sample and species dominance. For example, if two samples (X and Y) have nearly identical faunal percentage compositions and 1000 specimens were picked from sample X, but only 100 from sample Y, cosine-theta will indicate that those two samples are similar. Manhattan distance and Euclidean distance will indicate highly dissimilar samples, but that dissimilarity is driven not by faunal composition, as much as by the differences in the numbers of specimens. The effects of such additive and proportional changes on the results from similarity/dissimilarity coefficients are discussed and illustrated by Romesburg (1984) and Krebs (1989, p. 386).

Coefficients that are based on the presence or absence (i.e. binary) of taxa is another option. Depending on the data sampling methodology employed and other factors, this may be more appropriate for the data set or provide additional insights. Currently, we use the Ochiai similarity coefficient (Ochiai 1957; see also Romesburg 1984; Shi 1993), although other binary coefficients may be added in the future. See Shi (1993) for a detailed review of 39 association coefficients; Shi considered Ochiai's coefficient to perform amongst the best:

$$O_s = a\,[(a + b)\,(a + c)]^{-1/2} \qquad (4)$$

where a is the number of common taxa in two samples, and b and c are the number of taxa in the first and second samples, respectively. The Ochiai coefficient ranges from zero (i.e. no similarity) to one (i.e. identical samples), but sample size does have a moderate effect on this metric. The Ochiai coefficient is comparable to the cosine-theta coefficient, except that the input data are binary values; if a taxon is present in a sample, then '1' replaces its abundance, if it is absent, then it is assigned a value of '0'. All taxa occurrences (rare, common, etc.) will have equal weight and as such the resolution within the calculated matrix will be reduced. In practice, we have found the presence/absence metric to be most useful in the identification of dramatic, episodic discontinuities in faunal/floral assemblages, which can help to establish correlation tie-points.

In most petroleum geology applications, the actual differences in the raw faunal/floral abundances are not reliable due to a lack of a rigorous sample protocol. Therefore, we can seldom assume that the number of specimens in a sample accurately reflects the true absolute faunal abundance, but instead may be related to factors like: (1) dilution by volumes of sediment; (2) differences in sample volume; (3) taphonomic processes; or (4) differences in the number of specimens picked in the laboratory. One of the most commonly employed transformations to counter the effects of such factors is a *percentage transform*. This transformation is implemented by simply dividing the number of occurrences of each taxon in a sample by the total number of specimens in the sample. The percent transformation is also referred to as a *constant row–sum* transformation, because the sum of each sample (i.e. across rows) is 100 per cent. Stated mathematically, where S_i is a vector containing the total number of specimens in sample i, and X_{ij} is the abundance of taxon j in sample i the constant-sum transformed values for each taxon $j = 1, 2, \ldots n$, may be calculated as:

$$K_{ij} = 100\left(\frac{X_{ij}}{S_i}\right). \qquad (5)$$

In the absence of a strict sampling procedure in which volumes or counts are standardized, the percentage transformation is critical if one wishes to use either of the dissimilarity coefficients outlined above: Euclidean distance or Manhattan distance. If such a row normalization is not used with these two indices, the distances calculated may be largely an artefact of the sampling procedure. The percentage transformation is also important if one decides to use a homogeneity-of-variance transform, such as the range transform.

In most palaeontological data sets, there is a strong relationship between the mean value of a taxon and its variance, and taxa present in trace amounts almost always exhibit smaller variance than those measured at much higher concentrations. Therefore, without some sort of homogeneity-of-variance transform, the variables with the highest mean and variance usually have the greatest influence on the analysis. This is true for advanced multivariate methods such as principal components analysis and for relatively simple indices, such as Euclidean distance. However, taxa with low abundance can be tremendously important for geological understanding. As such we do not want to neglect the information that they may contain, simply because they have low variance. Analysis of geological data almost always requires some sort of homogeneity-of-variance transformation.

The homogeneity-of-variance transformation most commonly employed in this regard is the autoscale or Z transform, which sets the mean at zero and standard deviation at one for every taxon. For geological data, however, the less commonly used ‘range-transformation’ is often a better choice (Miesch 1976), where the (constant sum) matrix is denoted by K, the range-transformed matrix as R, and i and j indicate the sample and taxon, respectively, this transformation is calculated as follows:

$$R_{ij} = \frac{(K_{ij} - Kmin_{ij})}{(Kmax_{ij} - Kmin_{ij})}. \qquad (6)$$

This results in a matrix where the minimum value in each column equals zero and the maximum equals one. Unlike the Z transform, which results in absolute homogeneity of variance, the range transformation produces variances that are approximately homogeneous. The range transform has two additional advantages: (1) it does not force the data to conform to a theoretical distribution that might not be applicable, such as the standard normal distribution in the case of the Z transform; and (2) because transformed values are within the range zero to one, all values are non-negative, like the original data. The default, recommended transformation combination for most data sets is the percentage transform, followed by the range transform. Given these transformations, we have found the most robust of the aforementioned similarity coefficients to be cosine-theta.

Two-well similarity plots

The core of the slotting program is the search algorithm that finds a path through a similarity/dissimilarity matrix based on a series of locally optimal paths. A graphic expression of the similarity matrix is shown in Figure 1 where the 13/24a-6 well has 50 samples and the 13/24a-4 well has 59 samples, and each element represents a sample pair from the two wells. Sample pairs with similar faunal compositions will plot as pixels in the yellow to red range of the similarity plot.

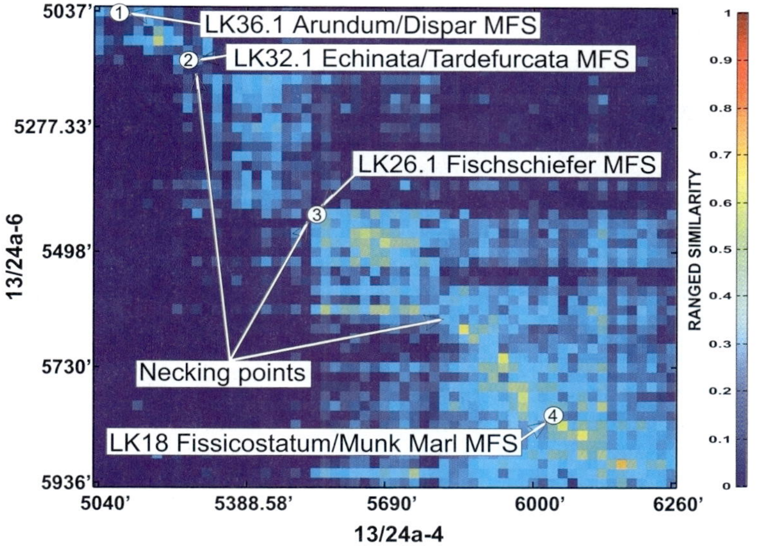

Fig. 1. Graphical representation of a two-well similarity matrix (Blake Field data using constant-sum and range transformations, and calculated using cosine-theta). Note that samples, whatever the inter-sample spacing, are plotted as if they are immediate neighbours (i.e. a non-linear depth scale). Hard tie-points (maximum flooding surfaces identified using the dinocyst dating of Duxbury 2001) are used as constraints by the slotting algorithm. Note that these MFS tie-points coincide with distinctive architectural elements within the matrix that are related to marked biofacies changes referred to as 'necking points'.

Correlation-path search method

The general approach to finding a correlation between the two wells shown in the similarity/dissimilarity plot is straightforward: given a matrix **C** (illustrated graphically in Fig. 1), we wish to find a path through the matrix where the similarities are highest. This is similar to the objective of finding a route on a topographic map that traverses the highest ridges and peaks. Therefore, we employ a gradient search routine with the heuristic that from the current element (i.e. current state) the path moves to an adjacent element (i.e. new state) with the highest average, local path. Highest average, local path is defined as the highest averaged similarity or lowest averaged dissimilarity for all possible paths considered from the current state. After an adjacent element is selected it becomes the current state and the procedure is repeated until the goal state is attained. The goal state is the path reaching the deepest sample in both wells. To accommodate basic stratigraphical principles, the valid paths are constrained to the elements $e_{i+x,j+y}$, where $e_{i,j}$ is the current state or element, $i \leq x \leq i+5$ and $j \leq y \leq j+5$. These constraints limit the search to the 'five-ahead' elements and only those samples that are of equal or deeper depths than the samples of the current element. The capability to 'look ahead' allows the path-finding algorithm to better bridge minor anomalies in the matrix without assistance.

The user can further constrain the path-finding heuristic by incorporating *a priori* information (i.e. stratigraphically useful events that are discrete and synchronous/isochronous in nature, such as maximum flooding surfaces, tuff beds, magnetostratigraphical dates, radioisotope dates, seismic picks, wireline log events, oxygen isotope events, etc.) that indicate sample pairs known to be correlated, and therefore, the slotted path can be forced to go through the known correlation tie. In the case shown in Figure 1, we happened to have several maximum flooding surfaces that could serve as hard constraints for the slotting algorithm. Forcing sample pairs to be included in the correlation path also allows for defining path starting or ending points when the wells do not represent identical intervals.

Given these objectives and constraints, the slotting algorithm is illustrated in Figure 2. For simplicity, this graphic represents a 'three-ahead' rather than a 'five-ahead' search method. The algorithm assumes that the first samples

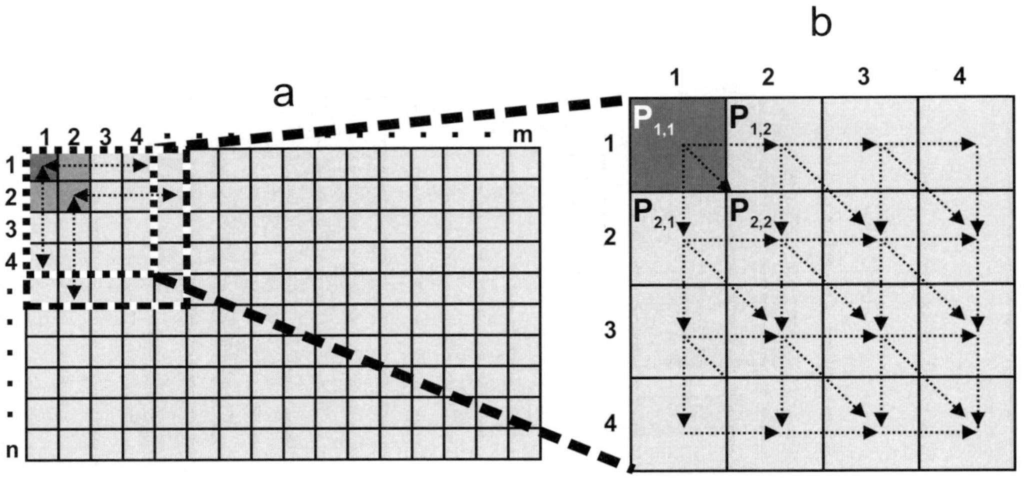

Fig. 2. Graphic illustration of the 'three-ahead' slotting method. Looking three matrix elements ahead the slotting algorithm determines the 'best' path and then establishes the next element of the path in the adjacent pixel corresponding to that path.

and last samples are correlated between the two wells, thus there is always a slotting 'tie-point' at the top left, and bottom right corners. Starting with the top left pixel, ($\mathbf{P}_{1,1}$), the algorithm searches all local path permutations through a subset of the full similarity/dissimilarity matrix. For a 'three-ahead' search that honours stratigraphic constraints the possible paths are limited, beginning with the current element ($\mathbf{P}_{1,1}$), to the three elements to the right and the three down (Figure 2b) resulting in 81 possible path permutations.

The method currently incorporated into the software uses a 'five-ahead' algorithm that results in 243 stratigraphically valid paths. After all permutations are evaluated, the path that yields the highest *average* path is chosen, and whichever adjacent pixel that the chosen path goes through ($\mathbf{P}_{2,1}$ $\mathbf{P}_{1,2}$ or $\mathbf{P}_{2,2}$) becomes the new current element on the slotting path.

After the element corresponding to the optimal path is chosen, the algorithm is repeated, using the newly chosen element (i.e. current state), for the next iteration. If for example, the best path in the example shown in Figure 2 went through $\mathbf{P}_{2,2}$ (rather than $\mathbf{P}_{1,2}$ or $\mathbf{P}_{2,1}$) then the next path evaluation would be based on a subset of the pixels whereby $\mathbf{P}_{2,2}$ was the top left pixel (i.e. current element) of the 4×4 matrix subset (Fig. 3). Figure 4 shows the slotting path using a 'five-ahead' algorithm, and forced to go through the maximum flooding surface tie-points shown on Figures 1 and 4.

Investigating subsets of the biostratigraphical data

Well data can be imported into the software from biostratigraphical database systems such as StrataBugs. StrataBugs exports what it terms 'category' information associated with each species in a dataset. This takes the form of a simple alphabetical code (e.g. dinocyst taxa have a DC code, acritarch taxa have an AC code, spore and pollen taxa have an SP code, etc.). The slotting software has the ability to read these codes enabling the user to select an individual category, or several categories together, and then calculate a similarity matrix using only those taxa in the selected category. It is also possible to manually select taxa and then calculate a similarity matrix using only those taxa. The software is constrained to using a minimum of two taxa in order to calculate a similarity matrix. There is no upper limit on the number of taxa or samples included in each stratigraphical succession. The software also has the ability to recognise StrataBugs qualifier codes that are appended to taxon sample records (i.e. reworked and caved taxa). Therefore, such taxon occurrences can either be ignored when undertaking a correlation or the possible presence of systematic stratigraphical reworking can be investigated.

Figure 5 compares similarity matrices calculated using acritarchs only, spores and pollen only, and dinocysts only with the matrix

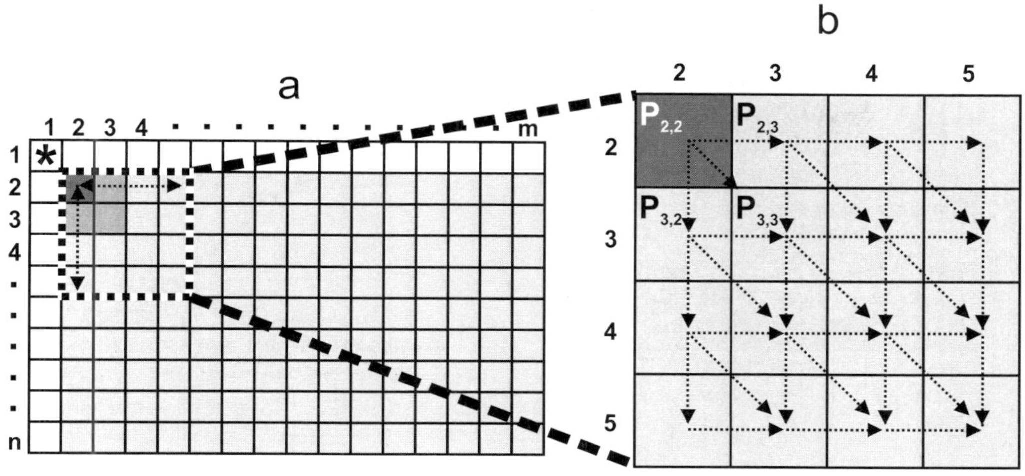

Fig. 3. Graphic illustration of the 'three-ahead' slotting method extended to the second slotting path link (assuming $\mathbf{P}_{2,2}$ was determined to be the best path in Fig. 2). Established elements on the chosen path are shown as asterisks (*).

calculated using the entire palynoflora. It is clear that the dinocysts offer the highest correlative potential in this instance and provide the dominant conditioning of the matrix calculated using the entire palynoflora. Though not all datasets exhibit such marked differences the ability to rapidly assess subsets of the data can help to focus the biostratigraphers attention upon the most stratigraphically useful subsets of taxa. Once a correlation has been defined using a specific subset of the dataset any tie points defined can be tested by viewing a similarity/dissimilarity matrix calculated from a different subset and the percentage abundance plots of taxa in that subset.

Species abundance curves

An important adjunct to the slotting algorithm described above is the graphical analysis of species abundance down-hole in multiple wells. This is not a new concept; species abundance is commonly used to provide more detailed biostratigraphical correlations, especially at the field scale. However, in this research, we have addressed two common problems typically associated with such an approach: (1) correlations based on species abundance are typically performed independently of mathematically based correlation methods, such as slotting algorithms; and (2) the species chosen typically require some prior user knowledge regarding which species are stratigraphically diagnostic.

The objective is to give the biostratigrapher the ability to evaluate individual species abundances, and to allow that analysis to be part of a more integrated and iterative inter-well correlation process. Therefore, we considered the following features to be important:

- Sample pairs can be graphically picked from an inter-well similarity plot (e.g. cosine-theta plot), and can be highlighted on the accompanying species abundance plots.
- After identification of the sample pair, an algorithm should determine which taxa contribute most to its similarity/dissimilarity. The user should be able to plot abundance curves for any taxon, but the taxa that contribute the most to the correlation should be made apparent to the user.
- The taxon abundance plots should show some indication of the degree of confidence the user should have in the abundance peaks. For example, an 'acme' of 5% of an assemblage that is calculated from a sample with 20 specimens carries little confidence and establishing a tie-point based on such an acme is questionable. Confidence would be considerably higher if that 5% 'acme' were based on a sample size of 1000 specimens.
- If the user deems the palaeontological evidence strong enough to warrant establishment of a hard tie-point, the user should be allowed to designate it as such, after which the slotting algorithm will be

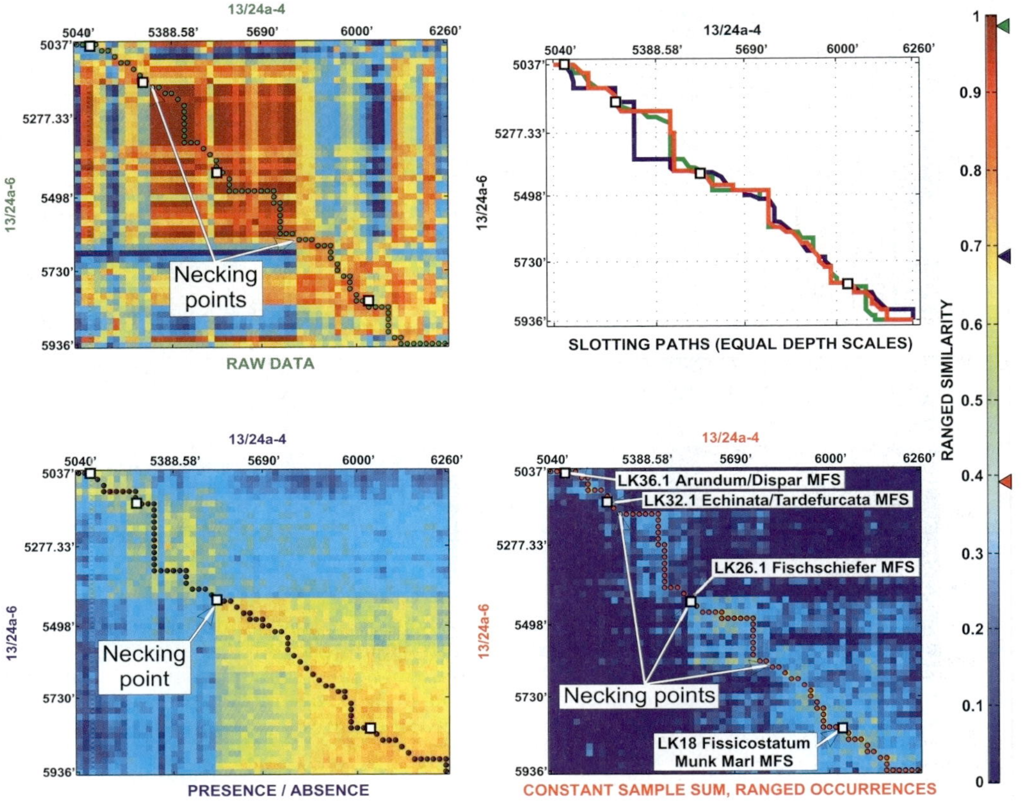

Fig. 4. Results of the 'five-ahead' slotting algorithm applied to the cosine-theta matrix for the Blake Field palynological data set. Similarity plots and lines of correlation are shown for each of three transformations: (1) raw data (top left); (2) presence absence (bottom left); and (3) per cent – range transformation (bottom right). The optimal slotting path is shown as a dotted line. Maximum flooding surfaces used as tie points are shown with the symbol □. The graph at the top right shows the slotting solutions, using a linear depth scale, calculated for each of the three transformations (the colour of each line coordinates with the colour of axis labeling of the three correlation matrices). Note the similarity in the positions of 'necking points' (major biofacies discontinuities) between the different matrices.

recalculated and constrained by that sample pair. The palaeontologist, not the algorithm, should make the final decision regarding the establishment of a tie-point. This process allows numerous qualitative and intangible considerations (e.g. taxonomic identification complications, differences in dissolution susceptibility, palaeoenvironmental considerations, etc.) to be weighed into the final decision.

An assumption of the method is that the primary stratigraphical utility of such species abundance curves will be at the field rather than a basinal scale. Regional isochronous events (e.g. maximum flooding surfaces, ash layers, LADs, etc.) can be imported into the application as tie-points prior to implementation of the initial slotting program. Species abundance curves are subsequently evaluated for use as higher-resolution tie points, between these first-order, hard correlation constraints. Finally, patterns formed by contrasts in the elements of the similarity/dissimilarity matrix can be used to infer depositional characteristics, such as continuous deposition or disconformities (Pielou 1979), and this information can be used to further constrain the correlation path.

If the biostratigrapher identifies a section where the slotting solution is problematic, the raw data can be interrogated by evaluating the relative abundance of species in both wells. In Figure 6, bisaccate pollen has the highest relative abundance in the selected sample pair

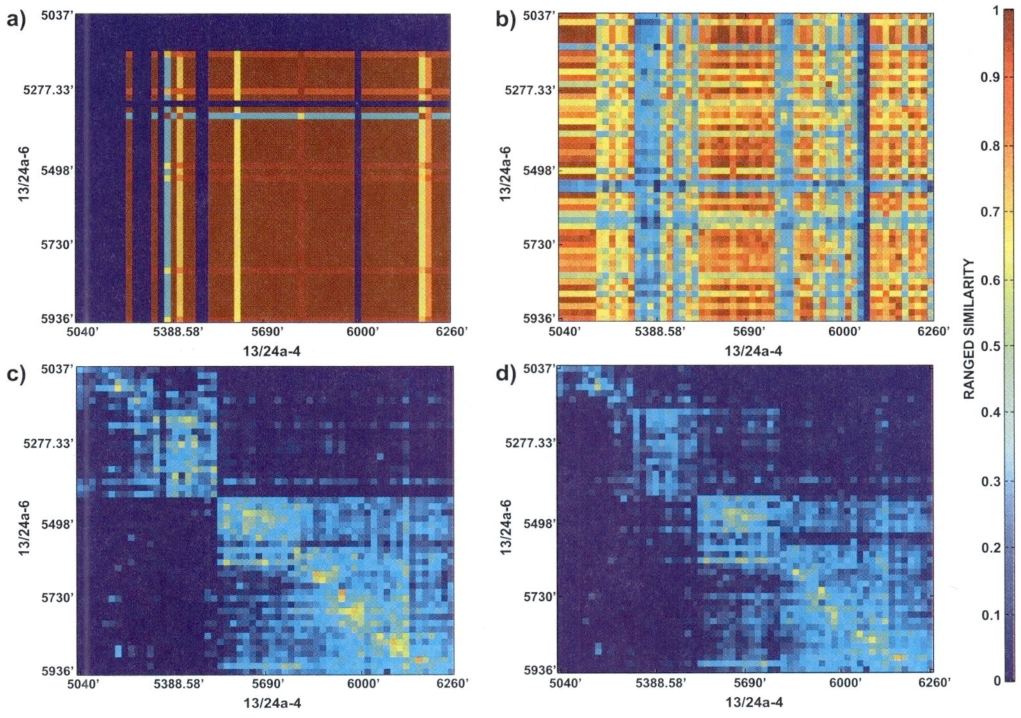

Fig. 5. Comparison of two-well similarity matrices calculated using subsets of the data (Blake Field data using constant-sum and range transformations calculated using cosine-theta); (a) acritarchs only, (b) spores and pollen only, (c) dinocysts only, and (d) the entire palynoflora.

relative to other taxa occurring in both samples. The two well depths being compared are indicated with a dashed line, and the hard constraints (tie-points) are indicated as solid lines. If, based on a review of the slotting algorithm results and individual species abundance curves, it is decided that these two samples are stratigraphically equivalent in the two wells, the software allows the biostratigrapher to designate that sample pair as a hard tie-point. If the user does so, the slotting algorithm is rerun, with the new tie-point being used as a hard constraint.

Confidence intervals on species abundance plots

A final consideration in the use of species abundance plots relates to the confidence placed in observed abundance events. As discussed above, it is considered that the species abundance plots should indicate the degree of confidence the user should have in the correlative utility of specific abundance peaks. Figure 7 provides an example of a real situation where one should be very concerned with such matters. The acme event that is being correlated here has a magnitude of just one-half of one per cent in both wells. The question one must ask before establishing a datum based on this event is: do we really trust that event? To answer that question, one must know how many specimens (n) that 0.5% calculation was based upon. Qualitatively, we would feel more comfortable if $n = 2000$ than if $n = 20$.

To allow a quantitative assessment of confidence in faunal percentage calculations, we use the concept of 'confidence interval' as defined in the classical statistics literature. As applied to palaeontological data, the approach we use is adapted from Hayek & Buzas (1997). Given a calculated faunal percentage p, taken from a sample with n specimens, the 95% confidence interval for that percentage p may be calculated as:

$$d = 1.96\left(\frac{pq}{n-1}\right)^{1/2} \tag{7}$$

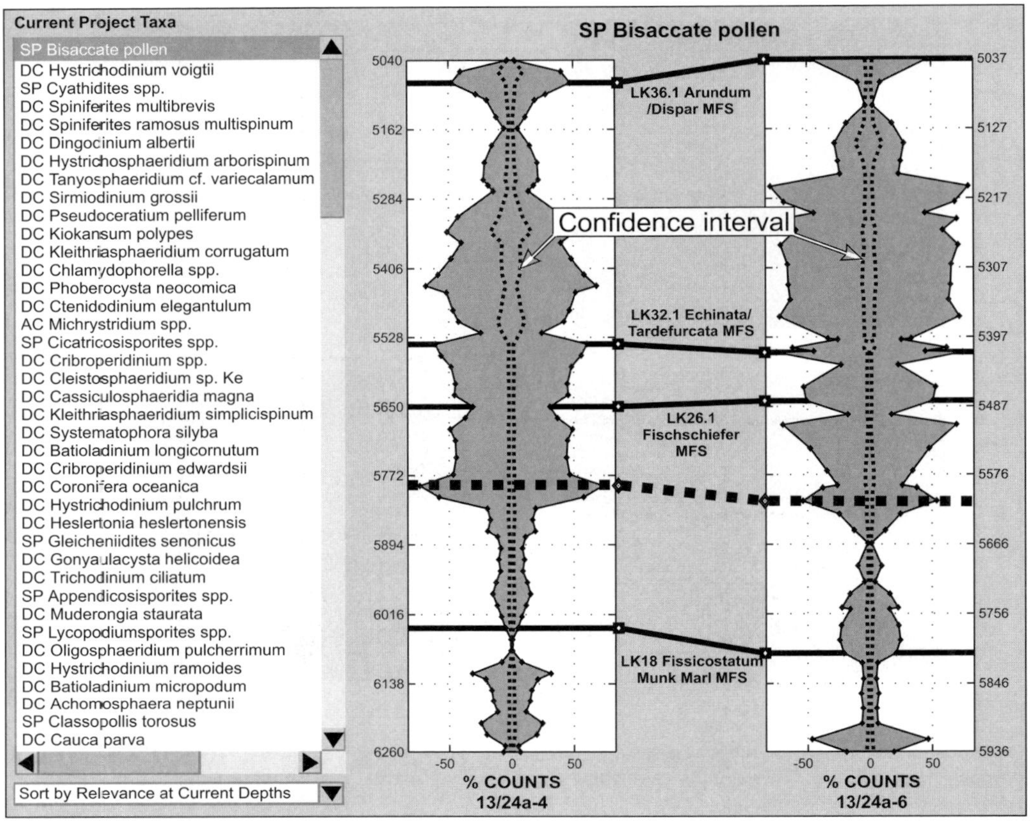

Fig. 6. Abundance plot for bisaccate pollen. The two samples connected by the thick dashed line correspond to the chosen pixel on the matrix similarity plot. The samples connected by the thinner black lines are the tie points (maximum flooding surfaces) that constrain the slotting solution.

where 1.96 is the Z value constant for 95% confidence interval, and $q = 1\text{-}p$. In terms of how we wish to use these confidence intervals, our primary concern comes in situations where $d > p$. That is, if the confidence interval is greater than the actual calculated faunal percentage, then we should not put much faith in that event. If $d < p$, then we can have more confidence in that event. The confidence interval is plotted as a thin dotted line on top of the butterfly plot (Fig. 7). Additional confidence in the determination of an abundance event tie point is obtained by comparing the abundance plots of several species at the same depth co-ordinates.

On these graphs, the confidence interval is the distance from the zero centerline to the thin dotted line. The correlated intervals are based on samples with large numbers of specimens (n = 1210 and 1424) and the confidence interval plots are within the confines of the shaded butterfly plot at the depths in question, therefore, these two acme events can be considered reliable. Note however, that in 13/24a-4, the highest stratigraphical occurrence of this species exhibits a confidence interval that exceeds the reported percentage of the taxon (i.e. the dotted line exceeds the bounds of the shaded butterfly plot). Therefore, we have less confidence in the reliability of this event. More to the point, we are not greatly concerned that this secondary event is not in evidence in the well shown on the right (i.e. 13/24a-6).

Summary

The methodology described in this paper was originally developed as a means to quickly achieve a higher degree of stratigraphical resolution than that provided by standard biozonation schemes. Our objective, to derive a

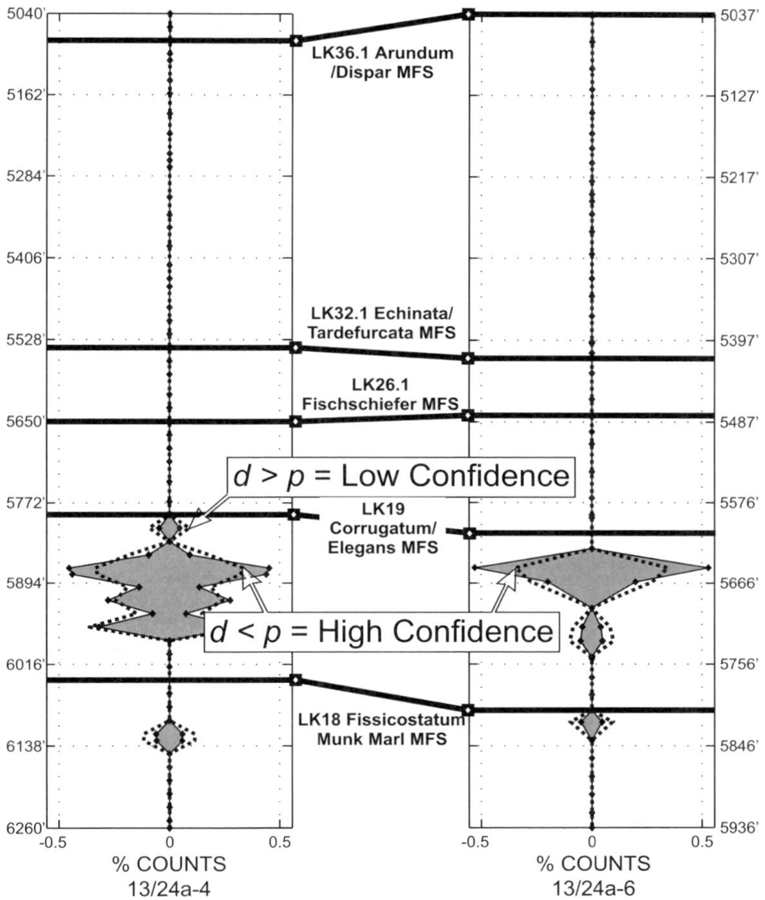

Fig. 7. Abundance plot for *Gonyaulacysta cladophora* sensu Duxbury. The confidence interval is the distance from the zero centerline to the thin dotted line.

viable stratigraphical correlation based on a similarity/dissimilarity matrix for two wells, is similar to the work of others (Clark 1985; Gordon 1973, 1980; Gordon & Reyment 1979; Thompson & Clark 1989, 1993; Swan & Sandilands 1995), with the primary difference being that we use a hill-climbing technique that composites a series of local optimal path solutions to derive the complete path. In contrast, the objective of others has been to determine a globally optimal path. An admirable characteristic of a global solution is that it provides a criterion by which one solution can be selected from many possible ones, although in the practice of stratigraphy there is no reason to assume that a mathematically optimal solution is necessarily the correct one. As a complement to a globally optimal approach, we use a local, hill-climbing technique that may better mimic how many experts would approach a stratigraphical problem – that is, solving it as a series of smaller problems. To date, there has been a surprising degree of correspondence between the independently derived solutions of experts and those derived using this methodology. Regardless, neither an optimal or local approach is likely to yield the most geologically reasonable solution without the guidance of some expert knowledge and, therefore, a critical aspect of our method is the capability to quickly relate the matrix and path-finding results to basic biostratigraphical data such as taxon abundance.

The system has now been used on many biostratigraphical data sets, by the authors, as well as by the industrial sponsors of the TACS program. Experience to date indicates that planktonic organisms, dinocysts and nannofossils in particular, rather than benthonic organisms, provide the highest stratigraphical resolution using this methodology. One of the surprises resulting from these applications has been the new insights gained from linking the results of the path-finding algorithm to basic biostratigraphical information (e.g. abundance plots). This combination increases the user's ability to 'discover' new information from a dataset. As discussed above, it is unlikely that an automated correlation method, operating independently of an expert's knowledge and experiences can result in a solution consistent with all the geological information available. Therefore, in order to take full advantage of both the algorithm and the biostratigrapher's knowledge, the user must be able to freely interact with a variety of data and results, which includes (1) specifying correlation ties between the stratigraphic sections that must be honoured and (2) examining the pathfinding results by looking at the absolute abundance or per cent occurrence of the taxa in common between a correlated sample pair. Our experience has been that users of the system tend to cycle quite quickly between the various results and data visualizations. This is a process of validating and then refining the path solution based on observations from the biostratigraphical data and the biostratigrapher's knowledge, during which the biostratigrapher gains new insights into the stratigraphical utility of various taxa. Users of this method have re-examined well-worked data sets, and in so doing, have found new, important stratigraphical information within hours. While such results can to some degree be credited to the pathfinding algorithm, the workflow of the method should also be credited. The manner in which biostratigraphers tend to use the system (i.e. rapid cycling between various results and data) suggests that the linking of the different types of information allows biostratigraphers to more effectively apply their expert knowledge.

The TACS group thank BG Group, BP, Chevron Texaco, ConocoPhillips, Shell and Unocal for funding this research. We thank BG Group and its partners Talisman Energy and Paladin Resources for permission to use palynological data from the Blake Field, North Sea, UKCS in order to demonstrate the technique development process, and G. Jones, B. O'Neill and I. Prince for their editorial reviews. MIW thanks S. Duxbury for many enlightening discussions on Cretaceous palynology. Contact ACG regarding availability of software.

References

BIRKS, H.J.B. 1979. Numerical methods for the zonation and correlation of biostratigraphical data. *In*: BERGLUND, B.E. (ed.). *Vol I. General Project Descriptions. Subproject B: Lake and Mire Environments. Project 158: Palaeohydrological Changes in the Temperate Zone in the Last 15,000 Years*. International Geological Correlation Programme. Lund, Sweden, 99–123 and Appendix 2.

CLARK, R.M. 1985. A FORTRAN Program for Constrained Sequence-slotting Based on Minimum Combined Path Length. *Computers & Geosciences*, **11**, 605–617.

CLARK, R.M. 1992. Sequence Comparisons and Sequence-slotting. *INQUA – Commission for the Study of the Holocene, Working Group on Data-Handling Methods, Newsletter*, **8**, 3–6.

CLARK, R.M. 1993*a*. Assessment of Sequence-slotting. *INQUA – Commission for the Study of the Holocene, Working Group on Data-Handling Methods Newsletter*, **9**, 5–10.

CLARK, R.M. 1993*b*. A New Version of PCSLOT. *INQUA – Commission for the Study of the Holocene, Working Group on Data-Handling Methods Newsletter*, **10**, 21–22.

CLARK, R.M. 1994. Corrections and Extensions to PCSLOT. *INQUA – Commission for the Study of the Holocene, Working Group on Data-Handling Methods Newsletter*, **12**, 21.

Clark, R.M. 1995. Depth-matching Using PCSLOT Version 1.6. *INQUA-Commission for the Study of the Holocene, Working Group on Data-Handling Methods Newsletter*, **13**, 11–13.

DAVIS, J.C. 1986. *Statistics and Data Analysis in Geology*. Wiley, New York.

DUXBURY, S. 2001. A palynological zonation scheme for the Lower Cretaceous – United Kingdom Sector, Central North Sea. *Neues Jahrbuch für Geologie und Paläontologie Abhandlungen*, **219**, 95–137.

GORDON, A.D. 1973. A sequence-comparison statistic and algorithm. *Biometrika*, **60**, 197–200.

GORDON, A.D. 1980. SLOTSEQ: A FORTRAN IV program for comparing two sequences of observations. *Computers & Geosciences*, **6**, 7–20.

GORDON, A.D. & REYMENT, R.A. 1979. Slotting of borehole sequences. *Mathematical Geology*, **11(3)**, 309–327.

HAYEK, L.C. & BUZAS, M.A. 1997. *Surveying Natural Populations*. Columbia University Press, New York.

JÖRESKOG, K.G., KLOVAN, J.E. & REYMENT, R.A. 1976. *Geological Factor Analysis*. Elsevier, Amsterdam.

KAUFMAN, L. & ROUSSEEUW, P.J. 1990. *Finding Groups in Data*. Wiley, New York.

KREBS, C.J. 1989. *Ecological Methodology*. Harper and Row, New York.

MAHER, L.J., JR. 1993. SLOTDEEP.EXE: Manual correlation using the dissimilarity matrix, *INQUA – Commission for the Study of the Holocene, Working Group on Data-Handling Methods Newsletter*, **9**, 21–26.

MAHER, L.J. JR. 2000. SLOTDEEP v. 1.8 Adds DC Profiles to its DC Map. *INQUA – Commission for the Study of the Holocene, Working Group on Data-Handling Methods Newsletter*, **18**, 4.

MIESCH, A.T. 1976. *Q-mode Factor Analysis of Geochemical and Petrologic Data Matrices with Constant Row Sums.* U.S. Geological Survey Professional Paper 574-G.

OCHIAI, A. 1957. Zoogeographical studies on the soleoid fishes found in Japan and its neighbouring regions II. *Bulletin of the Japanese Society of Fisheries Science*, **22**, 526–30.

PIELOU, E.C. 1979. Interpretation of paleoecological similarity matrices. *Paleobiology*, **5(4)**, 435–443.

RICH, E. & KNIGHT, K. 1991. *Artificial Intelligence*. McGraw-Hill, New York.

ROMESBURG, H.C. 1984. *Cluster Analysis for Researchers.* Lifetime Learning Publications. Belmont, CA.

SHI, G.R. 1993. A Comparative Study of 39 Binary Similarity Coefficients. *Memoir of the Association of Australasian Palaeontologists*, **15**, 329–341.

SWAN, A.R.H. & SANDILANDS, M. 1995. *Introduction to Geological Data Analysis*. Blackwell, Oxford.

THOMPSON, R. & CLARK, R.M. 1989. Sequence slotting for stratigraphic correlation between cores: theory and practice. *Journal of Paleolimnology*, **2**, 173–184.

THOMPSON, R. & CLARK, R.M. 1993. Quantitative marine sediment core matching using a modified sequence-slotting algorithm. *In*: HAILWOOD, E.A. & KIDD, R.B. (eds) *High Resolution Stratigraphy*. Geological Society, London, Special Publications, **70**, 39–49.

Non-acid wellsite palynology: widening opportunities

G. WILLIAMS[1], S. N. J. PAYNE[2], R. DYER[1], D. F. EWEN[2], N. PATRICK[1] & P. WATSON[1,3]

[1]*GeoStrat Ltd., Dalziel Street, Motherwell, Lanarkshire ML1 1PJ, Scotland, UK (e-mail: GeoStratUK@aol.com)*

[2]*BP Exploration Operating Company, Farburn Industrial Estate, Dyce, Aberdeen AB21 7PB, Scotland, UK*

[3]*Current address: Robertson Research International Ltd, Tyn Y Coed, Llanrhos, Llandudno LL30 1SA, Wales, UK*

Abstract: Wellsite palynology is a geoscientific tool, which involves real time preparation and analysis of subsurface samples. Analysis of palynomorphs during drilling is used to identify the location of the drill bit in the geological succession, critical for reservoir prediction, biosteering, casing, coring and terminal depth determinations to be made. Until recently, extraction of palynomorphs was only considered possible with the use of highly toxic and hazardous chemicals (e.g. hydrofluoric acid) and technical support in a mobile laboratory, resulting in high costs, logistical problems and potential Health, Safety and Environment (HSE) risk. The development of new processing techniques has removed the bulk of the chemical component in the sample processing and allow a primarily mechanical extraction technique to be employed at wellsite, representing a major breakthrough that reduces cost and logistical problems and virtually eliminates any chemical HSE risk. As a result, the application of wellsite palynology has expanded, operational costs at wellsite have been reduced, the dependence on a limited market place for the hire of specialized laboratory equipment is alleviated and the global reach of this technique is enhanced. The technique has been used successfully at wellsite for sediments of Middle Jurassic to Palaeogene age, on over 60 offshore wells in the UK, Norway and Faroes. Laboratory testing has shown wider geographical and age applicability. The technique has enabled horizontal biosteering of wells, the maximization of mature field recoverable reserves and the application of wellsite palynology to areas with logistical limitations, with significantly lower HSE risks.

Wellsite palynology is an established technique, which provides biostratigraphical information in real time at the site of drilling operations. It is used to provide stratigraphical information for the determination of casing points, coring points and terminal depths as well as to biosteer high angle and horizontal wells. This technique has been used extensively in a number of sedimentary basins worldwide.

The processing techniques normally employed to extract palynomorphs are based on acid maceration of rock samples and acid oxidation of the resultant organic residue. The acids used are normally a combination of hydrofluoric (HF), hydrochloric (HCl) and nitric acid (HNO_3). All of these acids are hazardous, potentially lethal and present a potential environmental risk. They require specialized laboratory facilities and medical backup.

Non-acid technique

A new, non-acid palynology processing technique has been developed to remove the problems associated with the use of acids at wellsite. The use of other chemicals is limited to relatively less hazardous substances, such as hydrogen peroxide (H_2O_2). The technique comprises a series of stages:

- The physical and chemical disaggregation of the rock sample is carried out using H_2O_2 at various strengths depending on lithology and palynomorph type. Typically, a 6% solution of H_2O_2 is required for mudstones and claystones buried to relatively shallow depths (less than 3000 m), whilst for siltstones and more deeply buried sediments, a 12% solution is preferable. The duration of this part of the process can vary from as little as one minute, up to 30 minutes depending on the rate of disaggregation of the lithology. As with all palynological processing techniques, the most suitable strength of solution and duration of process are determined through testing of samples. The optimum degree of disaggregation with minimum oxidation damage to the

From: POWELL, A. J. & RIDING, J. B. (eds) 2005. *Recent Developments in Applied Biostratigraphy*. The Micropalaeontological Society, Special Publications, 219–235. 1747-602X/$15.00

palynomorphs can then be achieved. The resultant residue is composed of mineral grains, calcareous and siliceous microfossils, disaggregated clays and released organic matter, including palynomorphs. This element of the process is the main chemical departure from conventional palynological processing.

- The sample is passed through a series of sieves, from 1 mm down to 63 micron mesh size, to separate the fines containing palynomorphs from the larger mineral and kerogen fragments.
- The organic matter is concentrated through a series of sieving and gravity separation techniques, including panning and settling. The resultant residue is then cleaned, using sieves and detergents, removing fine mineral and kerogen particles to leave a residue rich in palynomorphs and kerogen of similar size and specific gravity.
- Following final cleaning and concentration of the residue, it is mounted using coverslips and glass slides in the conventional manner.

Examples of typical recovery, using acid and non-acid preparations from a single cuttings sample from the North Sea, Sele Formation of earliest Early Eocene age, are provided in Figures 1 and 2 respectively. Both preparations show abundant palynomorphs, including dinoflagellate cyst marker taxa. Figures 3 and 4 illustrate non-acid preparations from Central North Sea (UK sector) sediments of Barremian and Early Oxfordian age respectively. Both samples show abundant and well-preserved dinoflagellate cyst recovery.

Effect of H_2O_2 on palynomorph recovery

The application of H_2O_2 to samples causes oxidation of organic material along with the breakdown of the sedimentary rock. Excessive oxidation of the palynomorph component is of significant concern as this will lead to a reduced assemblage and possible elimination of certain less robust taxa.

Hopkins & McCarthy (2002) showed oxidation effects on Quaternary palynomorphs of H_2O_2 in varying concentrations and periods of exposure. Their results showed progressive reduction of dinoflagellate cyst abundance over a six-hour period with certain taxa being less resistant than others. However, their results also show that dinoflagellate cysts will survive much more concentrated solutions of H_2O_2 (15% and 30%) for much longer periods than employed in the technique described here.

Variable percentage concentrations of H_2O_2 (3%–12%) have been tested for the technique described here and destruction of particular dinoflagellate cyst taxa (e.g. *Palaeoperidinium pyrophorum*) has been observed in certain samples. However, reducing the concentration of H_2O_2 or the length of exposure has resulted in satisfactory recovery in all cases.

Previous efforts to exclude acids

The vast majority of publications describing processing of palynological samples employ some variation of acid maceration using HCl and HF to remove carbonate and silicate material, for example Herngreen (1983), Doher (1980) and Batten (1999). This approach can remove most carbonate and silicate minerals from the residue if there are no time constraints in the laboratory. In the case of more rapid processing, as required for 'hot shot samples' or wellsite analysis, the length of time available for processing is reduced significantly and therefore the effects of these acids on the sample are greatly reduced. The result of this rapid processing is a less effective breakdown of the rock and only partial removal of mineral matter. The non-acid process described here has been compared to, and tested against, full laboratory processing and the rapid hot shot technique to ensure that best results are achieved (e.g. figure 6, Norwegian well 6707/10–1).

Certain authors have described palynological processing techniques, which reduce or remove the use of some chemicals in the process. Staplin *et al.* (1960), in a broad discussion of palynological techniques, noted that in rare cases sample treatment with core analysis fluid (Soltrol C), detergents, acetone and benzene over a lengthy period may result in a residue, which does not require further chemical processing. The time taken for the process, the chemicals involved and the apparently rare success of this method does not make it a practical technique for wellsite analysis.

Traverse (1978) briefly described a processing technique applied to Neogene DSDP samples from Leg 42B in the Black Sea. This technique employed HCl, Calgon detergent and zinc chloride heavy liquid ($ZnCl_2$) separation but avoided the use of HF. Whilst this technique removed the need for HF, it does involve hazardous HCl and $ZnCl_2$. Whilst successful in this instance, the technique was only applied to unconsolidated Neogene sediments buried to a maximum depth of 1000 m. For these reasons the technique is not considered to be a proven processing method for older, more consolidated sediments.

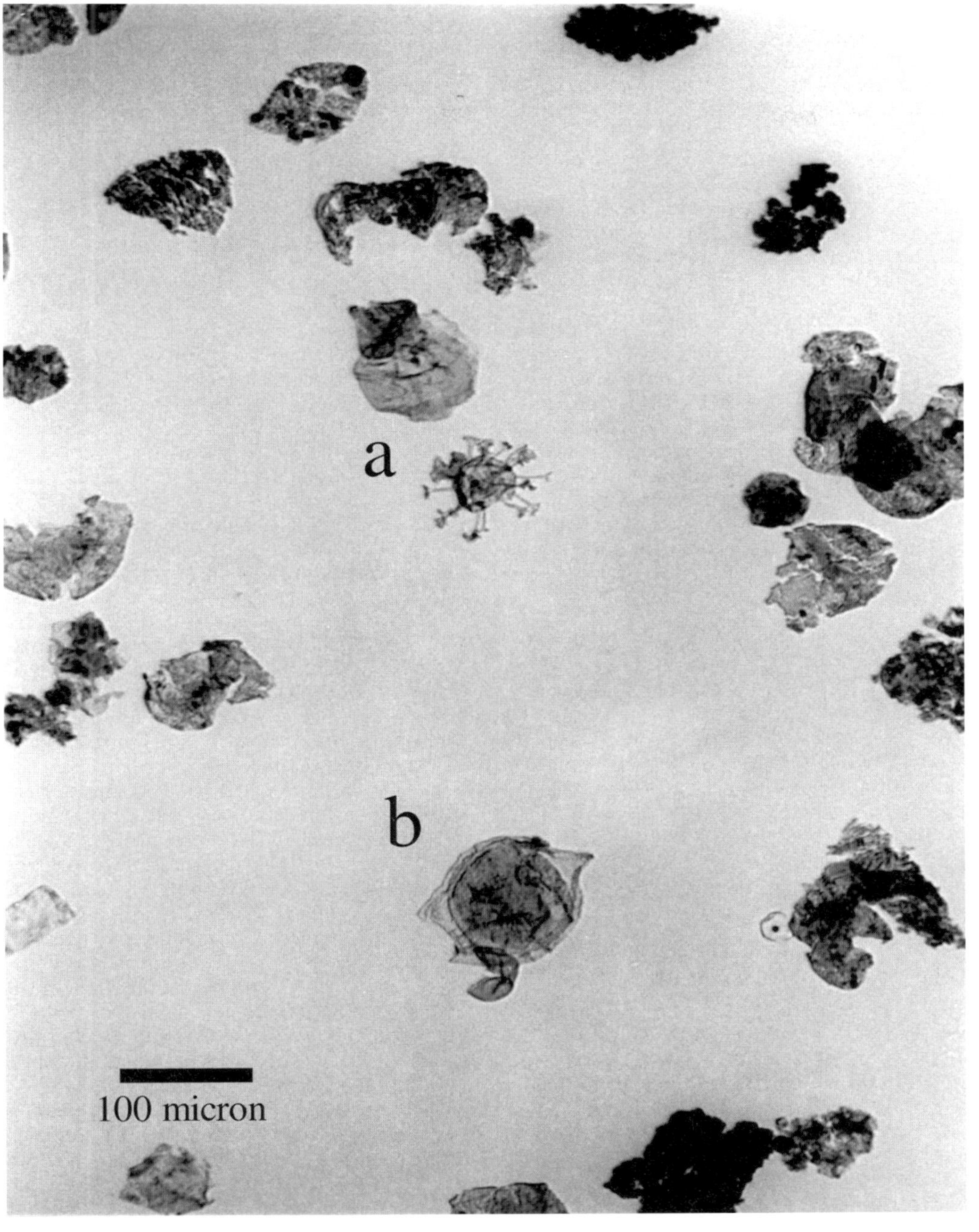

Fig. 1. Acid preparation, earliest Early Eocene, Sele Formation. (**a**) *Hystrichosphaeridium tubiferum*, (**b**) *Deflandrea oebisfeldensis*. (Well 22/9-2, 8300 ft.)

Development of the technique

The technique was first developed in 1990, when tests were carried out in the GeoStrat laboratory to establish whether the use of acids in palynology processing is essential. The motivation behind these tests was to provide a safer processing technique at wellsite. Initial results

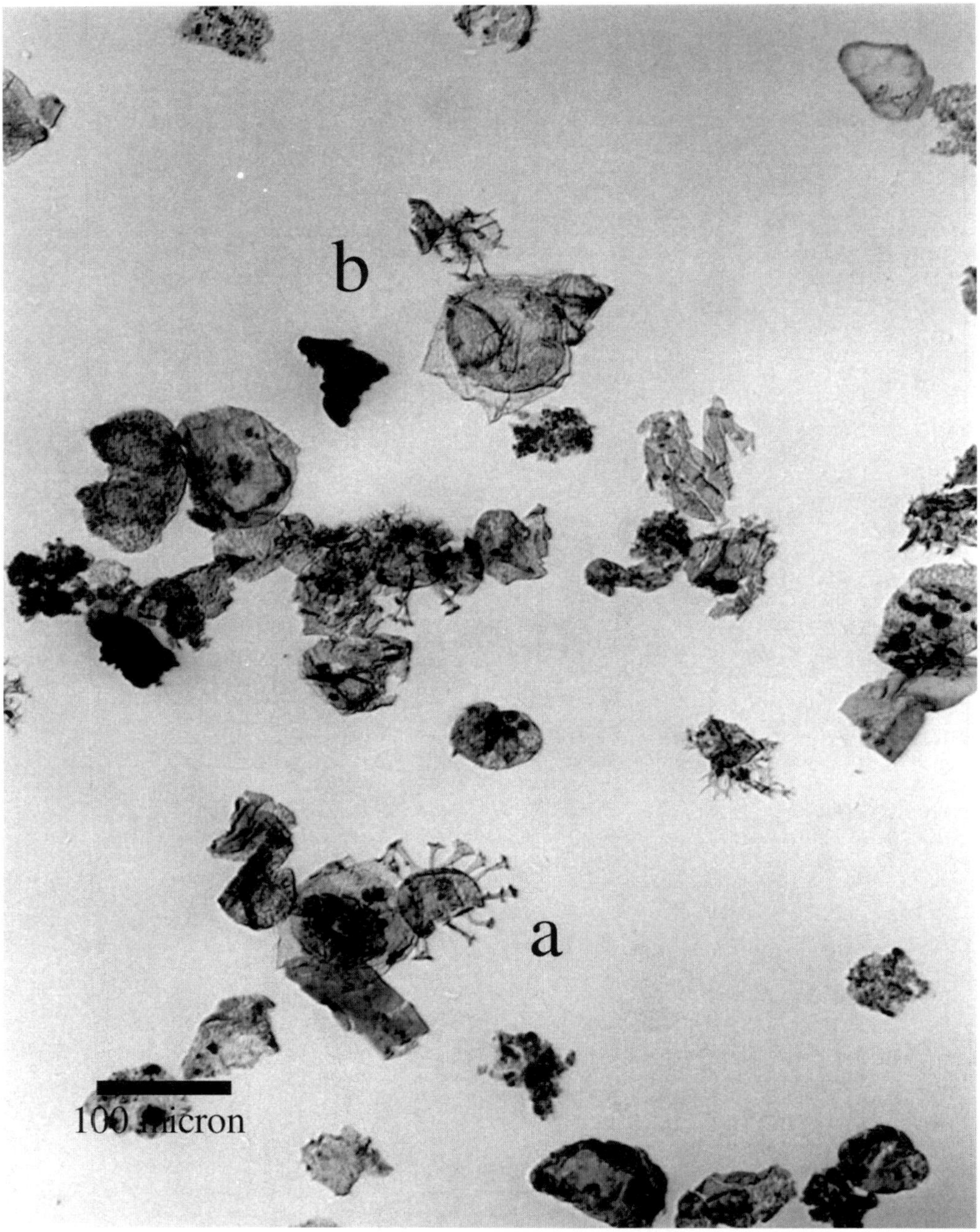

Fig. 2. Non-acid preparation, earliest Early Eocene, Sele Formation. (**a**) *Hystrichosphaeridium tubiferum,* (**b**) *Deflandrea oebisfeldensis.* (Well 22/9-2, 8300 ft.)

were favourable, leading to further testing on different lithologies and successions from the UK sector of the North Sea.

The technique was first used at wellsite in 1991. Following this successful application, a further 12 years of wellsite assignments, testing and additional research into additional refinements of the technique has resulted in a well

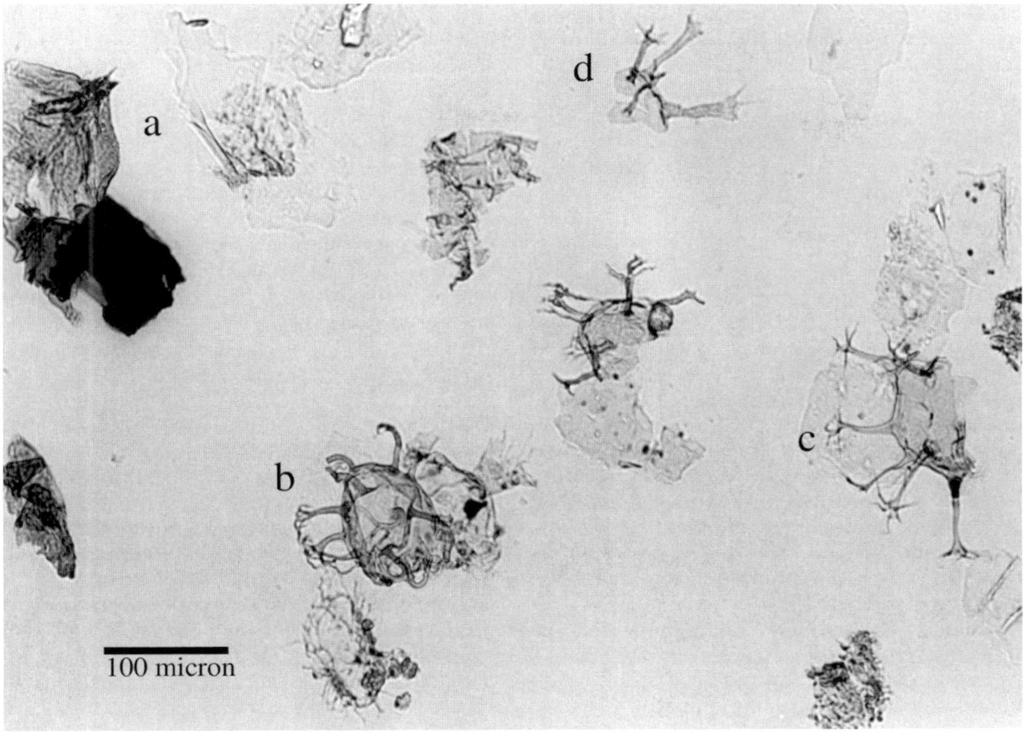

Fig. 3. Non-acid preparation, Barremian, Cromer Knoll Group. (**a**) *Rhynchodiniopsis* sp., (**b**) *Surculosphaeridium* sp., (**c**) *Oligosphaeridium asterigerum,* (**d**) *Oligosphaeridium complex.* (13/26a-2, 4100 ft.)

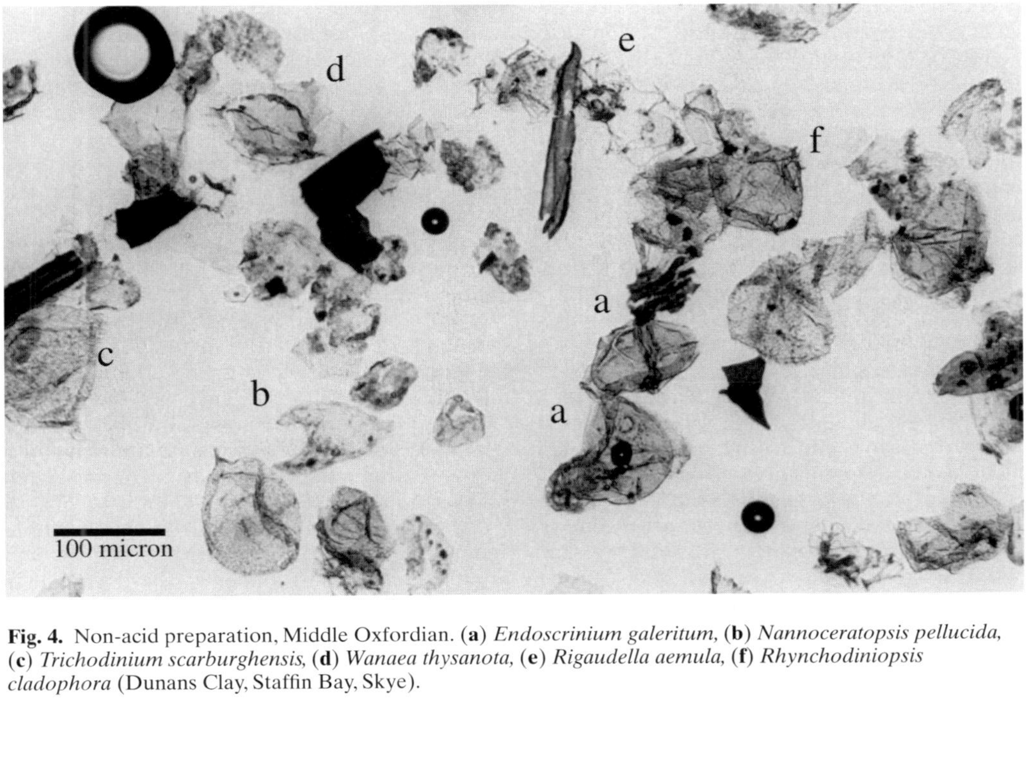

Fig. 4. Non-acid preparation, Middle Oxfordian. (**a**) *Endoscrinium galeritum,* (**b**) *Nannoceratopsis pellucida,* (**c**) *Trichodinium scarburghensis,* (**d**) *Wanaea thysanota,* (**e**) *Rigaudella aemula,* (**f**) *Rhynchodiniopsis cladophora* (Dunans Clay, Staffin Bay, Skye).

established process. The stratigraphical range over-which the technique has been applied ranges from Middle Jurassic to Palaeogene in the UK and Norwegian sectors of the North Sea, West of Shetland, the Faroes and Mid-Norway. Extensive testing of the technique has also been carried out to prove its application in other sedimentary basins.

Advantages

Health, safety and environment

The key advantages associated with the use of the non-acid technique are related to health and safety. The transportation, storage, handling and disposal of the acids used in conventional palynology processing all pose health, safety and environmental risks, which if not managed properly are significant. The removal of extremely hazardous acids from the process and replacement with the significantly less hazardous H_2O_2, results in a major mitigation of health and safety risk.

Potential environmental contamination from spillage of chemicals or waste is largely removed by the non-acid technique, as H_2O_2 is not classified as a potential marine pollutant.

Flexibility, transport and personnel

As a result of the less hazardous nature of the chemicals used in this process, a number of key advantages ensue. The transport of chemicals and equipment is made much easier by the elimination of acids and specialist equipment. All offshore work using acids requires a dedicated pressurised unit containing a fume hood or totally enclosed fume cupboard system. In contrast the non-acid technique requires approximately 50 kg of equipment and chemicals to complete a typical assignment. This difference has clear benefits in mobilization lead times and costs. No setting-up time is required for the non-acid technique and the palynologist can be operational within a matter of hours of arriving at the wellsite.

Similar advantages can be identified for the use of this technique in more remote regions where access to chemicals and delivery of equipment can pose significant logistical problems.

The use of the conventional acid technique requires a technician to carry out processing for the palynologist in almost all circumstances. Whilst this can be advantageous using the non-acid technique too, the nature of the non-acid processing technique entails that a palynologist can often prepare and analyse samples alone, or as a team of two palynologists working alternate shifts. This flexibility in manpower can have a significant impact in allowing the use of wellsite palynology where accommodation space is limited. This also provides a high degree of flexibility at the start of an assignment where the mobilization of a single palynologist can give initial cover to be followed by a full team once the interval of interest is penetrated. As a result of these benefits, wellsite palynology can be spread into areas where palynology has not been used previously.

Case studies

The following case studies have been selected to demonstrate the non-acid technique used for a range of wellsite tasks. All of the examples given were carried out by GeoStrat for BP on the UK and Norwegian continental shelf between 1996 and 2001. The examples range from exploration monitoring in a wildcat setting, through conventional exploration to production biosteering in mature field wells. All examples were carried out using a single palynologist at wellsite without a technician. Figure 5 shows the locations of the four case studies discussed.

6707/10-1: Mid Norway

This well provides an example of monitoring a wildcat exploration well over an extensive stratigraphical section in a relatively remote area (Fig. 5). The well was monitored using the non-acid technique for 114 samples over 2800 m of section.

A number of samples were processed onshore after the completion of the well using conventional acid preparation techniques. This allowed close comparison to be made between wellsite hotshot preparations using the non-acid technique and rapid analysis with standard laboratory processing and analysis. The two types of analyses are compared in Figure 6. The assemblages recovered comprise abundant and diverse marine dinoflagellate cysts with common pollen. The variation in recovery between acid and non-acid preparations is no greater than the range of variation seen between individual acid preparations.

The sections illustrated from this example comprise Late Cretaceous (Maastrichtian and Campanian) aged sediments. These sediments were studied mainly using drill cuttings with a limited amount of conventional core. The

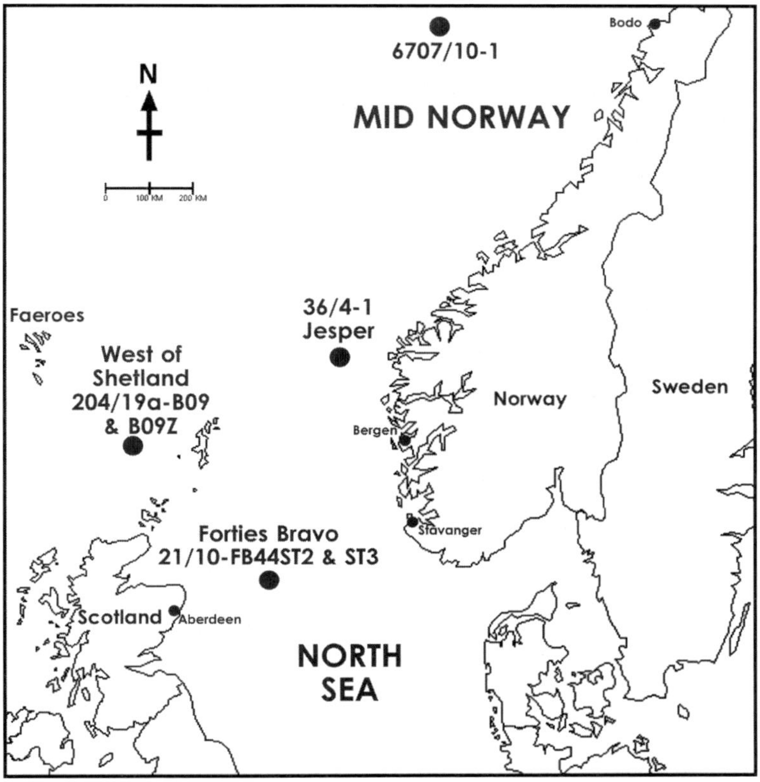

Fig. 5. Case study locations.

assemblages contain the expected abundance and diversity of marine dinoflagellate cysts and allowed for the recognition of the key stratigraphical markers through this section. Recovery was comparable to that achieved during the acid-based laboratory processing in all cases.

Jesper well 36/4-1: Norwegian North Sea

This well provides an example of the technique being used over an extensive range of lithostratigraphical units of Late Cretaceous to Middle Jurassic age. Recovery from all intervals was successful using the non-acid technique. Late and Early Cretaceous sediments of the Shetland and Cromer Knoll groups yielded typically abundant and diverse marine dinoflagellate cyst assemblages, with good representation of key stratigraphical events (Figs 7 and 8). All analyses were carried out at wellsite.

Late Jurassic sediments of the Viking Group, Heather Formation yielded abundant pollen and miospores with diverse dinoflagellate cysts and algae (Fig. 9). Recovery from Middle Jurassic sediments exhibits abundant pollen and miospores with abundant but low diversity dinoflagellate cysts, reflecting a marginal marine environment of deposition (Fig. 9).

Foinaven 204/19a-B09 & B09Z: UK West of Shetland

This case study illustrates the use of the non-acid technique during development drilling to recognise sequence stratigraphical events defined by biostratigraphy within Upper Paleocene sediments. The recognition of these events allowed the monitoring of a pilot hole, prior to making a casing point placement and biosteering a horizontal sidetrack (Fig. 10) across a section of 1200 m.

A series of unpublished BP sequence stratigraphical events from T38 to T28, including the

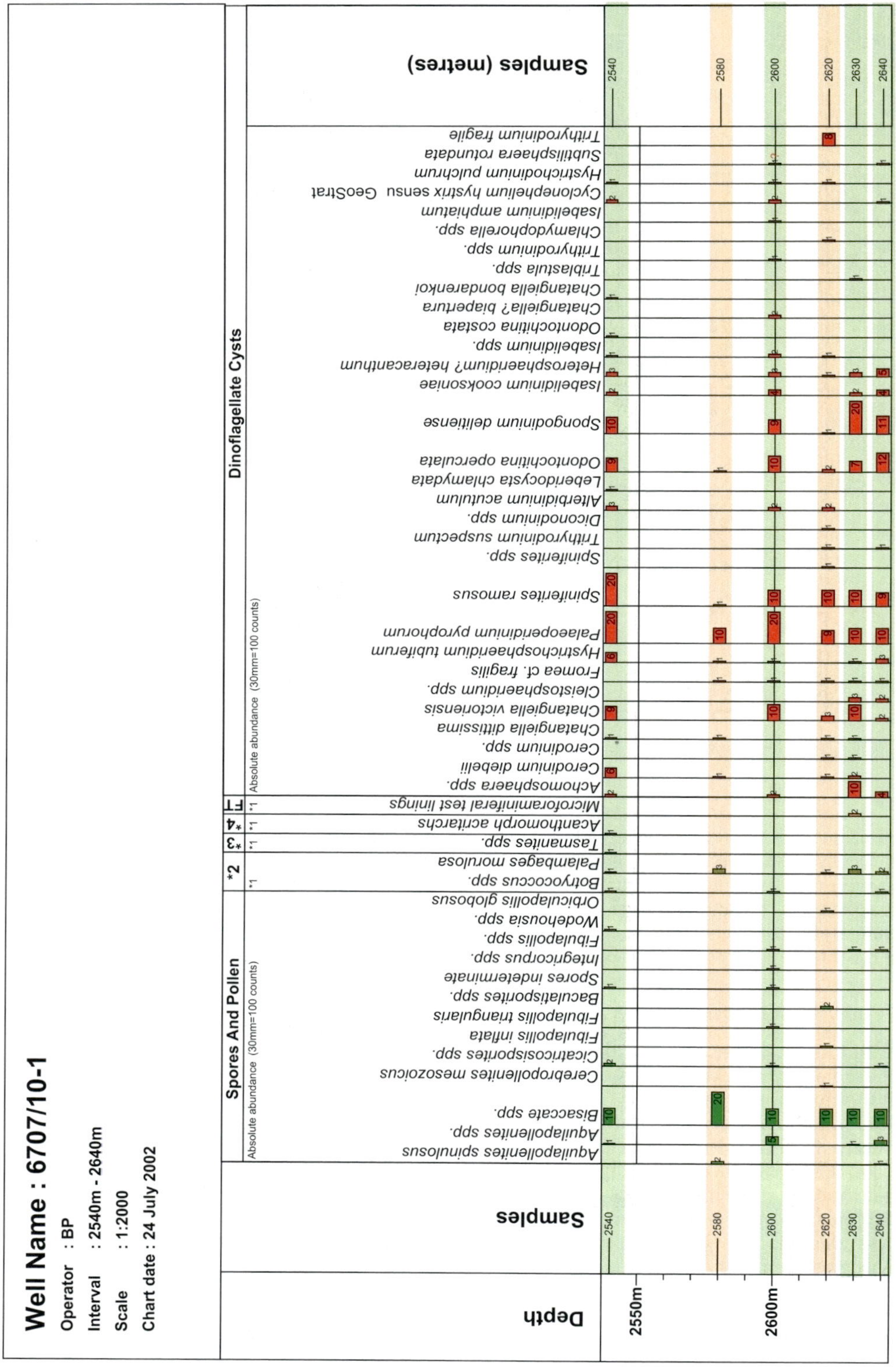

Fig. 6. (**a**) Well 6707/10-1 cuttings sample data from Maastrichtian and Campanian sediments. Green highlight: non-acid, red highlight: acid.

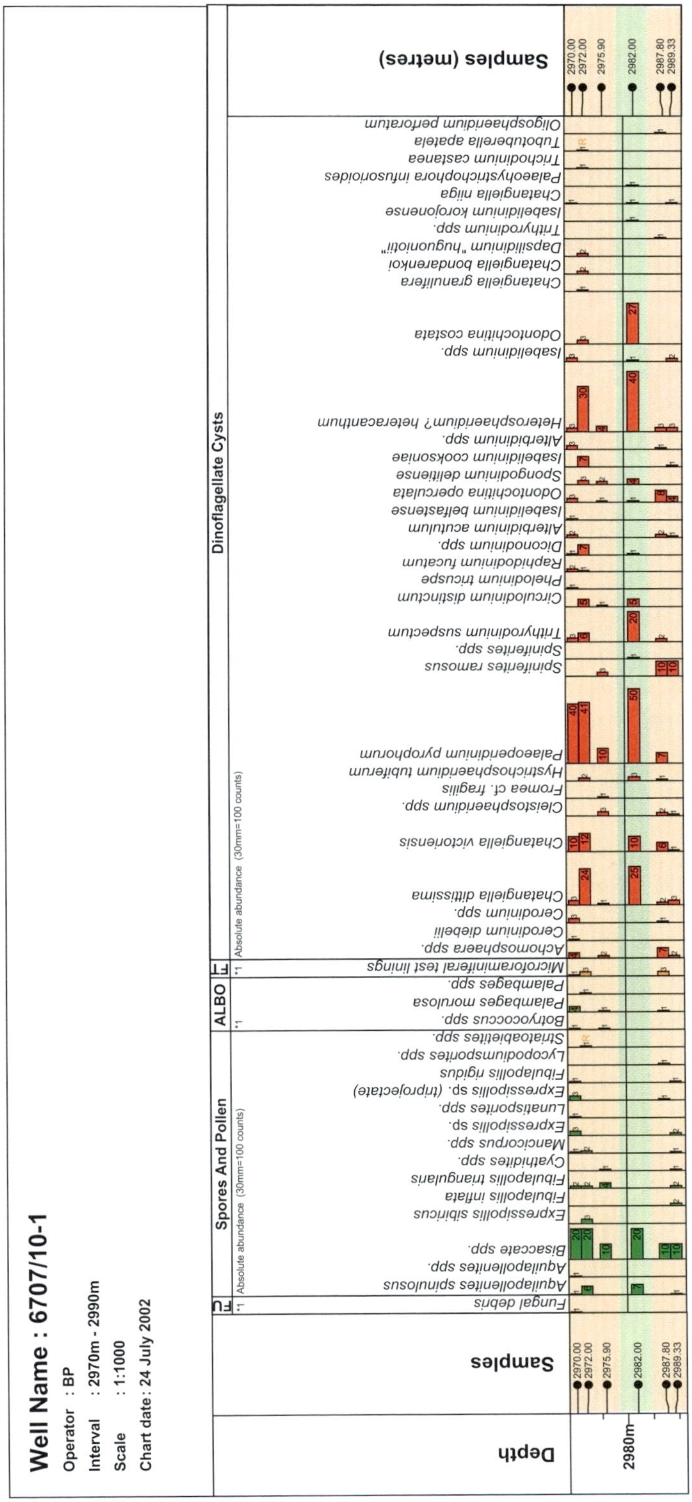

Fig. 6. (**b**) Well 6707/10-1 conventional core sample data from Campanian sediments. Green highlight: non-acid, red highlight: acid.

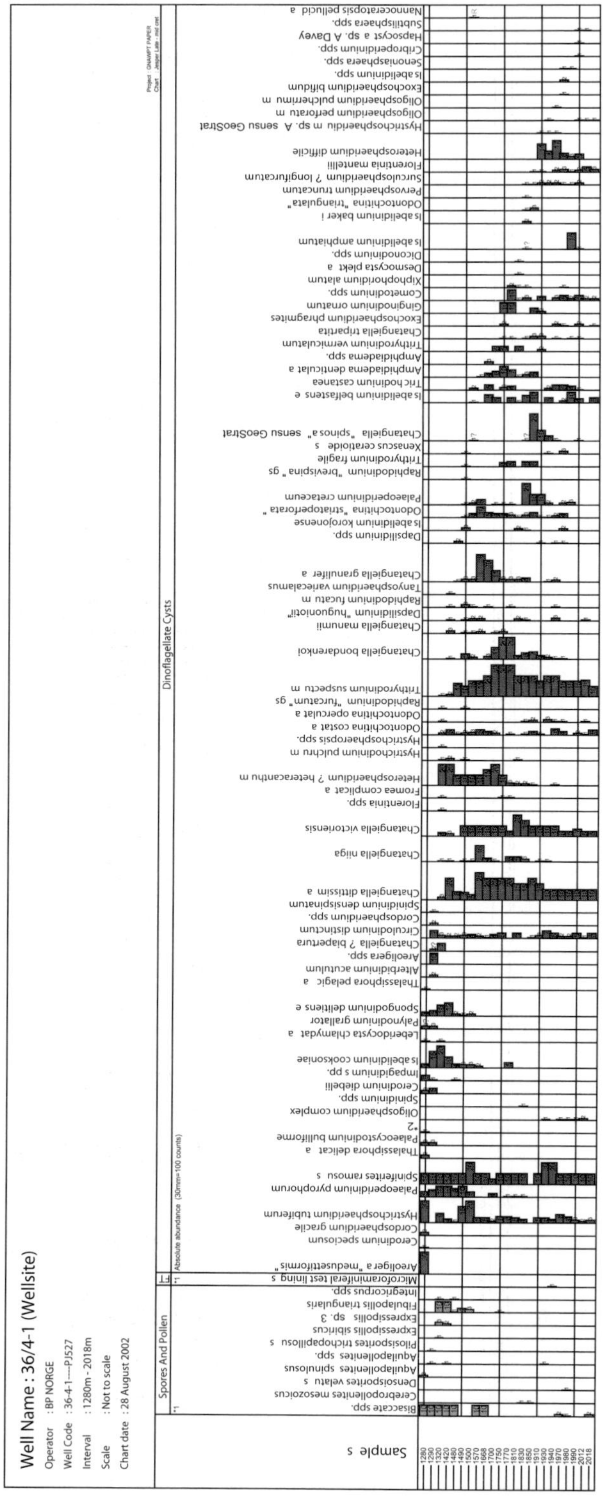

Fig. 7. Well 36/4-1, Maastrichtian to Turonian sediments of the Shetland Group. All samples processed using the non-acid technique.

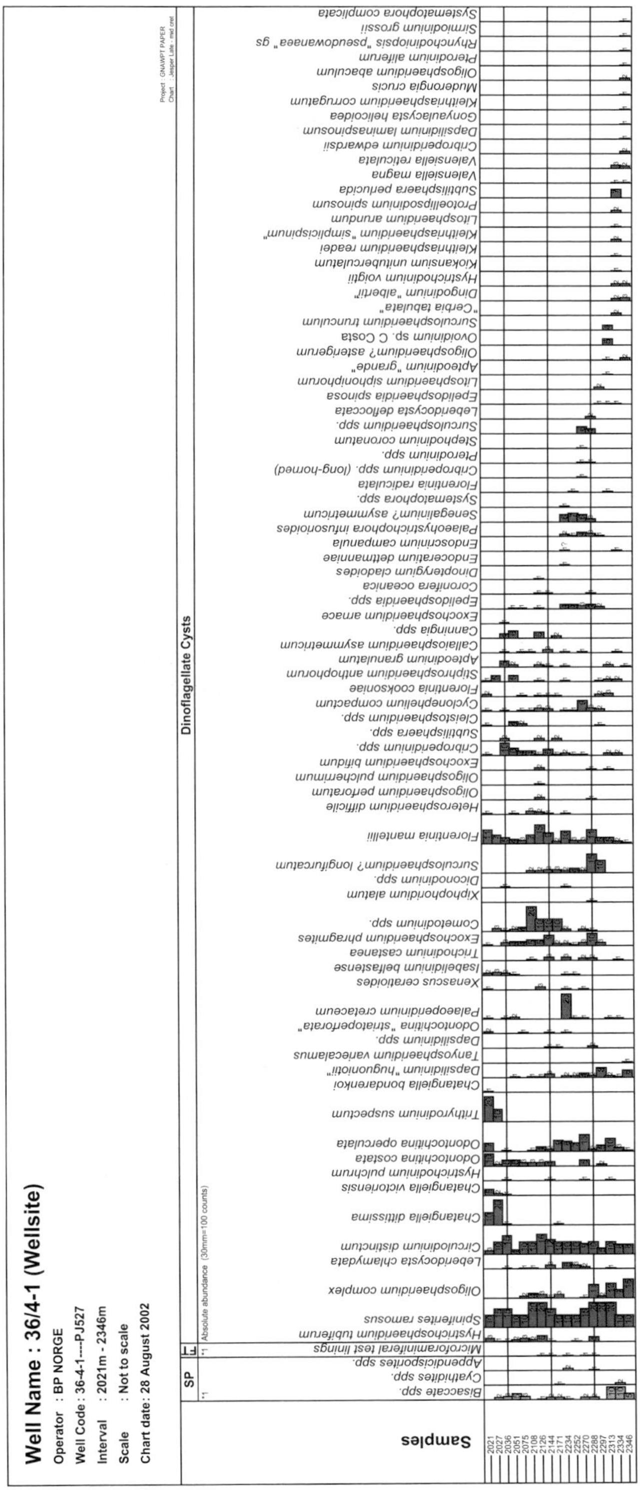

Fig. 8. Well 36/4-1, Cenomanian to Barremian sediments of the Shetland and Cromer Knoll groups. All samples processed using the non-acid technique.

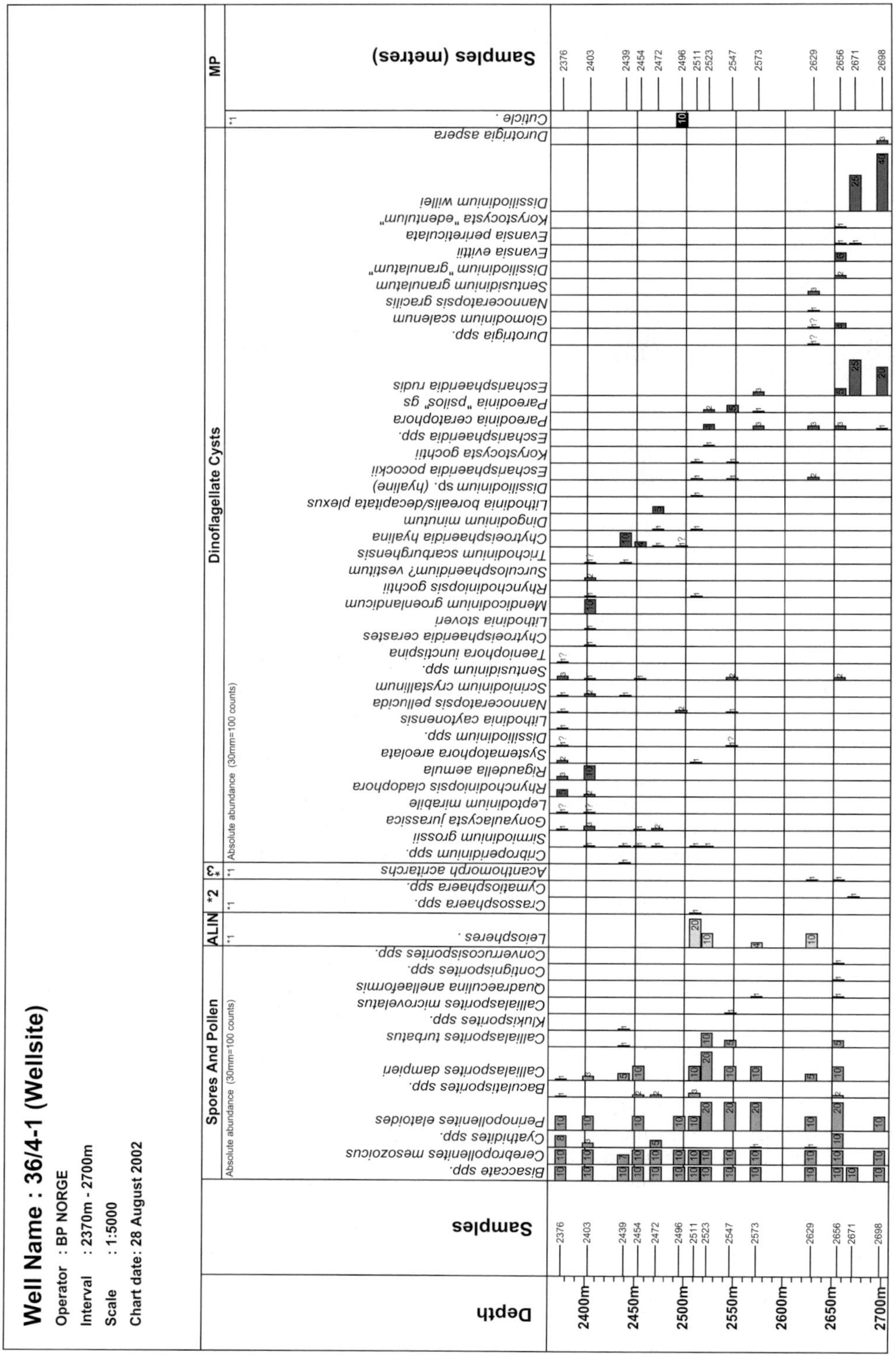

Fig. 9. Well 36/4-1, Oxfordian to Bathonian sediments of the Heather Formation. All samples processed using the non-acid technique.

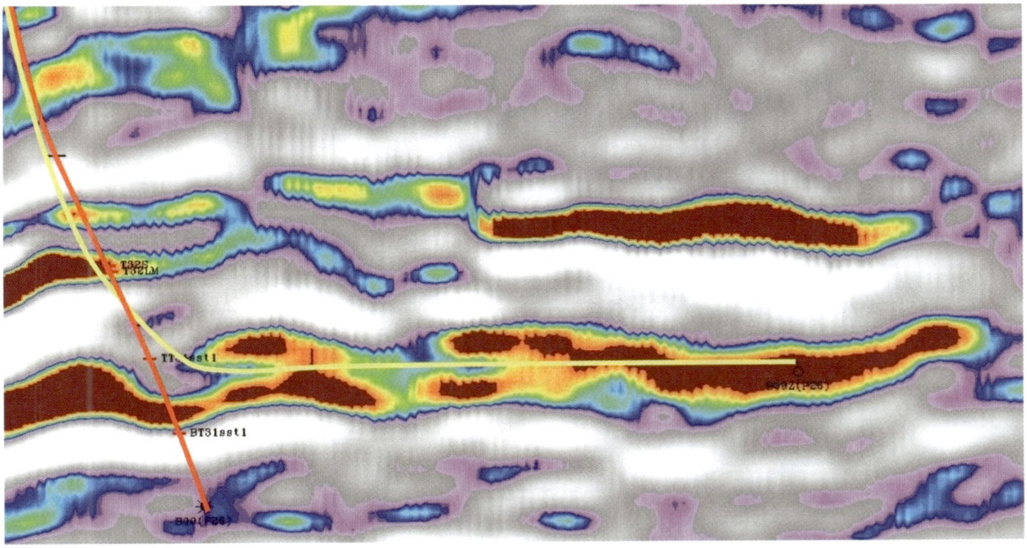

Fig. 10. Foinaven West of Shetland wells 204/19a-B09 and B09Z seismic section, pilot hole (red) and 1200 metre horizontal producer (yellow).

T31 target sand (equivalent to the Lambda Formation, Kettla Member and Vaila Formation of the Faroe Group), were recognized in well 204/19a- B09. The recognition of these sequences is based on a series of marine dinoflagellate cyst influxes, which are highlighted on Figure 11.

During the drilling of well 204/19a-B09Z, the same series of events were recognized, with the successful positioning of the 9⅝ inch casing shoe at the top of the T31 Sand Unit. Drilling of the T31 Sand was monitored to identify any departure from the planned well path and to identify the up-sequence exit of the T31 sand at TD (Fig. 12). All of these tasks were successfully completed.

Forties Bravo 21/10-FB44 ST2 & ST3: Central North Sea UK sector

These two wells are sidetracks from an earlier production well drilled from the Forties Field Bravo platform. The wells illustrate a drilling challenge on a mature field, the task being to access by-passed attic oil accumulations within the Late Paleocene Forties Sandstone Member. These reserves had not been drained by earlier drilling phases which concentrated on lower parts of the Forties Sand Member reservoir. The two wells were drilled from below through reverse stratigraphy, into the attic oil accumulation and completed in the overlying Sele Formation claystone cap rock (Fig. 13).

The key characteristics of this case study are the use of local field events, slim hole drilling using 3¾ inch drill bits, penetration of a sand rich section and the terminal depth determination in an unstable cap rock. All of the characteristics affect the use of palynology at the wellsite and provided a significant technical challenge for the non-acid technique. The combination of slim hole drilling in a sand rich section resulted in cuttings samples comprising mainly small quantities of shattered quartz grains with little argillaceous material to offer good palynomorph recovery. The need to recognize relatively subtle local field bioevents (Fig. 14) and rapidly identify penetration of the incompetent Sele Formation claystone cap rock to allow drilling to stop prior to any hole stability problems, provided additional technical tests for the non-acid technique (Fig. 15).

Conclusions

The technique discussed here provides a practical and proven approach to the processing of palynological samples at wellsite. The technique has been tested and successfully applied to a wide range of lithologies, depths of burial and age ranges in the UK, Norwegian and Faroes sectors of the North West European continental shelf. Conventional acid processing carried out

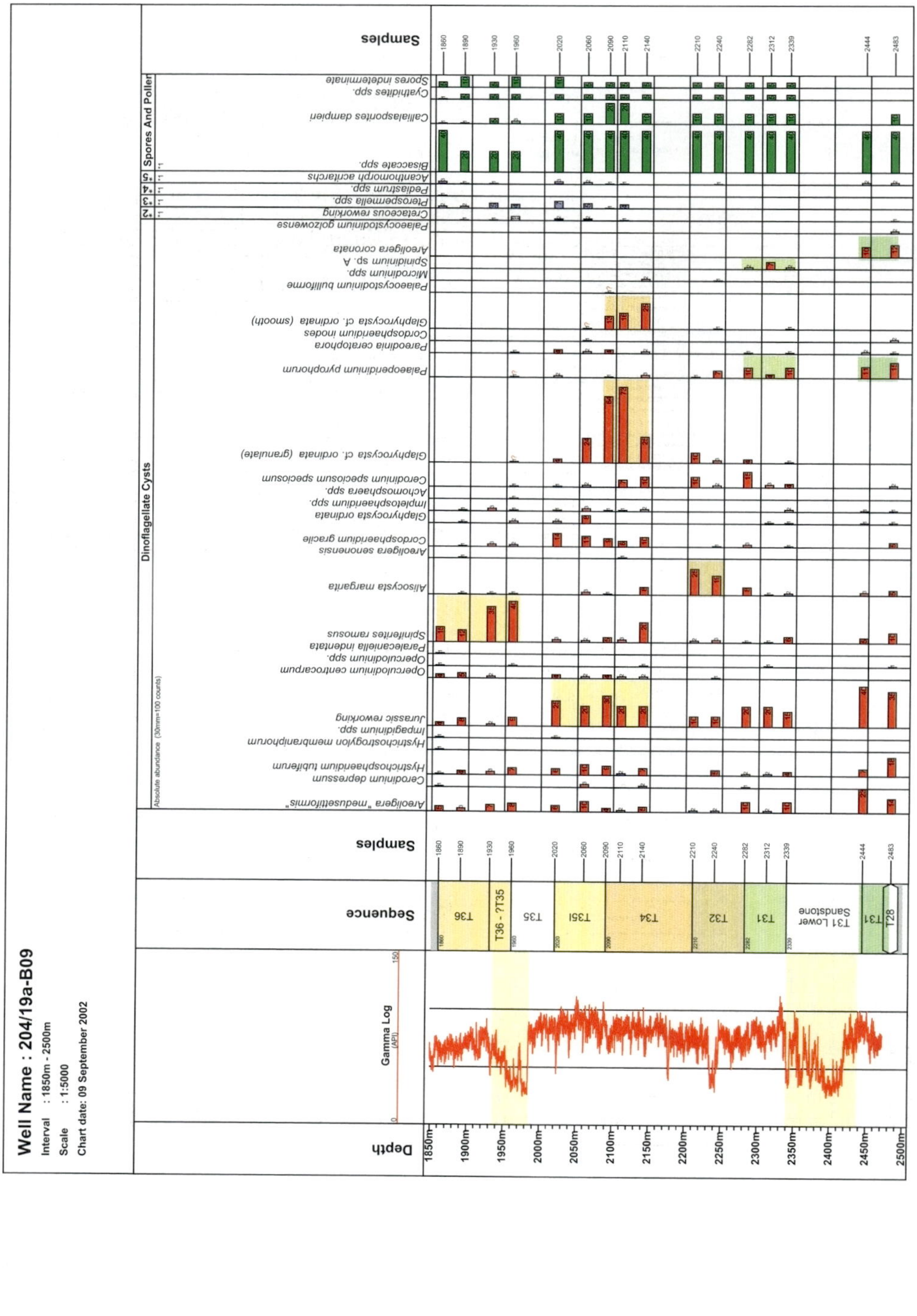

Fig. 11. Well 204/19a-B09 pilot hole. Sequence stratigraphical units and corresponding dinoflagellate cyst events are highlighted.

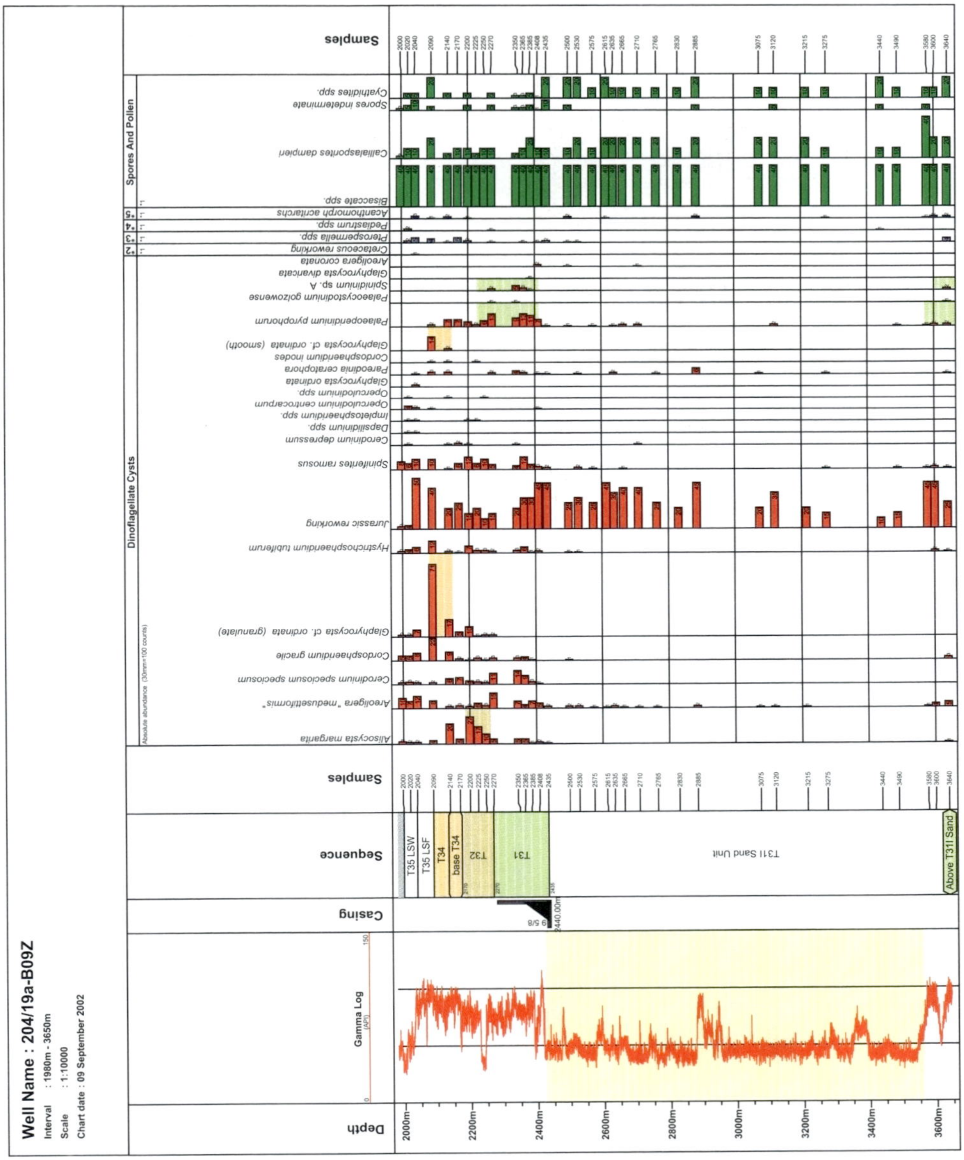

Fig. 12. Well 204/19a-B09Z horizontal producer. Sequence stratigraphic units and corresponding dinoflagellate cyst events are highlighted.

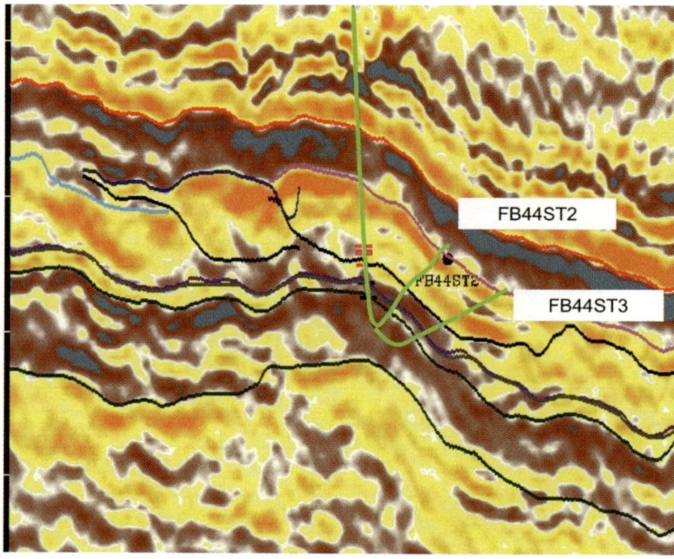

Fig. 13. Forties Field wells FB44ST2 and FB44ST3 and seismic section. Well paths: green, top reservoir: pink.

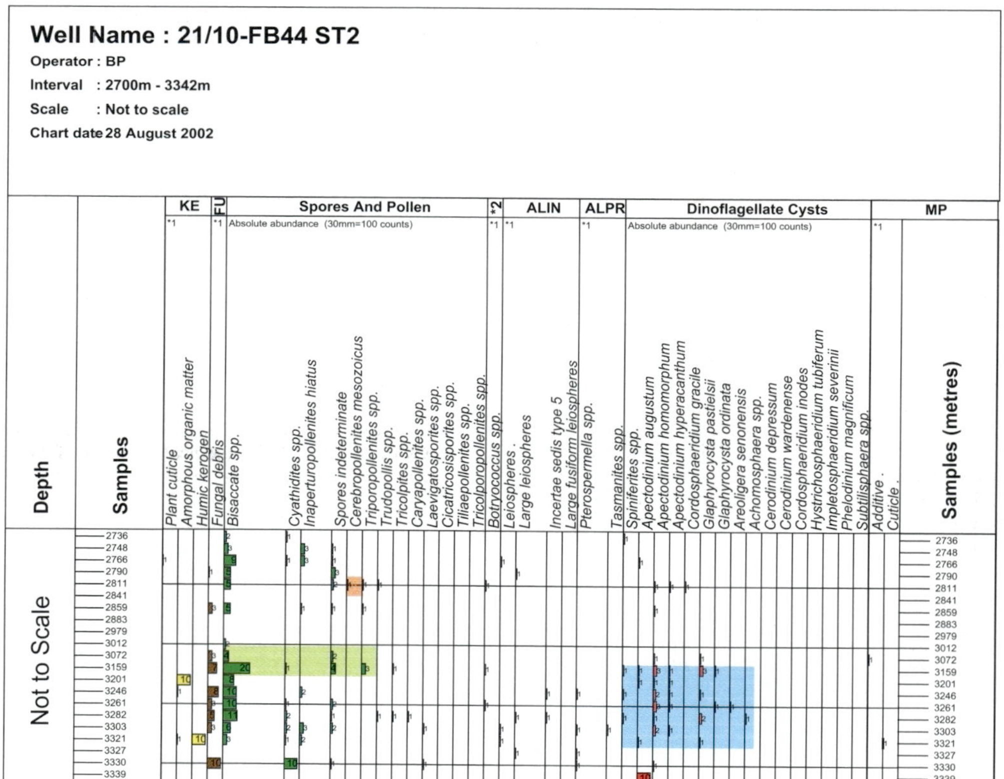

Fig. 14. Forties Field well FB44ST2. Key events are highlighted.

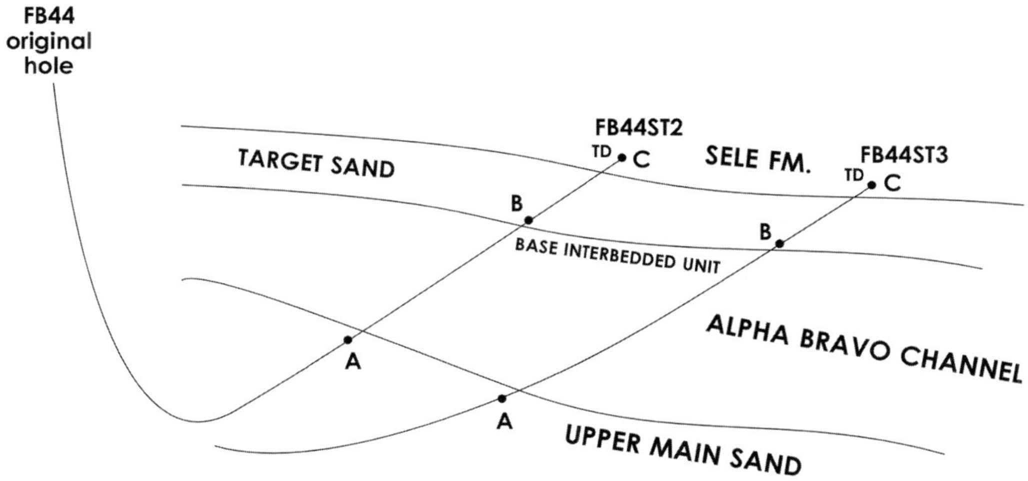

Fig. 15. Stratigraphical interpretation of Forties Field wells FB44ST2 and FB44ST3 following wellsite palynology.

in an onshore laboratory has been used to validate the results obtained at wellsite using this technique.

The key benefits of this technique are reduced health, safety and environmental risks, flexible logistics, rapid mobilization and reduced cost. It is hoped that these benefits can be extended to wider geographical areas in the future.

The authors wish to thank BP Exploration Ltd and GeoStrat Ltd for permission to publish. The input of BP geoscientists Roger Anderton, Andy Carter, Ewan Laws and Paul Milner is acknowledged. The approval of 36/4-1 well partners Idemitsu Petroleum Norge AS, Norsk Hydro Production AS, Petoro ASA and Statoil ASA is gratefully acknowledged. Likewise partners on well 6707/10-1, Esso Exploration and Production Norway A/S, Norsk ConocoPhillips AS, and Statoil ASA are thanked. Foinaven Field partner Marathon Oil (UK) Ltd and Forties co-venturers Agip (UK) Ltd, North Sea Ltd, EDC (Europe) Ltd, ExxonMobil International Ltd and Shell UK Exploration and Production Ltd are acknowledged.

References

BATTEN, D.J. 1999. Small palynomorphs. *In*: JONES, T.P. & ROWE, N.P. (eds) *Fossil Plants and Spores: Modern Techniques*. Geological Society, London, Miscellaneous Titles, London, 5–19.

DOHER, L.I. 1980. *Palynomorph Preparation Procedures Currently Used in the Paleontology and Stratigraphy Laboratories*. U.S. Geological Survey Circular **830**.

HERNGREEN, G.F.W. 1983. Palynological Preparation Techniques. *In*: COSTA, L.I. (ed.) *Palynology – Micropalaeontology: Laboratories, Equipment and Methods*. Norwegian Petroleum Directorate-Bulletin **2**, 13–35.

HOPKINS, J.A. & MCCARTHY, F.M.G. 2002. Post-depositional palynomorph degradation in Quaternary shelf sediments: a laboratory experiment studying the effects of progressive oxidation. *Palynology*, **26**, 167–184.

STAPLIN, F.L., POCOCK, S.A.J., JANSONIUS, J. & OLIPHANT, E.M. 1960. Palynological Techniques for Sediments. *Micropalaeontology*, **6**, 329–331.

TRAVERSE, A. 1978. Palynological analysis of DSDP Leg 42B (1975) Cores from the Black Sea. *In*: *Initial Reports of the Deep Sea Drilling Project*, **42(2)**, 993–1015, US Government Printing Office, Washington.

Index

Page numbers in *italic* indicate figures. Page numbers in **bold** indicate tables.